Brock/Springer Series in Contemporary Bioscience

Plankton
Ecology

Brock/Springer Series in Contemporary Bioscience

Series Editor: Thomas D. Brock
University of Wisconsin–Madison

Ulrich Sommer (Ed.)

Plankton Ecology

Succession in Plankton Communities

With 104 Figures and 12 Tables

Springer-Verlag
Berlin Heidelberg New York
London Paris Tokyo

Library of Congress Cataloging-in-Publication Data

Plankton ecology : succession in plankton communtities / Ulrich Sommer
(ed.).
    p.   cm. – (Brock/Springer series in contemporary bioscience)
Bibliography: p.
Includes index.
ISBN 0-387-51373-6 (U.S.)
    1. Plankton–Ecology. 2. Ecological succession. I. Sommer,
Ulrich, 1952-   . II. Series.
QH90.8.P5P53   1989
574'.92–dc20   89-11455
                                                                                    CIP

ISBN 3-540-51373-6 Springer-Verlag   Berlin Heidelberg New York
ISBN 0-387-51373-6 Springer-Verlag   New York Berlin Heidelberg

Production and editorial supervision: Science Tech Publishers

Cover art: Kandis Elliot

Printed in the United States of America
10 9 8 7 6 5 4 3 2 1

Contents

Contributors

W.R. DeMott   Department of Biological Sciences and Crooked Lake Biological Station, Indiana University-Purdue University at Fort Wayne, Fort Wayne, IN 46805, USA

Z.M. Gliwicz   Department of Hydrobiology, University of Warsaw, Nowy Swiat 67, 00-0046 Warsaw, Poland

H. Güde   Institut für Seenforschung und Fischereiwesen, D-7994 Langenargen, FRG

C. Pedrós-Alió   Instituto de Ciencias del Mar, Consejo Superior Investigaciones Cientificas, Paseo Nacional s/n, 08003 Barcelona, Spain

J. Pijanowska   Department of Hydrobiology, University of Warsaw, Nowy Swiat 67, 00-0046 Warsaw, Poland

C.S. Reynolds   Freshwater Biological Association, Windermere Laboratory, Ambleside, Cumbria, GB-LA22 OLP, UK

U. Sommer   Max-Planck-Institute für Limnologie, P.O. Box 165, 2320 Plön, FRG

R.W. Sterner   Department of Biology, UTA Box 19498, University of Texas, Arlington, Arlington, TX 76109, USA

E. Van Donk   Provincial Waterboard of Utrecht, P.O. Box 80300, 3508 TH Utrecht, The Netherlands

Preface

"The lake as a microcosm", the title of one of the earliest papers in limnology (Forbes, 1887), could be the motto of this book. This book was written by limnologists—especially plankton ecologists—for limnologists, marine biologists, and general ecologists. It is centered around the mechanisms that drive succession in plankton communities. There is no other field of limnology that has been more influenced by population ecology. The concepts of population ecology have been the skeleton to which plankton ecologists have attached the empirical flesh. Now it is time to make the balance of trade even, and to export the results of our work to the general ecological community. This book is especially timely because many of the classic concepts of theoretical ecology have recently come under criticism within the ecological community. We, the plankton ecologists, have the empirical data, the experiments, and the case studies that give life to the concepts of the theorists. Indeed, we conceive of lakes, and especially their planktonic communities, as "microcosms", i.e., as smaller models of the larger world of ecological interactions. This idea will be further developed in Chapter 1 of this book.

Descriptions of individual successions and phytosociological classification of succession patterns are of little interest for the nonplanktologist. However, mechanisms that cause species replacements are of interest because plankton do not differ fundamentally from other organisms in their requirements for growth and survival. Only their need to remain in suspension is unique. Otherwise, they face the same problems of fulfilling their nutritional demands for growth and reproduction and of avoiding predation. As in any other community, species-specific differences in the balance of reproduction and mortality cause changes in species composition.

The requirements for nutrition, the consequences of being eaten, and the functional division of the plankton community into phytoplankton, zooplankton, and bacterioplankton determine the order of this book. Chapters 2 to 5 deal with phytoplankton. In Chapter 2, the physical conditions (temperature, turbulence, stratification, and light) necessary for their growth and survival are explored; the stage is defined on which the successional play takes place. Chapter 3 discusses the nutritional requirements of phytoplankton and the successional consequences of competition for resources. The next two chapters deal with the successional consequences of being food for others. Chapter 4 discusses the impact of zooplankton grazing, and Chapter 5 concerns the impact of fungal parasitism. Chapters 6 and 7 cover the most important mechanisms of zooplankton succession. Chapter 6 explores the role of food limitation and competition for food, and Chapter 7 examines the impact of predation by higher trophic levels.

The chapters on bacterioplankton are a bit different from the rest of the book because the concept of succession as a series of species replacements cannot—at least not yet—be applied to bacteria. Chapter 8 discusses the nutritional control of succession, metabolic guilds substituting for species. Growthforms substitute for species in the treatment of predation on bacteria in Chapter 9.

The authors who have contributed to this book share the viewpoint that succession in plankton can best be approached by understanding the adaptive characters of the players and appreciating the importance of biotic interactions that center around eating and being eaten. None of these views is universally accepted and the debate continues. Hopefully, this book will provide an up-to-date and significant contribution to ongoing discussions of plankton ecology.

Acknowledgements: This book would not have been written without the invitation by T.D. Brock to do so. This invitation was the stimulus that I needed to summarize my own thoughts and work of the recent years and to invite the other authors to do the same. I also want to express my gratitude to all those with whom I have discussed plankton ecology during the recent years, and who have helped me to develop my own view of the world of plankton. In addition to those who are also contributors to this book, I want to acknowledge especially S.S. Kilham, P. Kilham, W. Lampert, M. Tilzer, W. Geller, and K.O. Rothaupt.

Ulrich Sommer

1

Toward a Darwinian Ecology of Plankton

Ulrich Sommer

Max Planck Institute for Limnology
Plön, Federal Republic of Germany

The time now is ripe to summarize current views and hypotheses about the mechanisms which drive succession in plankton communities, not only for the sake of "understanding" plankton, but also in order to use the world of plankton as a model for other kinds of ecological communities. Articles about the seasonal changes in species composition of phyto- and zooplankton abound in the literature. Unfortunately, they are mostly either descriptive or quite speculative. Although the patterns of seasonality within individual bodies of water can be quite regular, comparison between water bodies often leaves the impression of chaos. Some authors draw a pessimistic conclusion and consider the understanding of the mechanism of species shifts as an intractable problem (Harris, 1986). If taken to the extreme, this point of view rejects the idea that physiological ecology can be a legitimate enterprise in science. However, other authors are still optimistic and continue to try the traditional approach of the natural sciences to uncover causal relationships by the combined efforts of theoretical, experimental, and observational research. The intellectual debate between the proponents of these two attitudes has become increasingly hostile (cf. Peters, 1986; Lehman, 1986) and has reached the point where getting a referee from the opposing camp may seriously impair the publication chances of a manuscript or the funding chances of a grant proposal.

The authors of this volume, however, belong to the group of those who have not given up. We still believe that it is possible to understand the mechanisms driving species replacements. We derive this optimism from the substantial progress in this area that has been made during the last decades (Kerfoot, 1980; Reynolds, 1984; Lampert, 1985). This volume is intended to summarize and review this progress. Therefore, we will not present a collection of successional sequences or a typology of succession patterns for different types of water bodies. Instead, we will suggest some mechanisms that may drive species replacements in plankton communiities with the understanding that these mechanisms operate in all kinds of pelagic environments, although possibly with differences in relative importance. The reader is invited to compare our reviews with earlier ones (e.g., in Hutchinson, 1967; Smayda, 1980) to judge whether the progress since then justifies further hope for clarifying the mechanism of succession in plankton communities.

It is the common opinion of the contributors to this book that we have something to say not only to the plankton ecologists but also to ecologists in general. Many of the divisive issues in plankton ecology are also debated in general ecology (see Strong et al., 1983). The mechanisms discussed here are traditional topics of physiological and population ecology (physical control, nutrition, limitation of growth, competition, predation, parasitism). Many of the classic issues of theoretical ecology can be brought into the focus of experimental and field research more easily in plankton communities than anywhere else. Representative samples of planktonic organisms can be easily obtained. Estimates of abundance, biomass, and productivity are more reliable than in any other community, despite the increasing awareness of plankton patchiness. Cultivation of many plankton species is relatively easy and requires little space. Owing to their small body size, short generation times, and high growth rates, space and time requirements for experiments with them are modest, for example, in a 1-liter container, a phytoplankton competition experiment can reach steady state in one to two months while several years would be needed for herbs and several decades for trees. The small body size of plankton organisms makes it possible to use experimental populations of millions of individuals, which excludes any unwanted influence of inter-individual variability. The rapidity of population growth enables successional sequences to be produced within one season that contain as many stages as terrestrial vegetation succession does in the course of centuries. This means that the plankton ecologist can make repeated direct observations while the vegetation ecologist has to rely on indirect evidence from paleoecology or has to make the unlikely assumption that vegetation in different places represents different stages of the same sequence.

Beginning in the early days of limnology, the seasonal cycles of plankton abundances and species composition have been called "seasonal succession." The usage of the term "succession" has been quite loose and the analogies to the century-long, noncyclic process of terrestrial succession have rarely been explored (however, see Reynolds, 1980). The word "succession" can have two different meanings: the *loose concept* defines succession as a time series of species replacements; the *restrictive concept* reserves the term succession for those events in a chain of species replacements that are a consequence of preceding ones (Reynolds, 1980). Species shifts in direct response to external forcing factors do not qualify as succession. The restrictive definition of succession implies the concept of biotic interactions, i.e., that the presence and the activity of organisms have decisive influence on the living conditions of other organisms. For species shifts enforced by external perturbation (e.g., deepening of the mixed-water layer by wind and surface cooling) Reynolds reserves the terms "reversion" and "shift." Reversions are episodic and caused by unpredictable, short-term events; shifts are caused by regular, seasonally recurrent patterns of change in physical forcing factors.

1.1   Ecosystem Concepts

There are three different concepts which have been used to explain the functioning of communities and/or ecosystems. The *individualistic (Gleasonian) concept* assumes populations to respond independently to an external, physical, and chemical environment. Within this concept the restrictive definition of succession sensu Reynolds makes no sense. The Gleasonian point of view is quite common in phytoplankton ecology and quite rare in zooplankton ecology. The *superorganism (Clementsian) concept* views ecosystems as organisms of higher order and defines succession therefore as ontogenesis of this superorganism ("self-organization of an ecosystem," Margalef, 1978). Except for the work of Margalef, the explicit superorganism concept is nearly dead in scientific ecology but still very strong in political and ethical environmentalism. A more subtle form of Clementsianism has survived under the title "hierarchy theory" (Allen and Starr, 1982) which insists that the higher-level wholes (ecosystems, communities) have "emergent" properties that are independent of the properties of their lower-level components (e.g., populations).

The third concept of ecosystem and community organization is centered around the *nutrition* needed for survival and reproduction. The implication of this concept is that organisms have to compete for common resources and/ or to feed on each other. Interactions between populations are therefore primarily *negative* (competition, predation, parasitism), with symbiosis as a rather exotic case. Cooperation and division of labor for the sake of community stability or any other kind of community-level optimization have no place in this view of the world. The concept of negative interactions is already implicit in the third chapter of Darwin's *Origin of Species*. Therefore, I join Harper (1967) and call this approach to ecology "Darwinian." It is the classic domain of mathematical and experimental population biology (Lotka, Volterra, Gause) and has been disseminated among field ecologists primarily by Hutchinson and MacArthur. For mathematical ease, the emphasis has been put on equilibrium models for a long time.

Recently, dissatisfaction with the longstanding neglect of nonequilibrium states has led to a revival of Gleasonianism (e.g., Strong et al., 1983). I think, however, that most of the neo-Gleasonian critique originates in misunderstanding. The usual line of argument (cf. many chapters in Strong et al., 1983 and Price et al., 1984) runs as follows:

- First premise: Classic competition models have been equilibrium models.
- Second premise: Natural communities are normally not in equilibrium.
- Conclusion: Competition does not normally occur.

This syllogism unjustifiably identifies a hitherto overemphasized aspect of competition research with the process of competition as such. In the chapter on phytoplankton nutrient limitation and competition (Chapter 3) this point will be further developed in order to point out how competition can be an important causal factor in species replacements, even if competitive equilibrium is never fully achieved.

During the last decade a debate has broken out within the Darwinian camp as to whether "bottom-up" (limitation by resources) or "top-down" (control by predators) effects primarily control community dynamics. The chapters of this book are full of references supporting both of the two opposing views. Initially, many of the proponents of this debate assumed that "bottom-up" and "top-down" control were mutually exclusive alternatives. In many papers, the reader could identify the camp to which a researcher belonged just by reading the methods section. Top-down believers tended to prefer field experiments, often in enclosures where fish were added or removed. Bottom-up believers tended to prefer more traditional methods of comparative physiology with emphasis on culture experiments in the lab. Meanwhile, most researchers have realized that top-down and bottom-up control are not mutually exclusive, neither logically nor in reality. The selection of authors for this book clearly shows that I consider both to be potentially important. However, no attempt will be made to ascribe a certain percentage of relative importance to one of the two, as was done in a recent review by Sih et al. (1985). The rationale of such an approach lies in the assumption that trophic levels under heavy predation pressure are free of competition and vice versa. This view is deeply entrenched in ecological common sense, but is nevertheless false. As has been argued by Tilman (1983; see also Chapter 3) high mortality increases the resource requirements in order to maintain a reproductive rate in long-term balance with the mortality rate. Heavy predation thus alters the boundary conditions for resource competition, but does not switch off competition. Chapter 4 will discuss important indirect effects, such as predator-mediated alterations of the resource base, to which it is impossible to attach the labels bottom-up or top-down.

1.2 The PEG-model

In this book we will emphasize possible *mechanisms* of succession instead of *descriptions* of successional sequences. The examples used to illustrate the operation of these mechanisms will usually not be entire sequences of seasonal plankton succession but rather individual episodes, during which the mechanism under discussion is dominant and can most easily be demonstrated. In order to put these episodes into the context of a full seasonal sequence it may be helpful to describe the *PEG-model* (Sommer et al., 1986) of plankton seasonal succession. In several chapters of this book reference is made to the individual steps of this model. The PEG-model describes a model sequence of 24 steps of seasonal change in phytoplankton and zooplankton in an idealized lake. The steps are either necessary consequences of the previous ones (i.e., successional sensu Reynolds) or consequences of the regular seasonal change in external forcing factors (i.e., shifts sensu Reynolds). For each step a mechanistic explanation is provided.

The model emerged from a long-term discussion process among some 30 members of the Plankton Ecology Group (PEG) to which each participant contributed succession data from various lakes or ponds, including Lake Constance (Bodensee). In the course of the discussion the initial division between top-down and bottom-up views was overcome and a synthesis was reached. The model assumes certain boundary conditions: A fairly nutrient-rich, stratifying lake in which nutrients are depleted in the sequence P—Si—N. It is, however, intended to serve as a standard of comparison for a wider array of lakes. Comparison can be made in two ways. The *temporal sequences* can be compared and deviations can be detected and explained. If the explanation of deviation is found in the same mechanism as assumed in the model, but with different boundary conditions, this would not invalidate the model. Or, the temporal sequence as a whole can be neglected, and the individual steps of the model or a group of adjacent steps can be taken as *conditional statements* (if—then) about the mechanisms of species replacements in plankton communities.

The model sequence of 24 steps that comprise the PEG-model of plankton seasonal succession is as follows.

1. Toward the end of winter, nutrient availability and increased light permit unlimited growth of phytoplankton. A spring crop of small, fast-growing algae such as Cryptophyceae and small, centric diatoms develops.

2. This crop of small algae is grazed upon by herbivorous zooplankton species that soon become abundant due both to hatching from resting stages and to high fecundity induced by the high levels of edible algae.

3. Planktonic herbivores with short generation times increase their populations first and are followed by slower-growing species.

4. The herbivore populations increase exponentially up to the point at which their density is high enough to produce a community filtration rate, and therefore a cropping rate that exceeds the reproduction rate of phytoplankton.

5. As a consequence of herbivore grazing, the phytoplankton biomass decreases rapidly to very low levels. There then follows a "clear-water" phase which persists until inedible algal species develop in significant numbers. Nutrients are recycled by the grazing process and may accumulate during the clear-water phase.

6. Herbivorous zooplankton become food-limited and both their body weight per unit length and their fecundity decline. This results in decreases in their population densities and biomasses.

7. Fish predation accelerates the decline of herbivorous planktonic populations to very low levels and this trend is accompanied by a shift toward a smaller body size among the surviving crustaceans.

8. Under the conditions of reduced grazing pressure and sustained nonlimiting concentrations of nutrients, the phytoplankton summer crops start to build up. The composition of phytoplankton becomes complex due both to the increase in species richness and to the functional diversification into small "undergrowth" species (that are available as food for filter-feeders) and large "canopy" species (that are only consumed by specialist feeders such as raptors and parasites).

9. Then edible Cryptophyceae and inedible colonial green algae become predominant. They deplete the soluble reactive phosphorus to nearly undetectable levels.

10. From this time onward, algal growth becomes nutrient limited and this prevents an explosive growth of edible algae. Grazing by predator-controlled herbivores balances the nutrient-limited growth rates of edible algal species.

11. Competition for phosphate leads to a replacement of green algae by large diatoms that are only partly available to zooplankton as food.

12. Silica depletion leads to a replacement of the large diatoms by large dinoflagellates and/or Cyanophyta.

13. Nitrogen depletion favors a shift to nitrogen-fixing species of filamentous blue-green algae.

14.   Larger species of crustacean herbivores are replaced by smaller species and by rotifers. These smaller species are less vulnerable to fish predation and are less affected by interference with their food-collecting apparatus that can be caused by some forms of inedible algae. Accordingly, their population mortality is lower and their fecundity is higher than that of larger species.

15.   The small species of herbivores coexist under a persistent fish predation pressure and the increased possibility of food partitioning that is associated with the greater species complexity of the phytoplankton.

16.   The population densities and species composition of the zooplankton fluctuate throughout the summer, the latter being influenced by temperature.

17.   The period of autogenic succession is terminated by factors related to physical changes, including increased mixing depth resulting in nutrient replenishment and a deterioration of the effective underwater light climate.

18.   After a minor reduction of algal biomass, an algal community develops that is adapted to being mixed. Large unicellular or filamentous algal forms appear. Among them diatoms become increasingly important as autumn progresses.

19.   This association of poorly ingestible algae is accompanied by a variable biomass of small, edible algae.

20.   This algal composition together with some reduction in fish-predation pressure leads to an autumnal maximum of zooplankton including larger forms and species.

21.   A reduction of light-energy input results in a low or negative level of net primary production and an imbalance with algal losses which cause a decline of algal biomass to the winter minimum.

22.   Herbivore biomass decreases as a result of reduced fecundity due both to lower food concentrations and to decreasing temperature.

23.   Some species in the zooplankton produce resting stages at this time, whereas other species produced their resting stages earlier.

24.   At this period of the year, some cyclopoid species awake from their diapause and contribute to the overwintering populations in the zooplankton.

In the remaining eight chapters of this volume, various aspects of species succession in phyto-, zoo-, and bacterioplankton will be discussed and evaluated against the available data.

References

Allen, T.F.H. and Starr, T.B. 1982. *Hierarchy: Perspectives for Ecological Complexity.* University of Chicago Press, Chicago.

Harris, G.P. 1986. *Phytoplankton Ecology.* Chapman and Hall, 384 pp.

Hutchinson, G.E. 1967. *A Treatise on Limnology,* Vol. 2, *Introduction to Lake Biology and the Limnoplankton.* Wiley, New York, 1115 pp.

Kerfoot, W.C. (editor) 1980. *Evolution and Ecology of Zooplankton Communities.* University Press of New England, 793 pp.

Lampert, W. (editor) 1985. *Food Limitation and the Structure of Zooplankton Communities. Archiv für Hydrobiologie Beiheft* 21, 497 pp.

Lehman, J.T. 1986. The goal of understanding in limnology. *Limnology and Oceanography* 31: 1160–1166.

Margalef, R. 1978. *Ecologia.* Omega, Barcelona, 951 pp.

Peters, R.H. 1986. The role of prediction in limnology. *Limnology and Oceanography* 31: 1143–1159.

Price, P.W., Slobodchikoff, C.N., and Gaud, W.S. (editors) 1984. *A Novel Ecology.* Wiley, New York, 515 pp.

Reynolds, C.S. 1980. Phytoplankton assemblages and their periodicity in stratifying lake systems. *Holartic Ecology* 3: 141–159.

Reynolds, C.S. 1983. *The Ecology of Freshwater Phytoplankton.* Cambridge University Press, Cambridge, 384 pp.

Sih, A., Crowley, P., McPeek, M., Petranka, J., and Strohmeier, K. 1985. Predation, competition and prey communities: A review of field experiments. *Annual Review of Ecology and Systematics* 16: 269–311.

Smayda, T.J. 1980. Phytoplankton species succession. pp. 493–469 in Morris, I. (editor), *The Physiological Ecology of Phytoplankton.*

Sommer, U., Gliwicz, Z.M., Lampert, W., and Duncan, A. 1986. The PEG-model of seasonal succession of planktonic events in fresh waters. *Archiv für Hydrobiologie* 106: 433–471.

Strong, D., Simberloff, D., and Abele, L. (editors) 1983. *Ecological Communities: Conceptual Issues and Evidence.* Princeton University Press, Princeton.

Tilman, D. 1982. *Resource Competition and Community Structure.* Princeton University Press, Princeton, 296 pp.

2

Physical Determinants of Phytoplankton Succession

Colin S. Reynolds

Freshwater Biological Association
Windermere Laboratory, Ambleside
Cumbria, U.K.

Just as the composition of phytoplankton assemblages depends upon the presence and relative abundances of populations of individual species, so temporal changes in their composition are brought about by differences in the relative rates of augmentation and attrition of each population. These rates respond to a complex of interactions among various physical, chemical, and biotic environmental factors, operating at a variety of intensities and frequencies. This chapter addresses the impact of essentially physical variables on the population dynamics of individual species and it seeks to establish the particular properties of the organisms for which each selects. Factual information relating the performances of algae to quantifiable aspects of the physical environment is drawn largely from observations made in controlled laboratory experiments. Realistic potential combinations of the relevant physical factors are suggested in order to simulate the likely responses of specific populations in natural waters. The outcomes of such simulations are then compared with the PEG-model of phytoplankton succession (see Section 1.2) propounded by Sommer et al. (1986), which was originally elaborated to explain the pattern of seasonal change in species dominance, as regularly observed in Lake Constance (the Bodensee). A concluding section assesses the role of physical factors in regulating seasonal succession of phytoplankton generally. At the end of the chapter, beginning on page 52, there are three appendices. The first one, Appendix 2.1, defines the units used in this chapter. The second, Appendix 2.2, identifies the symbols used, and Appendix 2.3 explains the abbreviations used for algal names.

2.1   Relevant Physical Properties of the Aquatic Environment

Both because liquid water is a major constituent of all living organisms and because it is the medium in which pelagic organisms pass most of their lives, its physical properties are central to the physiological and evolutionary ecology of phytoplankton. Peculiarities in the structural chemistry of water molecules, notably their tendency to exist in "aquo" complexes having the general formula $(H_2O)_n$, determine that water is a relatively dense, viscous liquid at normal temperatures. Water also has a relatively high specific heat—that required to raise 1 g through 1 degree Kelvin (K)—poor thermal conductivity, and high latent heats of fusion and vaporization, as shown in Table 2.1. The mechanical support afforded to aquatic organisms and the substantial buffering against high-frequency fluctuations in external energy thus complement the solvent and transparent properties of water (see Section 2.2) as being beneficial to life in water.

Interactions between water bodies and their external environments nevertheless generate internal variability, at differing scales and intensities. Because these interactions constitute an essential component of the environment of phytoplankton, it is important to first establish the dimensions of their variability.

2.2   Physical Factors Impinging on Phytoplankton Dynamics

The principal role of physical factors in regulating the dynamics and structure of planktonic communities lies in the interactions of fluxes of radiant energy and the changes in heat storage with the dissipative forces of mechanical

Table 2.1   Some physical properties of pure liquid water

Property	Symbol	Value
Temperature	θ	273 to 373 K (0 to 100°C)
Coefficient of thermal expansion	γ	[a]1.6×10^{-5} K^{-1} [b]2.61×10^{-5} K^{-1}
Density	ρ_w	[c]999.972 kg m^{-3} [b]997.062 kg m^{-3}
Absolute viscosity	η	[a]1.5138 kg m^{-1}s^{-1} [b]0.8909 kg m^{-1}s^{-1}
Thermal conductivity	ζ	58.55 J m^{-2} K^{-1}s^{-1}
Specific heat	σ	4.186×10^3 J kg^{-1} K^{-1}
Latent heat of fusion	λ_f	3.33875×10^5 J kg^{-1}
Latent heat of evaporation	λ_e	2.243277×10^6 J kg^{-1}

[a] At 5°C.
[b] At 25°C.
[c] At 3.95°C.

kinetic energy (Reynolds, 1987). These processes are represented diagrammatically in Figure 2.1.

Surface heating, temperature, and buoyancy generation   In most water bodies, the primary source of heat derives from the sun. The flux entering a lake across its surface is liable to extensive modification with respect to direct solar energy (Figure 2.2). The reference point is the solar constant—the energy incident upon a notional surface outside the atmosphere and set perpendicular to the rays of the sun. Its maximal value has been approximated to be 1.94 cal cm^{-2} min^{-1} (Gates, 1962), or about 1350 W m^{-2}. The instantaneous value at a given location is a function of the elevation of the sun with respect to the surface of the earth and, hence, the potential daily radiation flux at that point is strongly dependent upon latitude and season (see Figure 2.2a). Even through a clear, dry atmosphere, substantial heat absorption occurs so that the maximum radiation flux on any lake (Q_s) (see Appendix 2.2) scarcely exceeds 900 W m^{-2} (Gates, 1972). Clouds and dust in the atmosphere absorb, scatter, and further deplete the heat income, although diffuse and reflected energy (Q_n) may contribute to the net flux, as does atmospheric long-wave radiation (Q_a). At the lake surface, some of the heat is immediately lost to the atmosphere through reflection ($-Q_r$), primarily as a function of the proportion and angle of incidence of direct radiation (Figure 2.2b), and through long-wave back radiation from the water ($-Q_b$). A potentially large proportion of the heat transfer across the water surface is consumed by evaporation,

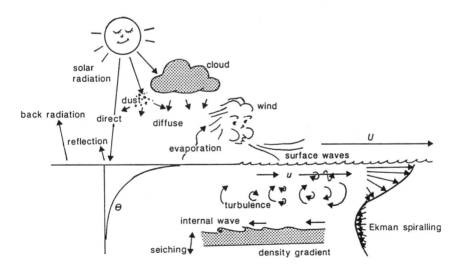

Figure 2.1 Diagram showing the major processes affecting the balance and distribution of heat within the upper water column of a lake. Based on figures in Harris (1986) and Spigel et al. (1986).

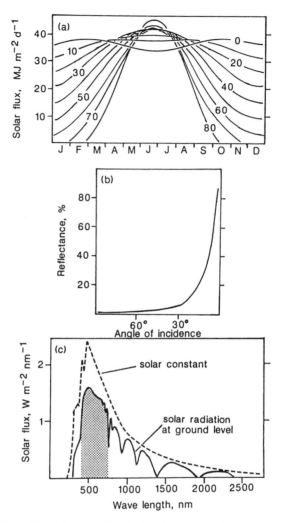

Figure 2.2   (a)The daily integrals of undepleted solar radiation received by a notional horizontal surface above the atmosphere, as a function of time of year and geographical latitude (northern hemisphere only shown); (b) reflection and backscattering of solar radiation at the surface of lakes, as a function of the angle of incidence; (c) the spectrum of the solar flux arriving at the lake surface, showing the major bands of atmospheric absorption; the visible spectrum is indicated by shading. Based on figures in Gates (1962) and information in Larkum and Barrett (1983).

$-Q_e$ ($= V_e \rho_w \lambda_e$ where V_e is the volume evaporated) and which should properly include the heat lost in vapor (Anderson, 1952). Evaporative heat losses are accordingly difficult to measure accurately or to model conveniently, as they depend upon differences in water and air temperatures, the saturation capacity of the air, and the rate of its renewal immediately above the water. $-Q_e$ can also be positive if there is condensation at the water surface. The signs of the terms are chosen in such a way that summation expresses either a net heat flux into the water (Q_T positive) or out of it (Q_T negative):

$$Q_T = Q_s + Q_n + Q_a - Q_r - Q_b \pm Q_e \pm Q_c \qquad (2.1)$$

Eq. 2.1 is a simplified version of Hutchinson's (1957) derivation which also included terms distinguishing components compounded here and also advective effects (ignored here). It is more explanative than deterministic, owing to inherent difficulties in precise measurement of the components. At a theoretical steady state, the gain and loss terms are balanced, $Q_T = 0$, and the water temperature is stabilized. In practice, the rate of change in the energy influx is too rapid for a short-term equilibrium to be established. Over a period of several days of similar conditions some stable, ambient temperature range will become manifest. Equally, the ambient range alters perceptibly in response to a change in diurnal integral of heat income. Such variations form the basis of annual budgets of heat exchange in lakes; for examples see Hutchinson (1957), Kuhn (1978), and Marti and Imboden (1986).

The net flux of solar radiation that crosses the lake surface is attenuated exponentially with depth, so that the heat gain tends to be confined to the near-surface waters. The heating at a given depth beneath the surface, z, approximates to:

$$Q_z = Q^*_T e^{-kz} \qquad (2.2)$$

where e is the base of natural logarithms and k is an exponent of the absorption of heat: Q^*_T is that fraction of the net heat flux, Q_T, which penetrates beyond the top few millimeters of the water column. Roughly, Q^*_T is about half the incoming short-wave radiation, Q_s, the remaining heat fluxes occurring to and from the surface-boundary layer. The heating below this surface layer, i.e., the heat absorption per unit volume per unit time is:

$$\partial Q = -\frac{dQ_z}{dz} + kQ^*_T e^{-kz} \qquad (2.3)$$

The absorption is not uniform across the spectrum, so the composition of the downwelling heat alters with respect to the radiation impinging on the surface (see Figure 2.2c). If it is assumed that the water column is uniformly mixed throughout its depth, H, and that at its lower boundary, the remaining heat flux, $Q_z(H) = Q^*_T e^{-kH}$ is negligible, then the rate of temperature change within the mixed layer will be:

$$d\theta/dt = Q^*_T (H\rho_w\sigma)^{-1} \qquad (2.4)$$

where σ is the specific heat of water (in J kg^{-1} K^{-1}) and ρ_w is its density (in kg m^{-3}).

The change in temperature causes a change in water density, viz.

$$d\rho_w/dt = -\gamma\rho_w \; d\theta/dt = -Q^*_T \; (H\sigma)^{-1} \tag{2.5}$$

γ is the coefficient of thermal expansion of water at the given temperature.

In many instances, however, the net heat flux, Q^*_T, raises the temperature and reduces the density of near-surface layers preferentially. Provided $\theta > 4°C$ (277 K; the temperature of maximum density), $\gamma > 0$ and $d\rho_w/dt < 0$, the water column will become *thermally stratified*. The same is true when $Q^*_T < 0$ and $\theta < 4°C$: $\gamma < 0$ but $d\rho_w/dt$ is < 0. Conversely, instability persists when $Q^*_T < 0$ but $\theta > 4°C$ and when $Q^*_T > 0$ while $\theta < 4°C$.

In a nonmixing column with $Q^*_T > 0$ and $\theta > 4°C$, the temperature change may be expressed:

$$d\theta/dt = \partial Q(\rho_w\sigma)^{-1} = k \; Q^*_T \; e^{-kz} \; (\rho_w\sigma)^{-1} \tag{2.6}$$

The corresponding change in density is:

$$d\rho_w/dt = (-\gamma k \; Q^*_T) \; e^{-kz} \; \sigma^{-1} \tag{2.7}$$

Starting from a homogeneous temperature profile, the change in density $\Delta\rho_w$ after time Δt (for otherwise constant conditions) approximates to:

$$\Delta\rho_w = -[(\gamma k Q^*_T) \; e^{-kz} \; \sigma^{-1}] \; \Delta t \tag{2.8}$$

The gradient of density thus generated is derived:

$$\Delta\rho_w/dz = d\rho_w/dz = (\gamma k^2 \; Q^*_T) \; e^{-kz} \; \sigma^{-1} \tag{2.9}$$

This gradient imparts stability to the water column, maintained by the buoyancy of the heated, near-surface water, relative to the denser, cooler water below. The opposite applies below 4°C (277 K). In waters with a high content of dissolved salts, density differentiation may owe more to a gradient of ionic strength than to temperature. In each case, however, the buoyant force imparted in the upper layers may be expressed as negative gravitational acceleration, $-g(\Delta\rho_w/dz)$ kg m^{-3}s^{-2}.

Mixing processes   Several mechanisms exist that generate motion within the water mass and can, either singly or in concert, directly oppose the stability forces. These include convectional cooling, wind-driven turbulence, and inertial effects induced by the earth's rotation (Coriolis' forces). Each has important separate effects which should be distinguished.

Convection   During the hours of darkness and at other times when the influx of solar radiation falls below the rate of efflux from the surface (i.e., $Q_T < 0$), heat is lost to the atmosphere from the surface of the water column. As the surface water cools, its density increases until it becomes heavier than the water immediately beneath it and, hence, unstable: the cooled water tumbles through the water column, accelerating as it does so, until it reaches a depth of approximate isopycny (i.e., where the descending water has the same density). In consequence, warmer water is displaced upward. The process continues so long as net heat imbalance between water and atmosphere persists, all the time deepening the layer of convectional mixing and altering the density gradient. The energy of this penetrative convection (B) is derived in eq. 2.10:

$$B = (g\gamma Q_T)(\rho_w\sigma)^{-1} \qquad (2.10)$$

The units are W m² s⁻² J⁻¹, which simplifies to m² s⁻³. The vertical velocity scale (in m s⁻¹) of convective mixing through a column h m in depth is equivalent to (B h m³ s⁻³). The examples of diurnal stratification and mixing by nocturnal, penetrative convection shown in Figure 2.3 are based on the data of Reynolds, Tundisi, and Hino (1983) for Lagoa Carioca, eastern Brazil (latitude 19°S); this is a small, tropical lake, well-sheltered from wind action. The overnight loss of heat from the water column (Figure 2.3a to Figure 2.3b) was equivalent to 14.6 MJ m⁻²; assuming the rate of heat loss to have been no less during the daytime period, a mean flux of 337 J m⁻² s⁻¹ (= W m⁻²) may be applied. Then the observed daytime heat gain (Figure 2.3b to 2.3c: 17.0 MJ m⁻² in 12 h, or 393 W m⁻²) is net of a simultaneous solar-energy income, averaging 730 W m⁻² over the daylight period. Solving eq. 2.10, the energy of night-time convection may be calculated at 2.36 × 10⁻⁷ m² s⁻³: putting h = 4.5 m, the velocity scale is found to have been 0.0102 m s⁻¹. In

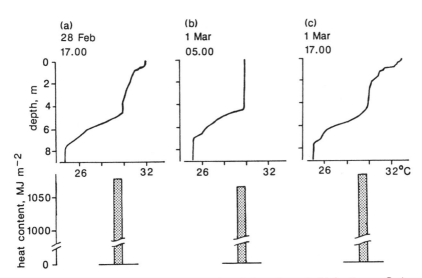

Figure 2.3 Vertical profiles of temperature in a sheltered, tropical lake (Lagoa Carioca, Brazil: 19°S 42°W) taken (a) in the evening, (b) just before dawn, and (c) in the evening of the next day, together with representations of the heat content of the water column, used in the calculation of the heat exchanges (see Section 2.2). Data of Reynolds et al. (1983).

this example, the daily heat income and loss are approximately balanced, suggesting that the temperature structure of the lake had achieved some seasonal steady state. The sharp temperature gradient (thermocline) below 4.5 m was, similarly, a seasonally persistent feature that developed during a previous phase of net warming ($Q_T > 0$) and liable to no serious erosion until Q_T fell to negative values during the winter months. As elsewhere, seasonal variations in radiation income and the ultimate imbalance between gains and losses underscore the basis of annual heat budgets, with characteristic periods of net warming, net cooling, thermocline formation, and erosion.

Wind-induced motion   Unlike the previous example of a small, sheltered, tropical lake, where water movements are dominated by large diel heat transfers and deep penetrative convection, most lakes are frequently subject to wind action on the surface. The movement of air over water is largely unimpeded but for the influence of adjacent topographical features, such as islands, mountains or forests, that may provide local shelter or exaggerate the wind action by funnelling. Away from the shores, the passage of the wind over the water surface nevertheless tends to generate a parallel drift of water immediately beneath the surface. Owing to its greater viscosity, however, the water moves rather more slowly than the parallel wind, setting up a force in the perpendicular plane. This force is known as shear, denoted by τ, and is proportional to the sharp gradient in the velocities of the air (U) and of the water (u) across the boundary. Thus,

$$\tau = -\eta \partial u / \partial z \tag{2.11}$$

where η is the viscosity in N m^{-2}s and the velocity gradient is measured in ms^{-1}m^{-1} (= s^{-1}). The shear force tends to accelerate the slower fluid (in this case, the water) and decelerate the faster (the air). Therefore, τ may also be interpreted as a transfer of momentum, ρu, through the velocity gradient $\partial u / \partial z$. The change in momentum can be represented:

$$\partial(\rho u)/\partial t = -\partial \tau / \partial z \tag{2.12}$$

The transfer of momentum may be characterized by a velocity, u^* (the friction, or shear velocity), and the densities of the adjacent media. For the water-side of the boundary,

$$\tau = (\rho_w u^*)(u^*) = \rho_w (u^*)^2 \tag{2.13}$$

Following similar reasoning in respect to the windspeed above the water,

$$\tau = \rho_a (U^*)^2 \tag{2.14}$$

where ρ_a is the density of the air (1.2 kg m^{-3}).

In this way, the stress applied across the boundary layer introduces a flux of kinetic energy (J_k) from the faster to the slower medium, which is itself scaled to u^*. Assuming J, U, and u to be aligned and $U > u$,

$$J_k = \tau(u^*) = \rho_w(u^*)^3 \tag{2.15}$$

So simple a formulation belies the complexity of the mechanisms of actual transfer of kinetic energy to the water. Above a critical windspeed ($U \geq 3.5$ ms^{-1}), the interface becomes broken into waves which, in turn, subject the surface drift to frequent accelerations and decelerations ($\pm u'$) about its mean downwind velocity, u, and introduce simultaneous rotational displacements in the vertical (z) plane, upward and downward, with velocities of $\pm w'$, and also in the perpendicular horizontal (y) plane, with velocities of $\pm v'$. This interaction between the horizontal and vertical velocity components of the surface drift provides the major source of turbulent kinetic energy (TKE) in the wind-stressed water column, so its quantitative evaluation is important. Modern approaches to the quantification of TKE are based on time-averages of fluctuations in instantaneous velocity components, within ranges ($u \pm u'$), ($0 \pm v'$) and ($0 \pm w'$). The arithmetic mean of a series of (e.g.) $+u'$ to $-u'$ tends to zero so the time-averaged components quickly approach steady states, (u, 0, 0, respectively). The squares of the series, however, are nonvanishing: $(u')^2$ is therefore a useful measure of intensity of turbulence along the axis, as is $(v')^2$ in the y direction and $(w')^2$ in the z direction. The mean turbulent velocity, u_t, is given by the time-averaged root of the horizontal and vertical velocity fluctuations,

$$u_t = [(u')^2 + (v')^2 + (w')^2]^{1/2} \tag{2.16}$$

The kinetic energy due to turbulent velocity fluctuation is equal to 1/2 (mass $\times$ velocity squared): thus,

$$TKE = 1/2\, \rho_w\, [(u')^2 + (v')^2 + (w')^2] \tag{2.17}$$

The scale of turbulent intensity is related to the time-averaged product of the horizontal and vertical components, usually written as $<u'\, w'>$. The friction velocity, u^*, is equivalent to the square root of the turbulent intensity at the surface; hence the turbulent intensity may be directly linked to the wind action on the surface, through the following relationships developed by Denman and Gargett (1983):

$$u^* = <u'\, w'>^{1/2} \tag{2.18}$$

$$\tau = \rho_a c U_{10}^2 = \rho_a(U^*)^2 = \rho_w(u^*)^2 \tag{2.19}$$

where τ is the work (i.e., mass $\times$ acceleration $\times$ distance) applied to the lake (kg m^2 s^{-2} m^{-3}), U_{10} is the wind velocity (m s^{-1}) measured 10 m above the survace and c is the coefficient of frictional drag of the water on the wind. This relationship between u^* and U_{10} is plotted in Figure 2.4a.

Dissipation of TKE   In fully developed turbulence, the kinetic energy introduced at the surface is progressively dissipated through a series of nested eddies ("the Kolmogoroff eddy spectrum") of diminishing size and velocity,

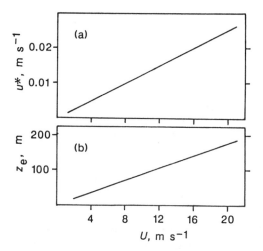

Figure 2.4 Relationship between (a) the friction velocity, u^* and (b) the depth of frictional resistance z_e at latitude 48°N assuming unconstrained mixing of a boundless water column, to wind speed, measured 10 m above the water surface (U_{10}). Plotted (respectively) from equations 2.19 and 2.23.

culminating at the molecular level (viscosity) with the discharge of the kinetic energy as heat. Nevertheless, while the driving energy persists, the flow is dominated by the largest eddies generated; near the surface, at least, smaller eddies are constantly liable to be displaced or re-entrained. A local "steady state" is struck when the input of TKE into the larger eddies is balanced by the rate of its dissipation, E, through the smaller scales. The depth of water actively mixed by the wind may be regarded as being subject to uniform stress and, like the wind field above the water, can be treated as a single boundary layer; its properties are directly related to the dimensions (l_m) and friction velocities (u^*) of the larger eddies generated. For instance, Stommel (1949); see also Hutchinson, 1957) has shown that turbulent energy varies as $l_m^{2/3}$ and the average velocity as $l_m^{1/3}$. Of particular interest to the present development is the relationship between the boundary-layer mixing to the eddy length, l. Provided that there is no physical barrier to mixing (i.e., the water is sufficiently deep to allow free turbulent penetration), smaller and smaller eddies can be propagated downward to a depth, z_m, representing the base of the boundary layer. As the flow becomes more viscous, so the velocity is also diminished exponentially. The depth of the mixed column, h_m, is directly scaled to l_m:

$$h_m = l_m/\kappa \tag{2.20}$$

where κ is von Karman's constant, the value of which has been consistently found to be 0.40. The vertical gradient of horizontal velocity through the mixed layer is then:

$$u/z = u^*/\kappa h_m \qquad (2.21)$$

Thus, a wind stress of τ kg m^{-1} s^{-2}, generating a turbulent intensity of $(u^*)^2$ (or τ/ρ_w m^2 s^{-2}), supplies TKE that is dissipated through a mixed layer with the horizontal velocity configuration denoted by $\partial u/\partial z$ m s^{-1} m^{-1}. The rate of dissipation is expressed:

$$E = (u^*)^2 \, (\partial u/\partial z) = (u^*)^3/\kappa h_m \qquad (2.22)$$

Inertial effects   Wind mixing at the scale of whole lake basins is compounded by inertial motion induced by the rotation of the earth (Coriolis' forces), which causes near-surface currents to be deflected, to the right of the wind in the northern hemisphere, and leftward in the southern. The extent to which the deflection becomes apparent depends upon the relationship between the energy of the wind mixing, $(u^*)^2$, and the downwind length of the lake, L, over which it operates. Inertial effects are well known in the sea but are said to be prevalent only in the largest lake basins (Hutchinson, 1957) and are often ignored in consideration of mixing in small lakes. However, they can be significant in the behavior of smaller lakes at low-wind stresses (see George, 1981). It is inevitable that a wind-driven surface current (u) must be compensated by a subsurface return: the forces determining the vertical location and direction of flow counterposed to the surface wind-drift are important to the consideration of boundary-layer mixing. In the simplest case, in a water body of unlimited depth and lacking any density gradient, the "depth of frictional resistance," z_e, where the current is directly opposed to the surface drift, is defined by:

$$z_e = \pi(C/\omega \sin \phi \cdot \rho_w)^{1/2} \qquad (2.23)$$

where ϕ is the geographical latitude and ω the angular velocity of the rotation of the earth; C is the coefficient of eddy viscosity or momentum transmitted through the velocity gradient at a given stress:

$$\tau = C\partial u/\partial z \qquad (2.24)$$

By rewriting eq. 2.24 and expressing τ in terms of u^* (eq. 2.19) and in terms of eddy size and velocity (eqs. 2.20 and 2.21) and then relating the wind stress τ back to the velocity of the wind (eq. 2.19), the following relationships may be derived:

$$C = \tau(\partial u \, z)^{-1} = \rho_w(u^*)^2 \, (\partial u/\partial z)^{-1} = \rho_w \, l_m^2 \, (\partial u/\partial z) \qquad (2.25a)$$

$$= \rho_w \, (u^*)^2 \, (u^*/l_m)^{-1} = \rho_w \, u^* l_m \qquad (2.25b)$$

$$= \rho_a \, c(U_{10})^2 \, (\partial u/\partial z)^{-1} \qquad (2.25c)$$

Solving for C yields a general value of $1.56 \times 10^{-3}\,U_{10}^2\,(\partial u/\partial z)^{-1}$ kg m^{-1} s^{-1} and the average velocity gradient that satisfies both eq. 2.21 and eq. 2.22 is 0.00363 m s^{-1} m^{-1}; C may be taken as being numerically equivalent to 0.43 U_{10}^2. Substituting into eq. 2.23 values for C in respect of windspeeds of 2–20 m s^{-1} and putting $\phi = 48°$N (the latitude of Lake Constance) and $\omega = 0.729 \times 10^{-4}$ radians s^{-1}, z_e may be evaluated at 17–177 m; h_e is the thickness of this layer (also known as the Ekman layer) from the surface to z_e (see Figure 2.4b).

The formation of density gradients   Above 4°C, thermal and kinetic energy fluxes counteract when net heat income and wind stress on the surface occur simultaneously. The turbulent motion distributes the incoming heat away from the immediate subsurface layer; at the same time, the downward propagation of turbulent eddies is increasingly resisted by buoyancy forces as their velocities weaken with depth. The relative magnitudes of the density and velocity components are instantaneously expressed by the gradient Richardson number, Ri_g. The following derivation owes to Smith (1975): The kinetic energy of a small volume of water, V', due to vertical velocity fluctuation (w') is equal to $0.5\,\rho_w\,V'(w')^2$; eqs. 2.18 and 2.21, by analogy, give $w' = l(\partial u/\partial z)$, so that the kinetic energy is equivalent to $0.5\,\rho_w\,V'\,l^2(\partial u/\partial z)^2$. The force (mass $\times$ acceleration) required to move the volume against a density gradient, $\partial \rho_w/\partial z$), is equivalent to $V'g\,l\,(\partial \rho_w/\partial z)$ and the work done (force $\times$ distance) is $0.5\,V'g\,l^2\,(\partial \rho_w/\partial z)$. If the energy available is insufficient to meet the work requirement, the turbulent motion subsides. Ri_g expresses the work requirement to the energy availability:

$$Ri_g = [0.5\,V'gl^2\,(\partial \rho_w/\partial z)]\,[0.5\rho_w\,V'l^2\,(\partial u/\partial z)^2]^{-1}$$
$$= (g/\rho_w)\,(\partial \rho_w/\partial z)\,(\partial u/\partial z)^{-2} \qquad (2.26)$$

$Ri_g > 1$ is indicative of zero turbulence but turbulent motion is substantially weakened in the range $0.25 < Ri_g < 1$. Within this range a water column becomes separated by a relatively steep transition between water that is actively mixed ("the surface-mixed layer") and a stable, deeper layer.

Thus defined, the surface-mixed layer may be regarded as an instantaneously discrete, though temporally variable, hydrological unit, functioning as a boundary zone between the surface, open to atmospheric exchanges of kinetic and radiant energy, and the relatively inert water at depth. Its properties have been extensively investigated in recent years (Dillon and Caldwell, 1980; Oakey and Elliott, 1982; Imberger, 1985; Spigel, Imberger, and Rayner, 1986). Motion in the surface mixed layer continues to be dominated by the flux of kinetic energy (J_k in eq. 2.15) and by its turbulent dissipation (eq. 2.22), while simultaneously distributing the net heat flux, Q_T, uniformly through its depth. Were these processes to reach equilibrium, the depth of the mixed layer would stabilize, with an approximately constant thickness, h_b. The absorption of heat into this layer causes a reduction in its density (see eq. 2.5) of $-\gamma\rho_w \cdot \partial\theta/\partial t$. As a result of this expansion, the center of gravity of the layer (at the center of its volume, $h_b/2$) is raised a short distance $\partial h_b/2$, with a concomitant change in its potential energy, J_{pot}:

$$\partial J_{pot} = g\rho_w h_b \cdot \partial h_b/2 = \tfrac{1}{2}\, g\gamma\rho_w (h_b)^2 \cdot \partial\theta \qquad (2.27)$$

Assuming the warming to be homogeneous, the temperature change per unit time $(\partial\theta/\partial t)$ is given from eq. 2.4 as $Q^*_T\,(h_b\rho_w\sigma)^{-1}$. Combination of these derivations yields an expression for the flux of buoyancy generation per unit mass per unit time. Thus:

$$J_{pot} = \tfrac{1}{2}\, g\gamma h_b\, Q^*_T \cdot \sigma^{-1} \qquad (2.28)$$

At the proposed equilibrium, J_k and J_{pot} are balanced. With appropriate evaluation of the components, calculation of the mixed layer depth is straightforward:

$$\tfrac{1}{2}\, g\gamma h_b\, Q^*_T \cdot \sigma^{-1} = \rho_w\,(u^*)^3$$

and solving for h_b,

$$h_b = 2\sigma\rho_w\,(u^*)^3\,(g\gamma Q^*_T)^{-1} \qquad (2.29)$$

The term h_b is an important theoretical quantity, known as the *Monin-Obukhov length*: it describes the depth to which the water column can be mixed under the influence of a given wind stress and heat influx. The value of h_b increases with increasing wind (U), through its relation to u^*, and/or decreasing Q^*_T. In reality, there is sufficient lag in the transfer of a dielly varying heat flux, with a superimposed variation due to changing cloud conditions and fluctuating wind velocity, for the true depth of the mixed layer to be often out of equilibrium with the governing forces. The ratio between the actual mixing depth, h_m, and the potential equilibrium depth, h_b, nevertheless predicts the mixed-layer behavior. If $h_m/h_b > 1$, buoyancy forces are dominant and h_m should decrease; if $h_m/h_b < 1$, turbulence dominates and h_m should increase.

The instantaneous balance between the buoyant resistance of a water column and the wind stress applied to its surface is also given by its bulk Richardson number, Ri_b (Phillips, 1977):

$$Ri_b = (\Delta\rho_w g h_m)\,[\rho_w (u^*)^2]^{-1} \qquad (2.30)$$

where $\Delta\rho_w$ is the jump in density between the surface-mixed layer and the underlying mass of water; the sharp density gradient (thermocline) between them, $\partial\rho_w/\partial z$, effectively prevents downward progress of turbulent eddies. The units of both the numerator and the denominator are $kg\ m^{-1}s^{-2}$, so the ratio is a dimensionless number with values ranging from 10 to > 1000. An increasing ratio is indicative of increasing stability (increasing buoyancy, decreasing turbulence). A decreasing ratio accompanies a net loss of heat and buoyancy or increasing wind stress, weakening the resistance of the density stratification to turbulent eddies. Those eddies with sufficient energy to strike the thermocline surface are quickly reversed but the impacts set up an oscillation. The stronger the stratification then the greater is the frequency of the oscillation. This frequency (also known as the Brunt-Väisälä or buoyancy frequency, N, in radians s^{-1}) is a useful measure of the stability of the stratification:

$$N = [(g/\rho_w)\,(\partial\rho_w/\partial z)]^{1/2} \qquad (2.31)$$

The period of complete oscillation (i.e., through 2π radians) is given by:

$$t_b = 2\pi/N \qquad (2.32)$$

Temporal variability in mixing   It has been shown that, in natural systems, heat flux and wind stress are highly variable, both at the diel and the seasonal scales, to which variability the depth of mixing is particularly sensitive. Increasing heat content and/or reducing wind stress permit fresh gradient formation nearer the surface, leading to the characteristic step-wise temperature structure of a well-developed metalimnion (Figure 2.5a). Reducing heat income or a net convectional heat loss weakens the buoyancy effect so the stratification is less resilient to wind action (Figure 2.5b). Increased wind stress raises the TKE input and larger eddies with greater vertical velocity assail the lower boundary (Ri_g and N decrease), while the horizontal return increases the shear stress across the thermocline, setting up internal waves. In this way, the deeper water is eaten away and entrained into the mixed-layer circulation (Blanton, 1973; Niiler and Kraus, 1977): the thermocline effectively migrates downward to a new equilibrium (Figure 2.5c).

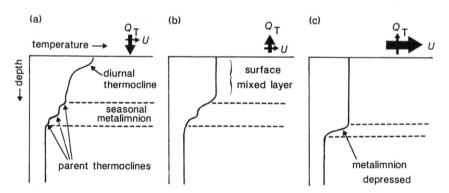

Figure 2.5   Diagrams representing various stages in the stratification of a lake: (a) Strong, stable stratification with a stepped metalimnion of thermoclines "accumulated" during net heating of the surface layer; (b) the effect of net loss of heat, with or without light wind mixing, from the water column is to establish a clearly defined, uniform surface layer which may be further deepened by more severe convectional heat loss and/or (c) wind stirring.

Seasonal differences in radiant energy income and the rate of heat exchange with the atmosphere are apparent in almost all lakes, the extent of the variability increasing with latitude (see Figure 2.2a). In addition, frequent variability in daily wind runs is a characteristic of most latitudes; in temperate latitudes—where the climate is dominated by unequal heating over land and sea, by the circumpolar rotation of air masses, by the passage of fronts, and by the attendant alternations in air pressure—cyclical weather events have an apparent periodicity of 4 to 11 days (Trimbee and Harris, 1983), though not necessarily with consistent intensity.

The outcome of this variability is that the thermal structure of lakes undergoes striking diurnal and annual oscillations. The general dynamics of mixed-layer variability are conveniently encapsulated in the dimensionless Wedderburn number (W), introduced by Imberger and Hamblin (1982). This relates the buoyant resistance of the surface boundary layer, $gh\Delta\rho_w$, to the force of wind acting thereon, $\rho_w(u^*)^2Lh^{-1}$, where L is the distance in m across the lake in the direction of the prevailing wind and Lh^{-1} is the aspect ratio of the mixed layer. Thus,

$$W = [gh^2\Delta\rho_w] [\rho_w(u^*)^2 L]^{-1} \qquad (2.33)$$

When $W = 1$, the counteracting forces are in equilibrium. For a given shear stress, net increase in buoyancy ($\Delta\rho_w$) must be compensated by a diminution in h_m, to maintain unity. Weakening of buoyancy, brought about by net cooling, causes W to fall below 1, leaving the surface layer liable to rapid deepening when wind action increases. The Wedderburn number is thus particularly useful in predicting the depth of the mixed layer in a given lake and variable capacity for buoyancy generation. Rewriting eq. 2.33, with $W = 1$,

$$h_m = \{[\rho_w(u^*)^2 L] [g\Delta\rho_w]^{-1}\}^{1/2} \qquad (2.34)$$

The caveat must be applied, however, that when solution of h_m exceeds either h_e or H, the depth of the entire water column, then the smallest value is indicative of the true depth of mixing.

Mixing times   The time scale for vertical movement of water (t_m) through a mixed-water column is given by the distance travelled divided by the average velocity. In uniform-stress layers, unbounded by a significant density gradient, the average passage time through the largest eddies corresponds to:

$$t_m = l_m u_t^{-1} \qquad (2.35)$$

The definition of u_t is given by eq. 2.16:

$$u_t = [(u')^2 + (v')^2 + (w')^2]^{1/2}$$

Equating eqs. 2.17 and 2.19, it follows that:

$$u_t = 2u^*$$

$$t_m = l_m (2u^*)^{-1} = 0.4\, h_m (2u^*)^{-1}$$

When h_m equals h_e, t_m equals $0.2\, h_e(u^*)^{-1}$. It should be noted in this case that while the vertical scale also increases with wind speed, the time scale for advection around the eddy remains constant. In the case of a mixing layer constrained by a basal density gradient, $h_m = h_b$,

$$t_m = h_b\, u_t^{-1} = h_b(2u^*)^{-1} \tag{2.36}$$

Here, the time scale around the eddy is inversely proportional to the wind speed. For a more complete discussion, see Denman and Gargett (1983).

Seasonal structure in Lake Constance   From these various theoretical considerations of the processes contributing to the physical environments of lakes, it is possible to construct a crude model of the seasonal variations in the physical structure of any given water body. An appropriate example is Lake Constance, as its phytoplankton periodicity has been selected as a model case (Sommer et al., 1986). Relevant limnological characters of the lake are listed in Table 2.2. Despite its central, continental location, the lake is sufficiently large and deep to store enough heat to avert the likelihood of winter ice-cover or even significant inverse stratification. Nevertheless, the water column becomes uniformly cold and approximately isothermal ($5 \pm 1°C$) during the winter months (Sommer, 1986). At this latitude, the solar-day length (T) in January is about 8 h and the undepleted solar constant is approximately 8.4 MJ m^{-2} d^{-1}, of which no more than 5.6 MJ m^{-2} d^{-1} penetrates the atmosphere to the lake surface under a clear, cloudless sky and perhaps only 0.3 of this (1.7 MJ m^{-2}d^{-1}) under overcast conditions. During the daylight period, then, the heat flux will average some 60–190 W m^{-2}. Given g equal to 9.81 m s^{-2}, then γ (at 5°C) equals 16×10^{-6} K^{-1}, σ equals 4186 J kg^{-1} K^{-1}, ρ_w equals 999.96 kg m^{-3}, and ignoring reflection and other heat losses, the kinetic energy necessary to dissipate the radiant energy income over the full depth of the lake ($H = 100$ m), may be calculated by rearranging eq. 2.29 and solving for u^*, as 0.00483–0.00709 m s^{-1}. The generating wind speeds, derived from eq. 2.19, are between 3.9 and 5.7 m s^{-1}. In reality, winds typically vary between 0 and 70 km h^{-1} (0–20 m s^{-1}), so that it is probable that the basin will remain liable to wind mixing throughout its depth, precluding

Table 2.2   Limnological features of Lake Constance

Geographical location: 48°N 9°E
Area (A): 539 km^2
Maximum depth (H_{max}): 250 m
Mean depth (H): 100 m
Maximum length (L): 67 km = 67×10^3 m
Physical classification: warm monomictic
Thermal stratification: April to January

formation of any density gradient. From eq. 2.23, the Ekman-layer thicknesses, h_e, associated with wind speeds of > 3.9 and > 5.7 m s^{-1} are estimated to be > 34.5 and > 50.5 m. The mean mixing time, t_m equals $0.2\, h_e(u*)^{-1}$ is similar in either case, about 23.5 minutes. The relevant quantities are summarized in Table 2.3.

Of course, there may well be days on which the mean wind velocity (U) is substantially less than the amount required to dissipate the net heat income but were the warming supplied by 5 MJ m^{-2} d^{-1} confined to the upper 10 m of the lake (i.e., 0.5 MJ m^{-3} d^{-1} or 500 J kg^{-1} d^{-1}), it would scarcely raise the temperature by more than 0.1 K or reduce its density by more than 0.003 kg m^{-3} each day. Any acquired structure must be regarded as being extremely fragile at this time of the year and quite unresistant to further episodes of full column mixing. Nevertheless, the entire lake would be acquiring heat at an accelerating rate: at the time of the spring equinox ($T = 12$ h), the daily solar radiation reaching the lake would fall in the range 5.0 to 16.8 MJ m^{-2} d^{-1}. Given two months during which a mean daily heat uptake was maintained (3.4 MJ m^{-2} d^{-1} in January, 10.9 MJ m^{-2} d^{-1} at the equinox), the total heat acquired by the 100-m water column (~ 430 MJ m^{-3}) would be sufficient to have raised its mean temperature by about 1 K (to 6°C).

Inferred equinoxial heat inputs (equivalent to 120–390 W m^{-2}) generate buoyancy at a rate of $4.5–14.6 \times 10^{-6}$ kg m^{-1}s^{-3}. The wind speed now required to bring about full column mixing is respectively > 6.1 and > 9.1 m s^{-1}. The kinetic energy flux associated with a wind speed of 4.2 m s^{-1} [$u*$ equals 0.0053 m s^{-1}; $\rho_w\,(u*)^3$ equals 146×10^{-6} kg s^{-3}], under the maximal equinoxial heat flux, would be sufficient to mix the water column only to a depth (h_b) of about 10 m. At first, the containment of the heat income within the layer (i.e., up to 1.68 MJ m^{-3} d^{-1}, or 1.68 kJ kg^{-1} d^{-1}, sufficient to raise its mean temperature by up to 0.4 K d^{-1}) imparts little stability to the stratification ($\Delta\rho_w \leq 0.0152$ kg m^{-3} d^{-1}). However, given 10–30 days of warming without serious disturbance by high winds, the temperature of the upper epilimnetic layer ($\sim 10°$C) would stand some 4°C higher than that of the unmixed, hypolimnetic water below ($\Delta\rho_w \sim 0.23$ kg m^{-3}) and the $W = 1$ condition becomes achievable at $U > 1$ m s^{-1} for the first time. Over the same period, Ri_b would have increased from zero to 800.

By the time of the summer solstice (T equals 16 h), daily solar radiation lies in the range 7.5–25.1 MJ m^{-2} d^{-1}; assuming a 10 m epilimnion throughout, a steady accumulation of an additional 58.6 MJ m^{-3} (586 MJ m^{-2}, or only 6.5 mJ m^{-2} d^{-1}) would be sufficient to have raised the epilimnetic temperature to 20°C. Neither assumption is realistic: there is considerable day-to-day variability in wind action and in solar heating: three of the combinations listed in Table 2.3 under June solstice illustrate interactions giving a 25-fold range of mixed depth (10.0–0.4 m). The Wedderburn numbers for these examples (respectively, 0.81, 0.042, and 0.005) give some idea of the sensitivity of these summer stratifications to poor irradiances and intense wind; in fact, a seasonal epilimnion of less than 11 m could not be considered stable on the basis of these three entries, while epilimnetic deepening to > 25 m in the face of wind speeds > 10 m s^{-1} may be predicted from eq. 2.34. After the summer solstice, the heat flux starts to decrease until the incoming energy and heat losses are approximately balanced. Eventually, the stratification is determined mainly by the net loss of heat and the fading resistance to wind: the epilimnion is deepened episodically during windy periods. Because on each occasion the increased epilimnetic circulation is diluted by cold hypolimnetic water, the temperature difference between them is lowered. The stability of the stratification is thus progressively weakened until it presents no resistance to the return of full isothermal conditions.

Putting W equal to 1 and ascribing values to $(u^*)^2$ for given wind speeds, eq. 2.33 may be solved to find $\Delta\rho_w$ and to define the extent of cooling required to bring about full column mixing. If U_{10} equals 5 m s^{-1}, $\Delta\rho_w$ equals 0.0267 kg m^{-3}, i.e., ρ_{wo} equals 999.915, equivalent to a surface temperature (θ_o) of 6.7°C. At 10 m s^{-1}, θ_o equals 7.6°C and at 20 ms^{-1}, θ_o equals 11.9°C. For completeness, the associated computations are included in Table 2.3. Given these levels of heat redistribution and loss required before full column mixing can be restored, it is scarcely surprising that isothermy is not generally restored until late autumn or winter.

Light   The optical properties of light (i.e., visible short-wave radiation), their behavior in water and their relevance to underwater photosynthesis are now well known (for comprehensive treatments, see Kirk, 1983; Larkum and Barrett, 1983). Photosynthetically active radiation (or *PAR*) almost coincides

Table 2.3   Modelled relationships among net heat flux (Q_T), wind velocity (U_{10}), density structure (ρ_w, $\Delta\rho_w$), predicted mixed depth (h_b, h_e), and mixing times (t_m) for different seasons in Lake Constance

Symbol	Unit	Mid-January		March equinox			Late April
θ_o	°C	5		6			10
ρ_w	kg m^{-3}	1000		999.9			999.7
Q_T	W m^{-2}	60	190	120	390	390	125
U_{10}	m s^{-1}	3.9	5.7	6.1	9.1		4.2
τ	kg m^{-1}s^{-2}	2.3×10^{-2}	5.0×10^{-2}	5.9×10^{-2}	1.3×10^{-1}	2.8×10^{-2}	2.6×10^{-2}
u^*	m s^{-1}	4.8×10^{-3}	7.1×10^{-3}	7.6×10^{-3}	1.1×10^{-2}	5.3×10^{-3}	5.1×10^{-3}
$(u^*)^3$	m^{3}s^{-3}	1.1×10^{-7}	3.6×10^{-7}	4.5×10^{-7}	1.5×10^{-6}	1.5×10^{-7}	1.4×10^{-7}
C	kg m^{-1}s^{-1}	6.5	14.0	16.0	35.5	7.6	7.3
h_e	m	34.5	50.5	54.0	80.4		
$\Delta\rho_w$	kg m^{-3}					0.015	0.23
h_b	m					10	10
t_m	s	1416	1416	1416	1416	948	980

Symbols defined in Appendix 2.2.

with the visible spectrum (wavelengths 380–750 nm), generally representing some 46–48 percent of global radiation reaching the earth's surface (see Figure 2.1c). Its maximum intensity is related to the solar constant, i.e., $0.47 \times 0.66 \times 1350 \simeq 420$ W m^{-2}. It is customary, however, to express the flux of light (irradiance) in terms of the photon flux density, where 6.023×10^{23} photons = 1 mol or 1 einstein ($\sim$218 kJ). Thus, the maximum intensity of incident *PAR* irradiance is 1.93 m mol m^{-2} s^{-1}. The actual instantaneous photon flux density impinging on the surface of a lake is dependent upon the angle of solar elevation, which varies between zero at sunrise and sunset, and a mid-day maximum that is governed by latitude and season. Moreover, cloud cover and atmospheric dust may significantly deplete and diffuse the photon flux density. Seasonal variations in the monthly mean irradiance on an English reservoir (52°N) and on an equatorial African lake are shown in Figures 2.6a and 2.6b, respectively.

Subject to backscattering by waves and surface reflectance, which is also a function of the solar elevation during the course of the day (cf. Figure 2.2b) but which is minimal under diffuse, cloud-occluded light, the fraction (generally < 0.9; or < 1.7 m mol m^{-2} s^{-1}) penetrating the lake surface (I'_o) becomes the *PAR* available to the phytoplankton. The absorption properties of water differ substantially from those of the atmosphere, such that the photon flux density is more or less sharply attenuated with depth, in accord with the Beer-Lambert law (see eq. 2.2):

$$I_z = I_o' \, e^{-\epsilon z} \qquad (2.37)$$

where I_z is the irradiance (units: mol photons m^{-2} s^{-1}) at a depth z and I_o' is the irradiance available immediately beneath the surface ($z = 0$ m); ϵ is the vertical extinction coefficient, expressed as a natural logarithm. Absorption is

Table 2.3   Continued

Symbol	Late April	June solstice				Autumnal mixing		
θ_o		20				6.7	7.6	11.9
ρ_w		998.2				999.9	999.8	999.5
Q_T	416	0	130	436	436	neg	neg	neg
U_{10}	4.1	6.2	10.0	4.5	2.3	5.0	10.0	20.0
τ	5.9×10^{-2}	1.6×10^{-1}	3.1×10^{-2}	2.4×10^{-2}	8.1×10^{-3}	3.9×10^{-2}	1.6×10^{-1}	6.2×10^{-1}
u^*	7.7×10^{-3}	1.2×10^{-3}	5.5×10^{-3}	4.9×10^{-3}	2.9×10^{-2}	6.3×10^{-3}	1.3×10^{-2}	2.5×10^{-2}
$(u^*)^3$	4.5×10^{-7}	2.0×10^{-6}	1.7×10^{-7}	1.2×10^{-7}	2.3×10^{-8}	2.4×10^{-7}	2.0×10^{-6}	1.6×10^{-5}
C	16.3	43	8.6	6.6	2.2	10.8	43.0	172.0
h_e							44.2	88.5	177.0
$\Delta \rho_w$	0.23	1.7	1.7	1.7	1.7			
h_b	10	25	10	2	0.4			
t_m	650	1000	896	205	70	1416	1416	1416

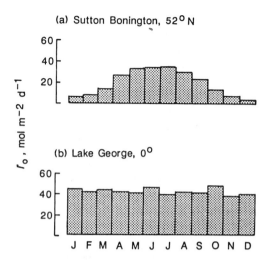

Figure 2.6   Mean monthly *PAR* income measured at sites close (a) to a small English reservoir (Sutton Bonington, UK; 52°N) and (b) to Lake George, Uganda (0°). Plotted from data presented respectively in Brierley (1985) and Ganf (1974).

not uniform across the visible spectrum but ranges from 0.005 to 1.0 m^{-1} according to wavelength, such that the spectral composition of the downwelling irradiance is altered with depth. Eq. 2.37 should strictly apply to irradiance of a given wavelength; extinction coefficients referenced to the visible spectrum are liable to errors and they usually differ from simultaneous values of the coefficient of heat flux (k in eq. 2.2). Subject to this caveat, pure water has a mean extinction coefficient of visible light extinction, ϵ, equal to 0.12 m^{-1}.

In natural lakes, extinction coefficients may be significantly enhanced by coloration of the water, whether due to dissolved organic "Gelbstoff" (Bowling, Steane, and Tyler, 1986), by the presence of suspended particles (e.g., of clay; see Allanson and Hart, 1979) or by the presence of planktonic organisms, especially algae (e.g., Talling et al., 1973). The photosynthetic pigments of planktonic photoautotrophs intercept light energy of specific wavelengths. Absorption is generally referred to the waveband of minimum extinction, usually that of green light, ~530 nm. The specific attenuation (ϵ_s) of chlorophyll a (the most ubiquitous of photosynthetic pigments among the algae), that attributable to a concentration of 1 mg of the pigment per m^3, ranges between 0.004 and 0.020 $m^2\ mg^{-1}$, depending upon the size and shape of the algae present (see Section 2.4). In a given water body, the attenuation coefficient represents the sum of the various components:

$$\epsilon = \epsilon_w + \epsilon_p + n\epsilon_s \tag{2.38}$$

where ϵ_w is the attenuation due to water and any colored solute, ϵ_p to suspended particulates ("tripton") and n is the concentration of the algal chlorophyll in mg m^{-3}.

The attenuation coefficient, ϵ, is an important property of a given water body since it responds to, and ultimately sets limits upon, the level of photosynthetic production that is sustained. At a given temperature, the chlorophyll-specific photosynthetic capacity, P_{max}, is saturated at irradiance levels significantly exceeding 0.2 m mol photon m^{-2} s^{-1} but low intensities limit the photosynthetic rate (P), such that it becomes a function of I. Plots of P on I thus take the general form of the examples presented in Figure 2.7a, which may be characterized by the values of P_{max} and the gradient (α) of light-limited P vs. I; where the slope α intersects the level of P_{max} corresponds to the theoretical point at which P is saturated by I and is called I_k. Provided $I'_o >$ I_k, the potential photosynthesis of a population evenly distributed along the light gradient is approximately described by the area of the trapezoid enclosed by the P vs. I plot, which corresponds to $P_{max} \times (I'_o - 0.5\,I_k)$.

Translated to the light field of the natural water column (Figure 2.7b) the potential production is given by the product $nP_{max} \times h_p$, where h_p is the length of the water column between the surface and $z_{0.5I_k}$, the depth at which I_z equals $0.5I_k$; from eq. 2.37,

$$h_p = \ln\,(I'_o/0.5I_k)\cdot\epsilon^{-1} \qquad (2.39)^*$$

Assuming I_k and ϵ fluctuate little at the scale of a few hours, h_p is an instantaneous function of the variable I'_o. Mathematical integration of the diel fluctuation in I'_o is not simple and is further complicated by the fact that $I'_o <$ I_k for part of the day. Planimetric determination of daily irradiance curves (Vollenweider, 1965) indicates that the mean daily light income into a lake, $\bar{I}'_o$, is 0.7 ($\pm$ 0.07) I'_o at the solar zenith and that $\bar{I}'_o$ typically exceeds I_k for 0.75 ($\pm$ 0.08) of the solar day (T). The daily integral of photosynthetic production (ΣnP) is given approximately by the relation:

$$nP = [nP_{max}\cdot0.75T\cdot\ln(\bar{I}'_o/0.5I_k)]\cdot\epsilon^{-1} \qquad (2.40)$$

It may be assumed that, as a consequence of net photosynthetic production, n and, therefore, ϵ increase, simultaneously raising in the water column the depth range of I_k; a uniformly distributed population now receives a lower irradiance and is, on average, exposed to a shorter period of light saturation. Catabolic processes (respiration) assume greater significance in the metabolic exchange of gases involved in photosynthesis (carbon dioxide and oxygen, in equimolar proportions). If R is the chlorophyll-specific respiration rate of phytoplankton in the water column under consideration and it is unchanged during light and darkness, then the integral rate of column respiration equals 24 nRh. When daily column respiration compensates daily photosynthetic

---

*Compare this equation with Talling, 1957.

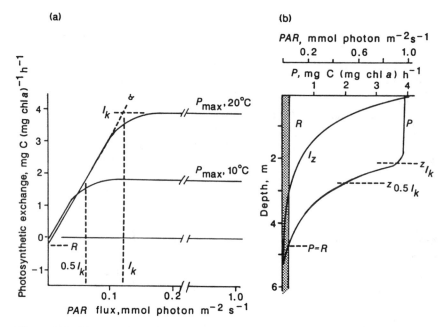

Figure 2.7 Photosynthetic behavior of *Fragilaria crotensis*, as an illustration of the quantities used in the text: (a) net photosynthetic rate, *P*, vs constant *PAR* flux (m mol photon m^{-2} s^{-1}), *I*, showing the onset of light saturation of photosynthesis (P_{max}) at I_k, at 20°C; the slope, α, of light-lighted photosynthesis vs irradiance; and the "negative photosynthesis," due to net respiration, *R*, at very low intensities of *PAR*; (b) the same quantities redrawn in the context of a water column with exponential attenuation of *PAR* (I_z) with depth: photosynthesis is still saturated at depths where $I_z > I_k$ but decreases at depth $z > z_{Ik}$ and is ultimately exceeded by *R*. The integral of photosynthetic production in the water column in b is equivalent to the product $P_{max} \cdot z_{0.5Ik}$, as shown in (a). Original data from Reynolds (1983).

production, further net increase is precluded. Compensation in a water column, h_m, is reached when:

$$\frac{nP_{max} \cdot 0.75 \; T \cdot \ln\,(\bar{I}_o{'}/0.5I_k)}{\epsilon \cdot 24\; nRh_m} = 1 \qquad (2.41)$$

Whence the compensation point is reached when:

$$n = \frac{nP_{max} \cdot 0.75T \cdot \ln\,(\bar{I}_o{'}/0.5I_k)}{24\; nRh_m} \times (\epsilon_w + \epsilon_p)\,\epsilon_s^{-1} \qquad (2.42)$$

By way of an example, eq. 2.42 may be solved for a water column in Lake Constance at the summer solstice and full radiation intensity (25 MJ m^{-2} over a 16-h day; i.e., 53.9 mol *PAR* m^{-2} or, on average, 0.94 m mol photon m^{-2} s^{-1}; if the fraction penetrating the surface is 0.9, then $I_o{'}$ equals 0.84 m mol photon m^{-2} s^{-1}) and which is simultaneously wind mixed to a depth of 2 m with a temperature of 20°C; assuming the population has the photosynthetic properties of the *Fragilaria* represented in Figure 2.7a, viz., P_{max} equals 3.9 mg C (mg chl a)$^{-1}$ h^{-1}, R equals 0.21 mg C (mg chl a)$^{-1}$ h^{-1}, and 0.5 I_k equals 0.063 m mol photon m^{-2} s^{-1}. Putting ϵ_s equal to 0.010 m^2 (mg chl a)$^{-1}$ and (ϵ_w + ϵ_p) equals 0.2 m^{-1}, then n approximates to 1183 mg chl a m^{-3}. In this extreme example, the extinction coefficient ϵ equals 12 m^{-1}, the mean light intensity at the base of the mixed layer, I_m, is about 3 × 10^{-11} m mol photon m^{-2} s^{-1}, and the depth of the productive portion of the water column (i.e., above the column compensation point) h_p equals 0.22 m. The point in the column where $P = R$ (Figure 2.7b; also called the euphotic depth) varies with I_o' through the course of a day, and its instantaneous value is not equivalent to the daily integrated compensation point of the column. Though the latter is perhaps a more relevant derivation, the euphotic depth is more commonly cited in studies of aquatic photosynthesis.

The compensation depth is especially relevant in relation to the mixing depth. In the above example, $h_p \simeq 0.11\ h_m$; were the wind mixing to reduce abruptly, relatively more of the new wind-mixed layer (h_m) would fall within the range of the light field (h_p) and algae entrained therein would receive significantly higher daily doses of irradiance. The mean irradiance experienced by algae in a mixed layer transported from the surface to its base is equivalent to $(I_m \cdot I_o')^{1/2}$. The daily integral of irradiance received by algae in a wind-mixed layer, I^*, is given by:

$$I^* = T \cdot (I_m \cdot I_o')^{1/2} \tag{2.43}$$

In the above example (h_m equal to 2 m), $I^* \simeq 0.3$ m mol photon d^{-1}. If only h_m is altered (to 0.4 m), then I^* for the population still entrained in the mixed layer rises to ~4.4 mol photon d^{-1} but the population below that depth is left in effective darkness.

The mean period (t_p) during which planktonic algae are exposed to irradiances that will support net photosynthesis depends on the relation between h_p and h_m. If $h_p > h_m$, then t_p approximates to T; if, however, $h_m > h_p$, then planktonic algae are liable to move with high frequency between full light to effective darkness. The mean time spent in the light is presumably equivalent to:

$$t_p = t_m \cdot (h_p \cdot h_m^{-1}) \tag{2.44}$$

and the daily sum of light periods Σt_p,

$$\Sigma t_p = T(h_p h_m^{-1}) \tag{2.45}$$

Applying the appropriate values of t_m from Table 2.3 to the above examples, it can be calculated that when h_p equals 0.22 m, h_m equals 2 m, and t_m equals 205 s, then t_p equals 22.5 s, i.e., 45 s in each full eddy cycle of 410 s. In the second case (h_m equals 0.4 m, t_m equals 70 s), t_p equals 38.5 s, or 77 s of each

140 s cycle. Given T equals 16 h, then $t_p \simeq 1.8$ h d^{-1} in the first instance and 8.8 h d^{-1} in the second. In clearer water, however, wherein $h_p > h_m$, Σt_p equals T. In the latter instance, net photosynthetic production is sustainable beyond the base of the mixed layer, i.e., at all depths where $\Sigma nP > 24\ nRh_m$.

The outcome of this empirical examination of physical factors is that lakes are subject to considerable seasonal variations and diel fluctuations in temperature, in the distribution of heating and cooling, and in the extent of wind mixing. Together, these factors make large impacts on the capacity of planktonic algae to increase their biomass. In the next sections of this chapter, the selective nature of this magnitude of physical variability is explored.

2.3 Dynamic Responses of Phytoplankton to Physical Variability

It must be stated at the outset of this section that the species composition of natural phytoplankton assemblages cannot be profitably correlated with the contemporaneous physical environment unless they are known to have been approximately constant for some considerable relative length of time. As Allen and Koonce (1973) stated succinctly, today's assemblage is not the product of today's environment but is rather yesterday's assemblage altered by a factor determined by yesterday's environment. Thus, the most significant advances in understanding the ecology of phytoplankton will come from a knowledge of how the rates of growth and attrition of individual species are affected by environmental variability. Equally, any understanding of seasonality in the composition of phytoplankton assemblages must be based on the knowledge of how environmental factors select in favor of or against the dynamic responses of particular species.

Replication rates   The appropriate point at which to begin an analysis of the responses to environmental limitations is the consideration of species-specific performances when all potential limiting factors and constraints are obviated. Such idealized growth is achieved only in monospecific cultures maintained under controlled laboratory conditions offering continuous light- and nutrient-saturation. Even then, there is no unique temperature at which the optimal growth of all species can be maintained: in most cases, temperature must be regulated within a range that is sublethal to cold-water stenotherms yet will still sustain significant growth of thermophilic species. For most comparative purposes, the relevant range is 18–22°C.

Indeed the data considered in this section pertain to the growth of cultures at 20°C. In this context, the term "replication rate" is preferred to "growth rate" to avoid any confusion with usage of the latter to imply apparent rates of population change, net of mortalities, for whatever reason. Here, the replication rate, r, is defined from the relation:

$$n_t = n_o \, e^{rt} \tag{2.46}$$

where n_o is the population or biomass of algae existing at a given time, n_t is the population after an interval, t, and e is the base of natural logarithms. Hence,

$$r = (\ln n_t - \ln n_o) \, t^{-1} \tag{2.47}$$

Various collections of data available in the literature (e.g., Hoogenhout and Amesz, 1965; Reynolds, 1984a) point to the existence of statistically robust relationships ($p < 0.05$) between the maximal exponential rates of replication of individual planktonic species at 20°C ($r_{20\,max}$) and the unit size (V) and the surface/volume ratio (SV^{-1}) of the units (Reynolds, 1984b and Figure 2.8). In fact, the line best fit to these data points is the log/log representation shown here in Figure 2.8b; the morphology (size and relative distortion of the spherical form) of phytoplankton units, whether free-living cells or, where appropriate, coenobial aggregates of individual cells, remains a powerful predictor of optimal growth performance. It might also be added that this relationship is much stronger than any distinction between phylogenetic groups (e.g., chlorophyta vs. cyanobacteria).

Effects of temperature upon replication rates   The rates of metabolic processes in algal cells are closely regulated by external temperatures though not all to the same extent: whereas the light-saturated rate of photosynthesis may be slightly more than doubled over the temperature range 10–20°C (see Figure 1.7a, $q_{10} > 2.2$), assimilation of photosynthate into proteins and new cell material is said to be more sensitive to temperature ($q_{10} > 2.5$; Tamiya et al., 1953). Maximal cellular replication rates are likely to respond to altered temperature by factors determined by the q_{10} of the rate-limiting anabolic process. In fact, several comparative studies (Foy, Gibson, and Smith, 1976; Konopka and Brock, 1978; Figure 2.8a) have indicated that the growth of larger, low- SV^{-1} algae tend to be limited more by relative rates of surface exchange and intracellular assimilation than by photosynthetic rate and, accordingly, the q_{10} values of their growth-rate responses to temperature variation tend to be > 2.2 (Harris, 1978; Reynolds, 1984b). Some of the available data pertaining to replication-rate responses in light- and nutrient-saturated monospecific cultures are assembled in Figure 2.9a to illustrate these interspecific differences in sensitivity to temperature. These plots also show interspecific differences in the temperature of optimal growth. For instance, the data for *Cryptomonas erosa* indicate a maximum growth rate (r_{max}) of 1.0 d^{-1} at 21°C, with a slope (β) of log r on A (temperature expressed on the Arrhenius scale) of -3.82 per 10^3A (equivalent to a q_{10}, between 10°C and 20°C, of 2.89); corresponding data for *Microcystis aeruginosa* (r_{max}: 0.8 d^{-1} at 27.5°C; the slope, β: -8.15 $(10^3A)^{-1}$; q_{10} between 10°C and 20°C: 9.6), and for *Synechococcus* (r_{max}: 7.97 d^{-1} at 41°C; β: -3.50 $(10^3A)^{-1}$; q_{10} between 10°C and

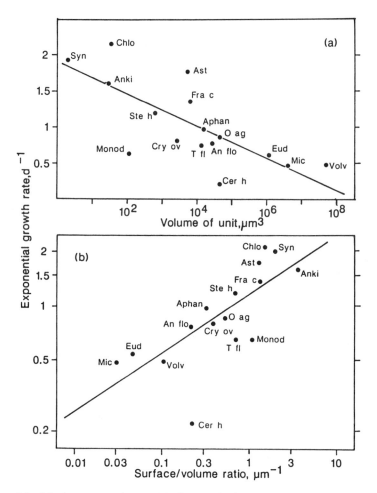

Figure 2.8   Maximum growth rates (r) of given planktonic algae in continuously light-saturated cultures at 20°C, plotted against (a) their mean volumes (in μm^3) and (b) surface/volume ratios (μm^{-1}), following the data set of Reynolds (1984a, 1984b) as appended in Reynolds (1988a). A key to abbreviations of algal taxa is given in Appendix 2.3. The least-square regressions fitted to the data are (a): $r = 1.916 - 0.223 \log V$; (b): $\log r = 0.058 + 0.325 \log (SV^{-1})$.

20°C: 2.6) emphasize the differing responses of "large" and "small" species, while the slope (β) for *Asterionella formosa*, a species of intermediate size but maintaining a relatively high SV^{-1} ratio, is -2.29 $(10^3 A)^{-1}$ (q_{10} between 10°C and 20°C: 1.9).

A statistically significant relationship ($p < 0.05$) between the slope (β) and logarithm of SV^{-1} for each of the species illustrated in Figure 2.9a is demonstrated in Figure 2.9c. Corresponding values for temperature-depen-

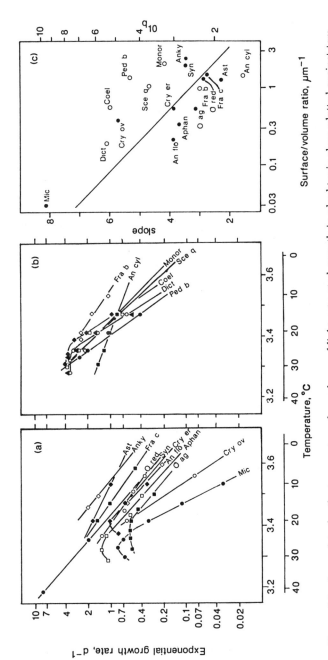

Figure 2.9   Some data for the temperature-dependence of light-saturated growth in planktonic algae plotted against temperature on an Arrhenius scale (1/K × 10³; equivalent customary temperatures shown below, in C), in respect of (a) data assembled in Reynolds (1984a, 1988a), and (b) the determinations of Dauta (1982). In (c), the slopes (β) of the regressions inserted for each species (a) and (b) are plotted against the approximate SV^{-1} ratios for the species concerned: filled circles, those used by Reynolds (1984a); open circles, those approximated from Dauta's (1982) illustrations. The regression inserted in (c) applies only to the former points; its equation is $β = 3.378 − 2.505 \log (SV^{-1})$. Abbreviations to species names are listed in Appendix 2.3. The right-hand scale of Figure 2.9c is expressed in terms of the q_{10} value for growth for the range 10–20°C; note that different solutions apply to other temperature ranges, since q_{10}, unlike β, depends on the choice of temperatures adopted.

dent growth rates of the organisms derived from the data of Dauta (1975; shown in Figure 2.9b), and relevant morphological data as approximated from his photographs, conform broadly to this relationship. It is apparent that the SV^{-1} ratio of a species gives an approximate guide to its likely replication-rate response to altered temperature.

Effects of mixed depth   The ability of autotrophic algae to maintain growth in their natural environments is also dependent upon an adequate income of *PAR*. Because of its attenuation with depth, most algae may be expected to benefit from mechanisms which prolong their residence within the upper, insolated water layers or at least to gain periodic access thereto. The problem of suspension faced by algae, having a higher intrinsic density (ρ_A) than the surrounding medium and, hence, a constant tendency to sink out of the surface waters, has been extensively investigated (see, for instance, Hutchinson, 1967; Smayda, 1970; Walsby and Reynolds, 1980); no more than the essential principles need be presented here.

A particle denser than water allowed to fall through a column of still water initially accelerates until it attains a constant settling velocity, s (positive downward), when the force causing the motion is equal to the drag force resisting the motion. For a sphere of diameter d, the force, F (in N = kg m s^{-2}), causing motion is

$$F = \tfrac{1}{6}\,\pi d^3\, g(\rho_A - \rho_w) \qquad (2.48)$$

Subject to limiting conditions that the particle is of a suitable small size (generally < 0.2 mm), and that the flow of water over it is laminar, then the drag force resisting motion has been shown on purely theoretical grounds (Stokes, 1851), to be:

$$D = 3\,\pi d\,\eta\,s \qquad (2.49)$$

where η is the viscosity of the liquid. The terminal velocity (in practice it is attained almost immediately) is derived from the equality of eqs. 2.48 and 2.49 and solved for s:

$$s = [\tfrac{1}{6}\,\pi d^3\, g(\rho_A - \rho_w)][3\,\pi d]^{-1} = \tfrac{1}{18}\, d^2\, g(\rho_A - \rho_w){\cdot}\eta^{-1} \qquad (2.50)$$

Because planktonic algae vary in size and density, especially when silicified diatoms are considered, settling velocities of algae potentially differ widely among individual species. In addition, shapes which are markedly subspherical generally sink more slowly than spheres of the same volume and density, by a factor referred to as the *form resistance*. Gas-vacuolate cyanobacteria may regulate their density, with little effect on overall unit size, and may become less dense than the water, such that ($\rho_A - \rho_w$) is negative, floating upward in still water (at a rate of $-s$). In the case of motile flagellates, the promoting forces are enhanced either to augment or offset gravitational acceleration; however, the drag forces continue to apply, so that the size of the alga continues to influence the swimming velocities attained.

Measured settling rates of various planktonic algae have been assembled in a recent review by Heaney and Butterwick (1985). The extremes of the reported ranges of still-water settling velocities and (where appropriate) swimming speed, assumed to apply in the upward vertical direction, are plotted in Figure 2.10a in log/log format, against the estimated volumes of freshwater species concerned. In spite of the several caveats indicated above, settling velocities are strongly correlated with unit volumes (coefficient of correlation = 0.84; $p < 0.05$). It is pertinent to note that the larger the unit size of a given nonmotile species, then the greater is the problem of near-surface residence. In contrast, the ability to float or swim against gravity not only obviates the problem but is itself enhanced by a large unit size.

It is also relevant that the fastest velocities attained by these algae (mostly < 1 mm s⁻¹) are slower than the turbulent friction velocities (u^*) generated by a wind (U_{10}) of 1 ms⁻¹ (eq. 2.19). Thus, even light winds (3–4 ms⁻¹) are likely to develop turbulence sufficient to exceed the intrinsic movements of most phytoplankton by one or more orders of magnitude and to keep them dispersed through the mixed layer. This deduction is convincingly verified by the data of George and Edwards (1976) pertaining to the vertical "patchiness" of buoyant *Microcystis* in a small reservoir, subjected to variable wind stress. Toward the base of the mixed layer, however, where turbulent velocities characteristically decrease, nonmotile algae are more likely to become "disentrained" from the mixed-layer motion and to be lost from the turbulent circulation. Thus, the critical factor maintaining phytoplankton in suspension is not the intensity of the turbulent mixing but rather the absolute depth of the mixed layer.

Smith (1982) showed that particles are lost exponetially from a constantly stirred layer according to a function analagous to dilution:

$$n_t = n_o\, e^{-t/t'} \qquad (2.51)$$

where n_o is the mixed-layer concentration of particles at a given point in time, n_t is the concentration after an interval, t, and t' is the time required by the last of an initially dispersed population of particles, having settling velocity s, to clear the same column in the absence of any mixing. If the column height is h, then the clearance time approximates to:

$$t' = h\, s^{-1} \qquad (2.52)$$

Thus, in the fully mixed column,

$$n_t = n_o\, e^{-(st\ h^{-1})} \qquad (2.53)$$

The exponential rate of sinking loss per unit time, (k_s), from a continuously mixed layer, h_m, is given by:

$$k_s = (\ln n_t - \ln n_o)\, t^{-1} = s\, h_m^{-1} \qquad (2.54)$$

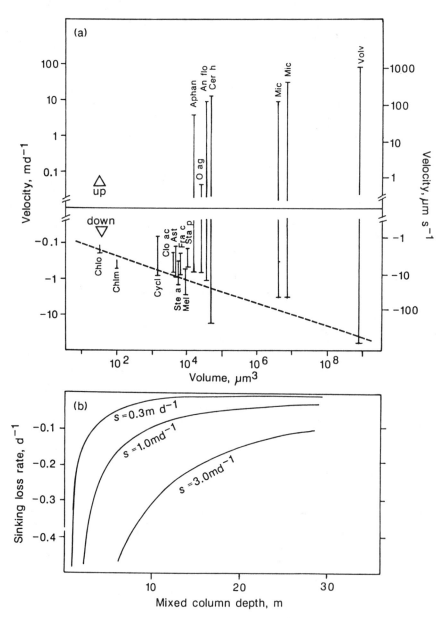

Figure 2.10   In (a) the ranges of sinking (s) and floating ($-s$) velocities of planktonic algae (mostly taken from Heaney and Butterwick (1985); additional data for *Microcystis* from Humphries and Imberger (1982) and for *Volvox* from Sommer and Gliwicz (1985); the equation of the line fitted to the maximum sinking rates is: log s = 0.319 log V − 1.268; the abbreviations to species names are set out in Appendix 2.3. In (b), the exponential loss rates ($-k_s$, d⁻¹) of algae of three different settling velocities (s = 0.3, 1.0, and 3.0 md⁻¹) are plotted against depth of fully mixed columns.

These relationships have been verified (Reynolds, 1984a) against experimental observations of the removal of *Lycopodium* spores dispersed into the surface circulation of a limnetic enclosure. For a population of nonmotile phytoplankton, sinking losses represent a drain on biomass which must be offset by growth. If population increase is to be maintained, $(r - k_s)$ must remain positive: clearly, the combination of small unit volume, affording a low intrinsic settling velocity, s, and a deep column mixing (h_m is large), is most advantageous in this respect. To demonstrate this principle, instantaneous values of k_s are plotted in Figure 2.10b against depth of the mixed layer, h_m, for three kinds of particles, each having a different settling velocity (0.3, 1.0, and 3.0 m d^{-1}). A near-surface diatom population replicating itself every 24 h ($r = 0.69$) and having a mean settling velocity, s, of 0.3 m d^{-1}, would fail to increase when $k_s \geq 0.69$ d^{-1}, i.e., when $h_m \leq 0.43$ m.

Effects of the underwater light field   Besides the critical role of the absolute depth of the mixed layer to the maintenance of nonmotile algae, its relation to the depth of effective light penetration, h_p, assumes considerable relevance to the growth of all planktonic photoautotrophs in lacustrine environments.

The principle role of the underwater light field in regulating the in situ replication rates of phytoplankton is mediated through the biomass-specific integral of photosynthetic production (see Section 2.2). However, it is rarely correct to regard net photosynthetic carbon fixation as more than the maximum capacity for growth; in fact, photosynthetic capacity generally exceeds the carbon requirements of replication rates maintained in natural environments. Equivalence between carbon-specific rates of growth and photosynthesis is attained in populations which are clearly light-limited (Reynolds, Harris, and Gouldney, 1985). Nevertheless, the same mechanisms representing adaptations to "harvest" light energy at low intensities and which are instrumental in sustaining photosynthetic rates in light-limited populations must also contribute to the maintenance of growth at low ambient levels of irradiation. These mechanisms include: the "antenna effect" of distributing pigment across the maximum area of interception, which may be achieved in markedly flattened (i.e., relatively shortened in one dimension) shapes or in attenuated (i.e., shortened in two dimensions) shapes, in order to enhance the ϵ_s value of the pigment; the increase of chlorophyll pigment per unit of cell biomass; and the increase of the spectral width of the light absorption by raising the concentrations of accessory pigments, especially phycobilins and xanthohylls (Larkum and Barrett, 1983; Kirk, 1983; Raven, 1984).

Some curves derived from the literature relating the replication rates (r) of cultures, maintained at various low levels of constant irradiance, to the photon flux are inserted on a common scale (I d^{-1}) in Figure 2.11a. This miscellany is exaggerated by the widely differing saturated growth rates of the various species. However, attention is usefully directed toward the initial

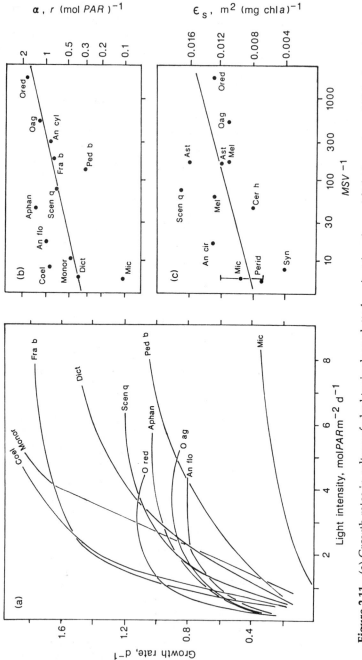

Figure 2.11   (a) Growth rates in culture of planktonic algae plotted against (continuous) light intensity at 20°C (data of Foy et al., 1976; Dauta, 1982; and Nicklisch and Kohl, 1983). In (b) values of the gradient of light-limited growth α_r ($=r/I$) are plotted against the approximate product MSV^{-1} for the algal unit. They follow a similar trend to the plot of ϵ_s, the chlorophyll-specific light absorbance, against MSV^{-1}, shown in (c). The equation of the regression in (b) is $\log(\alpha_r) = -0.590 + 0.236 \log(MSV^{-1})$; in (c), $\epsilon_s = 0.0062 + 0.0027 \log(MSV^{-1})$.

gradient (α_r) of r on I, where the replication rate is manifestly energy-limited. α_r is not significantly correlated with either unit size (V) or suface/volume ratio (SV^{-1}) of the individual algae but there is a significant log/log relationship ($p < 0.05$) between α_r and the product MSV^{-1}, where M is the greatest axial linear dimension of the typical algal unit (Figure 2.11b). As the independent variable, this product recognizes the relative inefficiency as light receptors of both small (minimal light interception) and large spheres (significant self-shading) as compared with flattened or attenuated shapes (simultaneously offering higher SV^{-1} and MV^{-1} ratios). Indeed, the chlorophyll-specific ϵ_s values of individual algae, which are customarily plotted against cell or unit volume (Harris, 1978; Reynolds, 1984a), are significantly correlated with log (MSV^{-1}), as shown in Figure 2.11c.

To translate these growth performances to the fluctuating light fields perceived by algae suspended in natural lakes is less than straightforward. The most convenient method would be to equate the daily integral "dose" of light, I^* (eq. 2.43) to I in Figure 2.11a, in order to predict r for a given species. As Gibson (1987) has pointed out in a thoughtful discussion of the concept of daily "light-dose," the integral includes periods of time in which I is definitely saturating and, hence, part of the integral I^* is "unused" by populations mixed through the daytime PAR gradient. Thus, dose (or "exposure," Gibson, 1987) is valid only under conditions of continuous light-limitation. Of greater potential use is the daily sum of light-periods, Σt_p, in which net growth is sustainable (eqs. 2.39 and 2.45), and which may be calculated, given $\bar{I}_o'$, T, h_m, and ϵ, as well as an I_k value derived as the quotient, r_{max} ($\alpha_r)^{-1}$, using components predicted from the morphology of individual species, according to the relationships shown in Figures 2.8b, 2.9c, and 2.11b.

2.4   Dynamic Responses of Phytoplankton to the Physical Environment

In the foregoing sections, sufficient information has been developed to enable the prediction of in situ growth rates of given species, determined by set combinations of quantitative characters of the physical environment. It must be emphasized that, in the following calculations, only the physical determinants of population change are considered; chemical limitations of growth rate and the effects of biotic factors (grazing, parasitism) are ignored. For present purposes, the removal of organisms in hydraulic outwash is also excluded (lakes are assumed to have an infinite retention time).

The relevant physical environmental determinants of in situ rates of population change are: water temperature (θ), mixed depth (h_m), the mean daily photon-flux (I_o'), over the daylight period (T), and the coefficient of its extinction with depth (ϵ). The relevant properties of specific algae are volume (V), surface area (S), and greatest axial dimension (M). Then the in situ rate of change for a given population in a given water column may be developed as follows:

The maximum replication rate at 20°C, under continuous light saturation is calculated as:

$$r_{20} \ d^{-1} = 1.142 \ (SV^{-1})^{0.325} \quad (2.55; \text{ from Figure 2.8b})$$

The maximum replication rate at another temperature is predicted from:

$$\log r_\theta \ d^{-1} = \log r_{20} + \beta \ [1000/(273+20) - 1000(273+\theta)] \quad (2.56)$$

where

$$\beta = 3.378 - 2.505 \log (SV^{-1}) \quad (2.57; \text{ from Figure 2.9c})$$

The modification of eq. 2.56 in respect of the daily *PAR* received over the course of 24 h in the mixed layer of a natural lake is given as a fraction of the 24 hours that is passed in the light:

$$r_{(\theta,I)} \ d^{-1} = r_\theta \cdot \Sigma t_p / 24 \quad (2.58)$$

where the daily sum of the photoperiods is

$$\Sigma t_p = T \ h_p \cdot h_m^{-1} \quad (\text{from eq. } 2.45)$$

the light-compensated column height is

$$h_p = \ln(\bar{I}_o' / 0.5 I_k) \cdot \epsilon^{-1} \quad (\text{from eq. } 2.39)$$

The onset of light saturation, I_k, in m mol photon $m^{-2}s^{-1}$, is expressed as the quotient of the light-saturated replication rate at the given temperature (r_θ) and the slope (α_r) of the light-limited portion of the r vs. I curve in Figure 2.11a, for example:

$$I_k = r_\theta \cdot \alpha_r^{-1}$$

and where

$$\alpha_r = 0.257(MSV^{-1})^{0.236} \quad (\text{From Figure 2.11b})$$

If, however, $h_p > h_m$, then $\Sigma t_p = T$.
By substitution in eq. 2.58,

$$r_{(\theta,I)} \ d^{-1} = [r_\theta T (24 \ h_m)^{-1}] \cdot \ln[2 I_o' \cdot 0.257(MSV^{-1})^{0.236} \cdot r_\theta^{-1}] \cdot \epsilon^{-1} \quad (2.59)$$

Or, when $h_p > h_m$,

$$r_{(\theta,I)} = r_\theta T / 24 \quad (2.60)$$

Finally, correction may also be made for sinking loss rates of nonmotile species:

$$(r_{(\theta,I)} - k) \ d^{-1} = r_{(\theta,I)} - sh_m^{-1} \quad (2.61)$$

where

$$s = 0.054V^{0.319} \text{ m d}^{-1} \quad (2.62; \text{from Figure 2.10a})$$

Following the sequence of eqs. 2.55 to 2.62, solved for the properties of the algal species in respect of set limnological conditions, it becomes possible to assemble provisional models of their potential rates of increase in situ through typical seasonal cycles of variation in the physical environment. In Figure 2.12, a series of realistic scenarios are set, each of which describes combinations of physical factors likely to obtain in Lake Constance, at certain times of the year. The model reconstructions of increase rate (replication rate, net of sinking losses but before any allowance is made for possible nutrient lim-

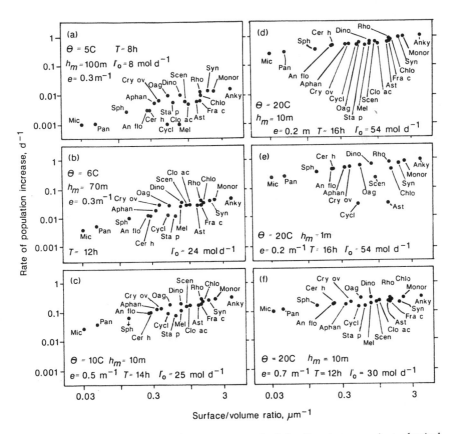

Figure 2.12 Modelled potential increase rates in Lake Constance, against physical conditions in the lake, set as appropriate to the time of the year (see also Table 2.3 and Section 2.2): (a) mid-January, water column uniformly cold and well mixed; (b) the vernal equinox, still cold and well mixed but light income increasing steeply; (c) late April, with selective surface warming and thermal stratification initiated; (d) the summer solstice, with stable thermal stratification and high water clarity; (e) the summer solstice, but with stratification from 1.0 m; (f) later summer conditions, with light limitation. Abbreviations for species are listed in Appendix 2.3.

itation or cropping rates) for each of a selection of common species are inserted to illustrate the sensitivity to the various physical limitations. The species chosen are not necessarily conspicuously present in Lake Constance but are included to extend the applicability of the model reconstructions to lakes elsewhere. The six panels (Figure 2.12a–f) are ordered in a sequence relating to the major seasonal changes in the physical environment but, for the purposes of this discussion, it is convenient to consider each of the limiting factors in turn.

The conditions most closely approaching those that might be furnished in laboratory cultures, whence the model equations are essentially derived, are advanced in Figure 2.12d. The criteria set are supposed to be the best for most species that can be derived from Table 2.3; these relate to the summer solstice (T equals 16 h) under continuously cloud-free skies, when the heat- and irradiance-fluxes into the water are maximal ($Q_T = 436$ W m^{-2}; $I_o' = 54$ mol PAR m^{-2} d^{-1}) and the surface temperature might easily surpass 20°C. The depth of the mixed layer (h_m) is set arbitrarily at 10 m, wherein the sinking loss rates of nonmotile algae are relatively suppressed, yet with sufficient optical clarity to ensure that, even at the bottom of the mixed layer, growth is light-saturated. Putting $\epsilon = 0.2$ m^{-1} (i.e., extinction is due neither to algae, nor to nonliving particulates in suspension), I_{10} ($\sim$7 mol PAR m^{-2} d^{-1}) significantly exceeds the saturation requirement during most of the solar day. In fact, such phases of high water clarity are prominent in the early stages of the summer stratification of many lakes, even of quite productive ones like Lake Constance (see Sommer, 1986). That the low planktonic biomass should coincide with near-optimal conditions for algal growth in situ recalls the earlier distinction made between the responses of standing crops as opposed to their dynamics. Besides, the larger standing crops realized later in summer reflect a preceding period of conditions favoring their development. The modelled rates of increase (in the range 0.25–1.0 d^{-1}) show similar alignment to the regression fitted in Figure 2.8b, but the intercept is displaced downward, in respect of the truncated day length in the natural lake. Downward departures below this line apply to populations of nonmotile species, which continue to accommodate modest rates of sinking loss.

This dependence upon physical mixing is emphasized in Figure 2.12e where, as a consequence of either reduced wind stress on the lake or intensified near-surface buoyancy generation under conditions of high insolation, the mixed-layer depth has been reduced ($h_m = 1.0$ m) while all other components are unchanged. The rates of net increase of all nonmotile species are depressed by amounts proportional to the sinking rates of the species concerned. Several species for which the modelled sinking-loss rate from a one-meter column exceeds the modelled growth during 24 h are omitted altogether: *Fragilaria crotonensis*, *Melosira granulata*, and *Staurastrum pingue*. Though nonmotile, the modelled sinking loss of *Sphaerocystis* has not been incorporated: this large mucilage-bound colonial alga maintains a density close to that of water and its velocity derived as a function of unit size (eq. 2.62) is less realistic than the assumption of neutral buoyancy. However, in fact neither is correct.

The reconstruction of Figure 2.12f introduces a stronger bias of light limitation of growth rates in a water column mixed to 10 m ($I_o' = 30$ mol PAR m^{-2} d^{-1}; $T = 12$ h; $\epsilon = 0.7$m^{-1}) though still with a temperature of 20°C; this combination is representative of conditions that might obtain toward the end of summer when the day length is shrinking and the vertical extinction coefficient due to algae is near to a seasonal high. It is interesting to note that under these circumstances the modelled rates of increase (range 0.1–0.3 d^{-1}) of attenuated forms (having high MSV^{-1}; e.g., *Oscillatoria* and *Dinobryon* spp.) are comparable with those of *Rhodomonas* and *Ankyra*, for example, because more of the mixed layer is capable of saturating their relatively lowered growth threshholds predicted by eq. 2.59.

Prior to the summer solstice, water temperatures are rather lower than the model provision of $\theta = 20$°C, with less buoyant resistance to mixing as well as lower insolation. The reconstruction in Figure 2.12c assumes physical components that might obtain in late April, in the early stages of thermal stratification ($h_m = 10$ m; $\theta = 10$°C), so that the differences with Figure 2.12f are mainly attributable to the lower temperature. The increased range of the modelled rates of increase (0.03–0.3 d^{-1}) is influenced by the relatively greater sensitivity to lowered temperature of the larger, low-SV^{-1} forms. Still earlier in the year, when the water is simultaneously cold and well mixed to depths of 50–100 m (Table 2.3), light limitation is especially severe. In Figure 2.12a, the physical limits set relate to conditions likely to obtain in Lake Constance in mid-January ($\theta = 5$°C; $h_m = 100$ m; $T = 8$ h); the potential rates of increase of the "larger" ($SV^{-1} < 0.3$) species are within the range 0.001–0.006 d^{-1}, whereas the modelled rates of most of those with $SV^{-1} > 0.3$ range between 0.005 and 0.012 d^{-1}. It may be deduced that under such extreme conditions of optical depth, light energy appears to be the factor limiting increase of the "smaller" algae and that, again, the disadvantage is least among the attenuated forms (e.g., *Oscillatoria*). Nevertheless, at such slow rates of growth, sinking-loss rates of diatoms, especially *Melosira* and *Cyclotella*, from even deeply mixed layers become significant. Accordingly, when light income is increased toward the equinox but with little change in either temperature or buoyant resistance to mixing (Figure 2.12b), it is the nonmotile, high-MSV^{-1} forms that experience the greater relative benefit in terms of accelerated rates of increase (5 to 15-fold; compare with *Microcystis*, 4-fold; *Ankyra*, 3-fold).

It may be seen that despite the inevitable coarseness of these various reconstructions, which ignore the wide confidence intervals that surround the regressions depicted in Figures 2.8 to 2.11 and which are based, in most cases, on limited factual data pertaining to the growth rates of named species, the main seasonal patterns of change in the dynamic performances of phytoplankton are realistically simulated. The depressed productivity in winter (biomass doubling times 45–700 d), the accelerated growth of species dominating the spring bloom period in Lake Constance (diatoms, cryptomonads: see Sommer, 1986), and the development of diverse populations of *Pandorina*, *Ceratium*, *Dinobryon*, and desmids through summer do not violate the provisions of the simulations in Figure 2.12. Moreover, the maximal rates of increase of each of the major dominants in the lake, reported by Sommer (1981), are reasonably predicted by the appropriate simulations. These deductions encourage an optimistic view of the substance of the models developed here and of their value in providing first approximations both of the role of physical environmental factors in influencing the structure and species composition of phytoplankton communities in lakes and of the direction of change in response to physical variability.

On the other hand, it is important to emphasize the shortcomings of a wholly physical model in predicting seasonal change in phytoplankton communities. One is that the populations of each species present at any given time are, at least in the short-term, independent of their rates of change. An established, slow-growing, or declining population may continue to dominate an assemblage even when physical conditions select in favor of other, more rapidly growing species present in only small numbers. A second shortcoming is that the comparative dynamics of individual species may well be simultaneously subject to other variables, which may enhance or override the species selectivity owing to physical factors. For instance, in situ replication rates may be subject to subsaturating availabilities of essential nutrients, some of which (e.g., silicon and nitrogen) may discriminate powerfully between the potential growth rates of individual members of the assemblage. Biotic variables, especially selective grazing by zooplankton and the incidence of host-specific parasites and pathogenic organisms, also contribute to the relative performances and representation of individual species. In many instances, the chemical and biological processes may interact directly with the physical limits set (e.g., temperature regulation of nutrient uptake kinetics; biomass-dependence of light penetration), or may be additive (e.g., grazing loss plus sinking losses), or may be alternatives (e.g., light limitation vs. nutrient limitation). Nevertheless, such factors are always likely to select only among those species favored by the contemporaneous physical conditions. This view concurs with Reynolds' (1984b) suggestion that the factors ultimately conditioning the structure of phytoplankton assemblages may be arranged in a hierarchical sequence, moving from interactions involving factors that are primarily physical (temperature, mixing, relative light penetration), to chemical (ionic environments, nutrient availabilities and relative gradients) and then to biotic (grazing, parasitism) factors. The effects of the superimposition of chemical and biotic constraints on the succession of phytoplankton species and on the temporal variation of community structure are explored in subsequent chapters of this volume.

2.5   Physical Determinants of Seasonal Succession

The present chapter has attempted to define, in empirical terms, the scales of variation in the physical environments of lakes and to demonstrate, also empirically, how morphological properties of individual species influence the dynamic responses of phytoplankton to the physical conditions obtaining. It has also been shown that certain of the primary components are seasonally predictable with a high probability—the day length; the heat balance; the incidence of thermal stratification—for a given latitude, local climatic conditions, and the morphometric properties of the lake concerned. It follows that the underlying patterns of seasonal change in the composition of limnetic phytoplankton—for instance, the prevalence of diatoms during the ascent of the spring bloom, and the restriction of the growth periods of larger, colonial species to the summer period—are broadly related to major changes in the physical environment.

The view that this is more than a coincidental association is encouraged by comparisons of the seasonal cycles of lakes either:

1. at differing latitudes and with differing patterns of seasonality (Reynolds, 1984a); or

2. in relation to extremes of interannual variability in the same lake (e.g., Reynolds and Reynolds, 1985); or

3. in artificially "engineered" systems in which the major alternations of stratification and mixing are either suppressed (e.g., Steel, 1976) or experimentally enhanced (Reynolds, Wiseman, and Clarke, 1984).

In each case, the predominant physical conditions seem to select consistently for the same species-association of algae. In the large tropical lake Lanao (Philippines) which may undergo several quite natural alternations between stable stratification and full- or partial-column mixing within a single calendar year, the same general responses of the phytoplankton are repeatedly reproduced (Lewis, 1978).

However, there are many kinds of community change (such as from dominance by colonial chlorophytes or chrysophytes to cyanobacteria or dinoflagellates, like *Ceratium* and *Peridinium* spp., or, in continuously mixed water columns, from diatoms to *Oscillatoria* spp.) that are scarcely influenced by variations in stratification and mixing, and in which changes in temperature and insolation are small or incidental (Reynolds, 1984b). These changes more often arise from the activities of the planktonic community itself, in partitioning nutrient and energy resources, in the diversification of niches, and in the effects of selective cropping by grazers. These internal, community-driven processes show many recognizable properties of true, ecological successions (following Odum, 1969) and it has been argued (Reynolds, 1988b) that the use of the word "succession" in the context of phytoplankton periodicity should be restricted only to those compositional changes which satisfy Odum's criteria. If this view is accepted, then, by definition, externally driven changes in the physical environment should not be regarded as determinants of successional pathways. This is very much in agreement with the PEG-model (Sommer et al., 1986). Of its 24 separate statements only five (numbers 1, 16, 17, 21, and 22) imply the intervention of physical controls on succession.

Rather, physical alterations should be regarded as perturbations or disturbance factors which—when imposed on an ecological succession (in ascending order of relative impact)—may either arrest its progress at a kind of "plagioclimax," or set it back to an earlier successional stage, or destroy much of the existing structure and initiate a new or "shifted" successional sequence. The difference between successional change and allogenic perturbations may be analogized to that between an untended field allowed to revert gradually toward supporting a climatic-climax vegetation (forest) and an adjacent field that may be mown, burned, or plowed at unpredetermined frequencies but left to recover naturally on each occasion.

For lakes, the frequency of storms, floods, and atmospheric changes and the extent to which they modify the environment are key determinants of the dynamics of individual algal species present and, while they persist, are able to discriminate among one or another of the broad categories of species. Which species within the selected category that will become abundant, either sooner or later, will be determined largely by internal, community-led processes and by the relative sizes of the populations present at the time. Put at its simplest, succession in phytoplankton communities is initiated by some physical, hydrographic event; its capacity for progress is also subject to physical controls but the direction and eventual outcome of the succession, so long as similar physical conditions persist, will be primarily influenced by interspecific interactions of the developing community. Although it is theoretically possible to generate equations to predict the impact of algal growth on the availability of chemical resources and their susceptibility to grazing animals—and such equations could take into account the predictable and stochastic fluctuations in the physical environment—to write such equations is clearly beyond the scope of the present chapter. The subsequent chapters of this volume, however, will furnish much of the information that will help to bring such workable simulation models closer to fruition.

References

Allanson, B.R. and Hart, R.C. 1979. Limnology of P.K. Le Roux Dam. *Reports, Rhodes University Institute for Freshwater Studies* 11 (7): 1–3.
Allen, T.F.H. and Koonce, J.F. 1973. Multivariate approaches to algal stratagems and tactics in systems analysis of phytoplankton. *Ecology* 54: 1234–47.

Anderson, E.R. 1952. Energy-budget studies: water-loss investigations. Vol. I. Lake Hefner Studies. *Technical Reports, U.S. Geological Survey* No. 229: 71–119.

Blanton, J.O. 1973. Vertical entrainment into the epilimnia of stratified lakes. *Limnology and Oceanography* 18: 697–704.

Bowling, L.C., Steane, M.S., and Tyler, P.A. 1986. The spectral distribution and attenuation of underwater irradiance in Tasmanian inland waters. *Freshwater Biology* 16: 313–335.

Brierly, S. 1985. The effects of artificial overturn on algal populations. *Research and Development Project Report* RP85-070. Severn-Trent Water Authority, Birmingham, U.K.

Dauta, A. 1982. Conditions de développement du phytoplancton: Étude comparative du comportement de huit espéces en culture. I. Détermination des parametres de croissance en fonction de la lumiére et de la température. *Annales de Limnologie* 18: 217–262.

Denman, K.L. and Gargett, A.E. 1983. Time and space scales of vertical mixing and advection of phytoplankton in the upper ocean. *Limnology and Oceanography* 28: 801–815.

Dillon, T.M. and Caldwell, D.R. 1980. The Batchelor spectrum and dissipation in the upper ocean. *Journal of Geophysical Research* 85: 1910–1916.

Foy, R.H., Gibson, C.E., and Smith, R.V. 1976. The influence of daylength, light intensity and temperature on the growth rates of planktonic blue-green algae. *British Phycological Journal* 11: 151–163.

Ganf, G.G. 1974. Incident solar irradiance and underwater light penetration as factors controlling the chlorophyll *a* content of a shallow equatorial lake (Lake George, Uganda). *Journal of Ecology* 62: 593–609.

Gates, D.M. 1962. *Energy Exchange in the Biosphere*. Harper and Row, New York.

Gates, D.M. 1972. *Man and His Environment: Climate*. Harper and Row, New York.

George, D.G. 1982. The spatial distribution of nutrients in the south basin of Windermere. *Freshwater Biology* 11: 405–424.

George, D.G. and Edwards, R.W. 1976. The effect of wind on the distribution of chlorophyll *a* and crustacean plankton in a shallow eutrophic reservoir. *Journal of Applied Ecology* 13: 667–690.

Gibson, C.E. 1987. Adaptations in *Oscillatoria redekei* at very slow growth rates—changes in growth efficiency and phycobilin complement. *British Phycological Journal* 22: 187–191.

Harris, G.P. 1978. Photosynthesis, productivity and growth: the physiological ecology of phytoplankton. *Ergebnisse der Limnologie* 10: 1–163.

Harris, G.P. 1986. *Phytoplankton Ecology*. Chapman and Hall, London.

Heaney, S.I. and Butterwick, C. 1985. Comparative mechanisms of algal movement in relation to phytoplankton production, pp. 114–134, in Rankin, M.A. (editor), *Migration: Mechanisms and Adaptive Significance*. University of Texas Press, Austin.

Hoogenhout, H. and Amesz, J. 1965. Growth rates of photosynthetic microorganisms in laboratory cultures. *Archiv für Mikrobiologie* 50: 10–25.

Humphries, S.E. and Imberger, J. 1982. *The Influence of the Internal Structure and Dynamics of Burrinjuck Reservoir on Phytoplankton Blooms*. Environmental Dynamics Report ED82-023. University of Western Australia, Nedlands.

Hutchinson, G.E. 1957. *A Treatise on Limnology, Vol. I*. Wiley, New York.

Hutchinson, G.E. 1967. *A Treatise on Limnology, Vol. II. Introduction to Lake Biology and the Limnoplankton*. Wiley, New York.

Imberger, J. 1985. The diurnal mixed layer. *Limnology and Oceanography* 30: 737–770.

Imberger, J. and Hamblin, P.F. 1982. Dynamics of lakes, reservoirs and cooling ponds. *Annual Review of Fluid Mechanics* 14: 153–187.

Kirk, J.T.O. 1983. *Light and Photosynthesis in Aquatic Ecosystems*. Cambridge University Press, Cambridge.

Konopka, A.E. and Brock, T.D. 1978. Effect of temperature on blue-green algae (Cyanobacteria) in Lake Mendota. *Applied and Environmental Microbiology* 36: 572–576.

Kuhn, W. 1978. Aus Wärmehaushalt und Klimadaten berechnete Verdunstung des Zürichsees. *Vierteljahrsschrift der Naturforschenden Gesellschaft in Zürich* 23: 261–283.

Larkum, A.W.D. and Barrett, J. 1983. Light harvesting processes in algae, pp. 1–219, in Woolhouse, H.W. (editor), *Advances in Botanical Research, Vol. 10.* Academic Press, London.

Lewis, W.M. 1978. Dynamics and succession of the phytoplankton in a tropical lake: Lake Lanao, Philippines. *Journal of Ecology* 66: 849–880.

Marti, v.D.E. and Imboden, D.M. 1986. Thermische Energieflüsse an der Wasseroberfläche: Beispiel Sempachersee. *Schweizerische Zeitschrift für Hydrologie* 48: 196–229.

Nicklisch, A. and Kohl, J.-G. 1983. Growth rates of *Microcystis aeruginosa* (Kütz.) as a basis for modelling its population dynamics. *Internationale Revue des gesamenten Hydrobiologie* 68: 317–326.

Niiler, P.P. and Kraus, E.G. 1977. One dimensional models of the upper ocean, pp. 143–172, in Kraus, E.G. (editor), *Modelling and Prediction of the Upper Layers of the Ocean.* Pergamon Press, Oxford.

Oakey, N.S. and Elliott, J.A. 1982. Dissipation within the surface mixed layer. *Journal of Physical Oceanography* 12: 171–185.

Odum, E.P. 1969. The strategy of ecosystem development. *Science* 164: 262–270.

Phillips, O.M. 1977. Entrainment, pp. 92–101, in Kraus, E.G. (editor), *Modelling and Prediction of the Upper Layers of the Ocean.* Pergamon Press, Oxford.

Raven, J.A. 1984. A cost-benefit analysis of photon absorption by photosynthetic cells. *New Phytologist* 98: 593–625.

Reynolds, C.S. 1983. A physiological interpretation of the dynamic responses of populations of a planktonic diatom to physical variability of the environment. *New Phytologist* 95: 41–53.

Reynolds, C.S. 1984a. *The Ecology of Freshwater Phytoplankton.* Cambridge University Press, Cambridge.

Reynolds, C.S. 1984b. Phytoplankton periodicity; the interaction of form, function and environmental variability. *Freshwater Biology* 14: 111–142.

Reynolds, C.S. 1987. Community organization in the freshwater plankton, pp. 297–325, in Gee, J.H.R. and Giller, P.S. (editors), *The Organization of Communities, Past and Present.* Blackwell Scientific Publications, Oxford.

Reynolds, C.S. 1988a. Functional morphology and the adaptive strategies of freshwater phytoplankton, pp. 388–433, in Sandgren, C.D. (editor), *Growth and Survival Strategies of Freshwater Phytoplankton.* Cambridge University Press, New York.

Reynolds, C.S. 1988b. The concept of ecological succession applied to seasonal periodicity of freshwater phytoplankton. *Verhandlungen der internationale Vereinigung für theoretische und angewandte Limnologie* 23: 683–691.

Reynolds, C.S., Harris, G.P., and Goudney, D.N. 1985. Comparison of carbon-specific growth rates and rates of cellular increase of phytoplankton in large limnetic enclosures. *Journal of Plankton Research* 7: 791–820.

Reynolds, C.S. and Reynolds, J.B. 1985. The atypical seasonality of phytoplankton in Crose Mere, 1972: an independent test of the hypothesis that variability in the physical environment regulates community dynamics and structure. *British Phycological Journal* 20: 227–242.

Reynolds, C.S., Tundisi, J.G., and Hino, K. 1983. Observations on a metalimnetic *Lyngbya* population in a stably stratified tropical lake (Lagoa Carioca, Eastern Brasil). *Archiv für Hydrobiologie* 97: 7–17.

Reynolds, C.S., Wiseman, S.W., and Clarke, M.J.O. 1984. Growth- and loss-rate responses of phytoplankton to intermittent artificial mixing and their potential application to the control of planktonic algal biomass. *Journal of Applied Ecology* 21: 11–39.

Smayda, T.J. 1970. The suspension and sinking of phytoplankton in the sea. *Annual Review of Oceanography and Marine Biology* 8: 353–414.

Smith, I.R. 1975. Turbulence in lakes and rivers. *Scientific Publications of the Freshwater Biological Association* No. 29: 1–79.

Smith, I.R. 1982. A simple theory of algal deposition. *Freshwater Biology* 12: 445–449.

Sommer, U. 1981. The role of r- and K- selection in the succession of phytoplankton in Lake Constance. *Acta Oecologica* 2: 237–242.

Sommer, U. 1986. The periodicity of phytoplankton in Lake Constance (Bodensee) in comparison to other deep lakes of central Europe. *Hydrobiologia* 138: 1–7.

Sommer, U. and Gliwicz, Z.M. 1986. Long range vertical migration of *Volvox* in tropical lake Cahora Bassa (Mozambique). *Limnology and Oceanography* 31: 650–653.

Sommer, U., Gliwicz, Z.M., Lampert, W., and Duncan, A. 1986. The P.E.G.-model of seasonal succession of planktonic events in fresh waters. *Archiv für Hydrobiologie* 106: 433–471.

Spigel, R.H., Imberger, J., and Rayner, K.N. 1986. Modelling the diurnal mixed layer. *Limnology and Oceanography* 31: 533–556.

Steel, J.A. 1976. Eutrophication and the operational management of reservoirs of the Thames Water Authority, Metropolitan Water Division, pp. J1–J12 in IPHE, *Eutrophication of Lakes and Reservoirs*. Institute of Public Health Engineers, London.

Stokes, G.G. 1851. On the effect of the internal friction of fluids on the motion of pendulums. *Transactions of the Cambridge Philosophical Society* 9 (2): 8–14.

Stommel, H. 1949. Horizontal diffusion due to oceanic turbulence. *Journal of Marine Research* 8: 199–225.

Talling, J.F. 1957. The phytoplankton population as a compound photosynthetic system. *New Phytologist* 56: 133–149.

Talling, J.F., Wood, R.B., Prosser, M.V., and Baxter, R.M. 1973. The upper limit of photosynthetic productivity by phytoplankton: evidence from Ethiopian soda lakes. *Freshwater Biology* 3: 53–76.

Tamiya, H., Iwamura, T., Shibata, K., Hase, E., and Nihei, T. 1953. Correlation between photosynthesis and light-independent metabolism in the growth of *Chlorella*. *Biochimica et Biophysica Acta* 12: 23–40.

Trimbee, A.M. and Harris, G.P. 1983. Use of time-series analysis to demonstrate advection rates of different variables in a small lake. *Journal of Plankton Research* 5: 819–833.

Vollenwieder, R.A. 1965. Calculation models of photosynthesis-depth curves and some implications regarding day rate estimates in primary production. *Memorie dell'Istituto italiano di Idrobiologia* 18 (Suppl.): 425–457.

Walsby, A.E. and Reynolds, C.S. 1980. Sinking and floating, pp. 371–412, in Morris, I. (editor), *The Physiological Ecology of Phytoplankton*. Blackwell Scientific Publications, Oxford.

Appendix 2.1 Definition of units of measurement

Term	Name of unit*	Abbreviation	Equivalent unit
Mass	Kilogram	kg	
Length	Meter	m	
Area	Square meter	m^2	
Volume	Cubic meter	m^3	
Density	Kilogram per cubic meter	$kg\ m^{-3}$	
Time	Second	s	
	Hour	h	3600 s
	Day	d	86400 s
Velocity	Meter per second	$m\ s^{-1}$	
Acceleration	Meter per second per second	$m\ s^{-2}$	
Force	Newton	N	$kg\ m\ s^{-2}$
Work	Joule	J	$N\ m\ (= kg\ m^2\ s^{-2})$
Power	Watt	W	$J\ s^{-1}\ (= kg\ m^2\ s^{-3})$
Viscosity	Poise	P	$kg\ m^{-1}\ s^{-1}$
Temperature	Degree celsius	°C	
Absolute temp.	Degree kelvin	K	$\theta°C + 273$
Arrhenius temp.	1/absolute temp. in kelvins (herein expressed $\times\ 10^3$)		
Heat flux	Watts per square meter	$W\ m^{-2}$	$J\ m^{-2}s^{-1}$
Photon flux	Mol photon per square meter per second	mol photon $m^{-2}s^{-1}$	6.023×10^{23} photons $m^{-2}s^{-1} = 1$ Einstein $m^{-2}s^{-1} = 218\ k\ W\ m^{-2}$

* Wherever possible, S.I. units (or multiples thereof) have been used to express all quantities.

Appendix 2.2 Identification of symbols used

Symbol	Definition	Unit
A	Area (of a lake)	km^2
B	Energy of penetrative convection	m^2s^{-3}
C	Coefficient of eddy viscosity	$kg\ m^{-1}s^{-1}$
D	Drag force acting on a body settling through water	$kg\ m\ s^{-2}$
E	Rate of dissipation of turbulent kinetic energy	m^2s^{-3}
F	Force	$kg\ m\ s^{-2}$
H	Height of the full water column of a lake	m
I	Photosynthetically active irradiance, or photon fluence rate, in mol photon per unit area per unit time	$mol\ m^{-2}s^{-1}$
I_k	Photon fluence rate required to saturate the chlorophyll-specific photosynthetic capacity of phytoplankton	$mol\ m^{-2}s^{-1}$
I_m	Mean photon fluence rate at the base of a mixed water column	$mol\ m^{-2}s^{-1}$
I_z	Photon fluence rate at a specified depth, z, beneath the surface	$mol\ m^{-2}s^{-1}$
I'_o	Photon fluence rate immediately beneath the surface of a water body	$mol\ m^{-2}s^{-1}$

I^*	Daily integral of irradiance perceived by algae in a mixed layer as defined in eq. 2.43	mol d^{-1}
J_k	Flux of kinetic energy	W m^{-2}, = kg s^{-3}
J_{pot}	"Flux" of buoyancy generation (see eq. 2.28)	kg s^{-3}
L	Maximum dimension of a lake or longest dimension in the direction of the wind (note units in eqs. 2.33 and 2.34)	km or m
M	Maximum linear dimension of a single algal unit (unicell, colony, or filament)	μm
N	Buoyancy frequency (as defined in eq. 2.31)	s^{-1}
P	Chlorophyll-specific photosynthetic rate, generally expressed as mol carbon fixed or mol oxygen generated per unit of chlorophyll per unit time	mol (mg chl a)$^{-1}$ h^{-1}
P_{max}	Maximum rate of chlorophyll-specific photosynthesis attainable at the irradiance levels and temperature obtaining	mol (mg chl a)$^{-1}$ h^{-1}
PAR	Photosynthetically active radiation	(mol photon) m^{-2}s^{-1}
Q	Flux of heat on the surface of a lake. Various components are identified by subscripts, e.g., Q_T is the term introduced for the net change in storage	W m^{-2}
R	Chlorophyll-specific rate of algal respiration	mol (mg chl a)$^{-1}$ h^{-1}
Ri_b	Bulk Richardson number (defined in eq. 2.30)	dimensionless
Ri_g	Gradient Richardson number (see eq. 2.26)	dimensionless
S	Surface area of a cell. Used especially relative to cell volume, SV^{-1}, as an index of metabolic activity	μm^{-2}
T	Length of solar day (sunrise to sunset)	h
U	Wind velocity in horizontal direction (U_{10} denotes measurement 10 m above the surface of a lake)	m s^{-1}
V	Volume (e.g., of algal unit)	μm^{-3}
V_e	Volume of water evaporated	m^3
W	Wedderburn number (as defined in eq. 2.33)	dimensionless
c	Coefficient of frictional drag	dimensionless
d	Diameter of a spherical particle	μm
g	Gravitational acceleration (9.81 m s^{-2})	m s^{-2}
h	Height of a part of the water column subject to a particular physical process	m
h_b	Depth of surface mixed layer constrained by a density gradient (the Monin-Obukhov length)	m
h_e	Thickness of the Ekman layer, from the surface to the depth of frictional resistance, z_e	m
h_m	Water column height subject to boundary layer mixing	m
h_p	Water column height within which net photosynthesis can be maintained, "the euphotic depth"	m
k	Exponential coefficient of attenuation of vertical heat transfer	m^{-1}
k_s	Exponential rate of sinking loss of settling particles from a constantly stirred layer	d^{-1}

l_m Length scale of mixing eddy   m

n Algal population, cited as cells per unit volume or by analogy to its chlorophyll a content m^{-3}

p Statistical probability that a result was obtained by chance, rather than as a consequence of the experimental design   fraction of 1

q_{10} Relative increase in the rate of a temperature-dependent process over a range of $10°C$. For $10–20°C$, $q_{10} = 10^{-\beta(1000/283-1000/293)}$ $(10 \text{ K})^{-1}$

r Exponential coefficient of replication, as in $N_t = N_o e^{rt}$ d^{-1}

s Settling velocity of a particle $m\ d^{-1}$

t Time unit (in s unless otherwise stated) for a given process to occur

t_b Time for oscillation of a thermocline   s

t_m Time taken for water to mix through a layer of given thickness   s

t_p Mean time, or photoperiod, in which an alga is exposed to light during its transport through the fully mixed water depth   s

u Velocity of water current in the horizontal x-axis $m\ s^{-1}$

u_t Mean turbulent velocity $m\ s^{-1}$

u^* Friction velocity of turbulent motion

$(u^*)^2$ Square of friction velocity of turbulent motion, $\tau \rho_w^{-1}$ $m^{-2}s^{-2}$

u' Velocity fluctuation in horizontal (x) plane $m\ s^{-1}$

v' Velocity fluctuation in horizontal (y) plane $m\ s^{-1}$

w' Velocity fluctuation in vertical (z) plane $m\ s^{-1}$

x Denotes a coordinate along the horizontal, long axis of the lake

y Denotes a coordinate along the horizontal axis perpendicular to the x axis

z Denotes a coordinate along the vertical axis (i.e., depth beneath the surface; positive values increase downward)

z_e Depth of frictional resistance   m

z_m Depth at the base of the mixed layer   m

Δ Difference; e.g., $\Delta\rho_w$ is the difference in ρ_w between two water masses located at depths z_1 and z_2

Σ "Sum of," as in ΣnP, the daily integral of gross photosynthetic capacity of an algal population

α Slope of a correlated, dependent variable on a limiting, independent variable; used here to denote gradient of light-limited, chlorophyll-specific photosynthetic rate (P) against increasing irradiance (I); the gradient, α, is a measure of photosynthetic efficiency of the algal chlorophyll   mol C (mol photon m^{-2})$^{-1}$

α_r	Slope of light-limited replication rate (r) against irradiance (I)	(mol photon m^{-2})$^{-1}$
β	Slope of regression predicting response of maximum growth rate to change in γ	
γ	Temperature coefficient of thermal expansion of water; $= -(1/\rho_w)(d\rho_w/d\theta)$	K^{-1}
∂	Increment of a variable; $\partial x/\partial y$ is the gradient between two pairs of variable coordinates (x_1 y_1),(x_2 y_2)	
ϵ	Coefficient of extinction of *PAR* with increasing depth	m^{-1}
ϵ_p	Coefficient of extinction due to the presence of suspended particles	m^{-1}
ϵ_s	Chlorophyll-specific coefficient of extinction attributable to suspended phytoplankton	m^{-1}(mg chl a·m^{-3})$^{-1}$, = m^2 (mg chl a)$^{-1}$
ϵ_w	(Background) coefficient of light extinction due to water and solutes	m^{-1}
ζ	Thermal conductivity of water	$J\ m^{-2}K^{-1}s^{-1}$
η	Absolute viscosity of water	P, = kg $m^{-1}s^{-1}$
θ	Temperature	K, °C
κ	Ratio of eddy size to boundary layer height ($\sim$0.4), von Karman's constant	dimensionless
λ_e	Latent heat of evaporation	J kg^{-1}
λ_f	Latent heat of fusion	J kg^{-1}
π	Ratio of circumference to diameter of a circle	dimensionless
ρ_A	Density of an alga	kg m^{-3}
ρ_a	Density of air ($\sim$1.2 kg m^{-3})	
ρ_w	Density of water ($\sim$1000 kg m^{-3}, but see text)	
σ	Specific heat of water	J kg^{-1} K^{-1}
τ	Wind stress on a lake surface	N m^{-2}, = kg $m^{-1}s^{-2}$
ϕ	Geographical latitude	°N or °S
ω	Angular velocity of the rotation of the earth	radians s^{-1}

Appendix 2.3   Abbreviations for algal names used in Figures 2.8 to 2.12, with authorities

An cir	*Anabaena circinalis* Rabenh. *ex* Born. *et* Flah.
An cyl	*Anabaena cylindrica* Lemm.
An flo	*Anabaena flos-aquae* Bréb. *ex* Born. *et* Flah.
Anki	*Ankistrodesmus braunii* (Näg.) Collins
Anky	*Ankyra judayi* (G.M.Sm.) Fott
Aphan	*Aphanizomenon flos-aquae* Ralfs *ex* Born. *et* Flah.
Ast	*Asterionella formosa* Hass
Cer h.	*Ceratium hirundinella* O.F. Müll
Chlm	*Chlorococcum* sp.
Chlo	*Chlorella pyrenoidosa* Chick
Clo ac	*Closterium aciculare* T. West
Coel	*Coelastrum microporum* Näg
Cry er	*Cryptomonas erosa* Ehrenb.
Cry ov	*Cryptomonas ovata* Ehrenb.
Cycl	*Cyclotella* spp.
Dict	*Dictosphaerium pulchellum* Wood

Din	*Dinobryon divergens* Imhof
Eud	*Eudorina unicocca* G. Sm.
Fra b	*Fragilaria bidens* Heib.
Fra c	*Fragilaria crotonensis* Kitton
Mel	*Melosira granulata* (Ehrenb.) Ralfs
Mic	*Microcystis aeruginosa* Kütz. emend. Elenkin
Monod	*Monodus subterraneus* Boye Petersen
Monor	*Monoraphidium minutum* (Leslie) Komarkova-Legnerova
O ag	*Oscillatoria agardhii* Gom.
O red	*Oscillatoria redekei* Van Goor
Pand	*Pandorina morum* Bory
Ped b	*Pediastrum boryanum* (Turp.) Menegh.
Perid	*Peridinium willei* Huitfield-Kaas
Rho	*Rhodomonas minutus* var. *nannoplanktica* Skuja
Scen q	*Scenedesmus quadricauda* (Turp.) Bréb.
Sph	*Sphaerocystis schroeteri* Chodat
Sta p	*Staurastrum pingue* Teiling
Ste a	*Stephanodiscus astraea* (Ehrenb.) Grun.
Ste h	*Stephanodiscus hantzschii* Grun.
Syn	*Synechococcus* sp.
T fl	*Tabellaria flocculosa* (Roth) Kütz. var *asterionelloides* (Roth) Knuds.
Volv.	*Volvox aureus* Ehrenb.

3

The Role of Competition for Resources in Phytoplankton Succession

Ulrich Sommer

Max Planck Institute of Limnology
Plön, Federal Republic of Germany

3.1 Introduction

The present time is an exciting period to write a review article on resource competition among phytoplankton. For decades, competition has been one of the main topics of both theoretical ecology and evolutionary biology. About 25 years ago the topic was introduced into plankton ecology by Hutchinson's (1961) famous paper on the "paradox of the plankton." About ten years ago Tilman (1977) initiated a period of intensive experimental research with his chemostat competition experiments between the diatoms *Asterionella formosa* and *Cyclotella meneghiniana*. At about the same time, others began to develop the concept of "top-down" control of plankton communities (e.g., Porter, 1977); they assumed that herbivory controls both seasonal biomass patterns and species composition of phytoplankton. Further development showed that "bottom-up" (competition for resources) and top-down control are not mutually exclusive, because differential mortality can be incorporated into a competition model (see Figure 11 in Tilman, 1982; and eq. 3.5).

One attempt, among others, to give balanced attention to bottom-up and top-down control in the explanation of plankton seasonal succession is the PEG-model (Sommer et al., 1986). This model is the outcome of a long-term discussion project carried out by about 30 plankton ecologists (the "Plankton Ecology Group," PEG). The PEG participants compared seasonal chemistry, phytoplankton, and zooplankton dynamics of 24 different lakes. As a result of our discussion, we proposed a word model, describing in 24 sequential statements, a chain of successional events for a model lake (see Chapter 1).

The model was constructed based on succession in Lake Constance, but it was intended to provide a framework for the interpretation of seasonal plankton succession in all lakes. It should be possible to derive a chain of events for other lakes from the PEG-sequence by *causally explained* deletion of events, addition of new ones, and/or modification.

Several of the statements of the PEG-model assume that resource limitation and competition occur. On the basis of the situation in Lake Constance (Figure 3.1), no limitation by resources at the start of the vegetation period is assumed (see Chapter 1, statement 1). Phytoplankton populations are assumed to reproduce at the maximal rate sustainable under the ambient physical conditions (see Chapter 2). It is further assumed that the usually smaller algae that dominate such growth are highly edible for zooplankton (statement 2). The subsequent increase of zooplankton is assumed to be sufficiently rapid to culminate in an overexploitation of the algae before the latter become seriously nutrient limited. Overexploitation leads to a crash of algal biomass ("clear water phase," statement 5) during which nutrient concentrations may again increase. After a decline of grazing pressure, still high nutrient concentrations initially permit nutrient-saturated growth of early summer algae (statement 8). Before nutrients become limiting, Cryptophyceae and colonial Chlorophyceae are dominant (statement 9); depletion of phosphorus (P) and competition for P under rich silicon (Si) concentration favors transition to diatoms (statement 11); depletion of silicate under sustained P-limitation leads to a dominance of dinoflagellates and/or *Cyanophyta* (statement 12); among them, species of *Nostocales* become dominant if reduced nitrogen (N) also becomes limiting (statement 13). The sequence of summer dominants is considered to be a combined result of nutrient competition and grazing selectivity. Grazing is thought to determine the size structure (inedible "canopy" species, edible "undergrowth"; statement 8), competition is considered to select for taxa. Later in the year decreased external input of light (shorter day length, deeper mixing) is considered to decrease the importance of biotic interactions (statement 17). In Section 3.9, the predictions of the PEG-model will be examined in order to determine how they can be interpreted in terms of modern competition theory and how they must be modified for other lake types.

While plankton ecology was describing biotic interactions and gaining confidence in the possibility of causal explanation of species replacements, a new school of ecological thought was emerging (mainly at Florida State University, Tallahassee), which doubted the evidence of biotic interactions and conceived communities as being randomly assembled or being controlled, if at all, mainly by external physical factors (Strong et al., 1983). In phytoplankton ecology, this point of view found support from Goldman et al. (1979). They hypothesized that phytoplankton of the oligotrophic, tropical ocean are growing at maximal reproductive rates (i.e., without resource limitation).

For the purpose of this review, I will restrict myself to resource competition ("exploitative competition" in the terminology of the general ecol-

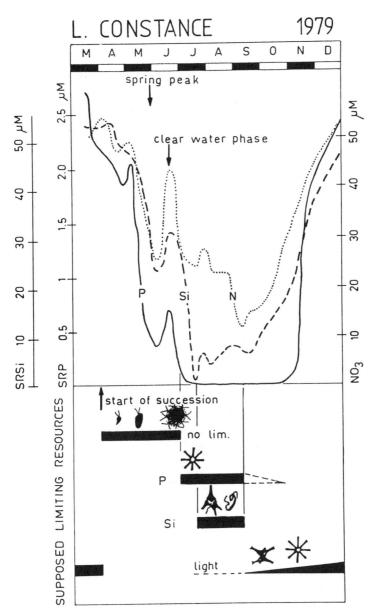

Figure 3.1   The resource-based predictions of the PEG-model for phytoplankton and the major nutrient data for the 0–6 m water layer for Lake Constance, 1979 (except for the duration of the P minimum in early autumn, other years are very similar; cf. Sommer, 1987a). The black horizontal bars in the lower portion indicate the presumed duration of limitation by a given resource. Phytoplankton species shifts are indicated by pictograms (*Rhodomonos, Cryptomonas, Pandorina, Asterionella, Ceratium, Anabaena, Staurastrum*).

ogist) and will not discuss allelopathy ("interference competition"). This restriction is not meant to imply that allelopathy plays no role in phytoplankton succession; however, at present, knowledge of phytoplankton allelopathy is so deficient that the only thing that can be stated with some confidence is: "more research is needed." In accordance with Tilman (1981), resource competition is defined as the depression of a competitor's reproductive rate by consumption of commonly needed resources. It follows necessarily that there can be no resource competition without limitation of the reproductive rate by resource availability. Resources are those factors which can control the reproductive rate of a population and which can be consumed. Unconsumable factors that may affect reproductive rates, such as temperature, salinity, and toxicity, are not—by definition—resources. Resources potentially limiting phytoplankton reproduction include light, inorganic carbon, those mineral nutrients which become scarce during the vegetation period (usually P, N, and—for diatoms—Si, perhaps also some trace elements) and, for some species vitamins. Unlike the chemical resources, light energy may be limiting prior to consumption by organisms (in the case of short daylength, deep mixing, high nonorganismic light attenuation). With the exception of extremely oligotrophic bodies of water, mineral nutrients are usually sufficiently abundant to sustain maximal reproductive rates of many algal species so long as most of them are freely available in the dissolved phase and relatively little is bound in biomass. For the majority of lakes, the statement "no nutrient competition without nutrient limitation" can be directly inverted into "no limitation without competition."

Because competition experiments of the last decade have concentrated on nutrient competition and because the role of light limitation in phytoplankton succession is covered by Reynolds (see Chapter 2), this chapter will mainly address the effects of competition for nitrogen, phosphorus, and silicon. These three are intended to stand as examples for any essential resource, including light. Conventionally, light limitation is described by different equations (Blackman-type) from those used to describe nutrient limitation (Michaelis-Menten-type). Nevertheless, the same competition models apply if light is treated as a (or the) limiting resource.

In this review, the concept and effects of nutrient limitation will be discussed (Sections 3.2 to 3.4). Section 3.5 is a short introduction to Tilman's (1982) *equilibrium theory of competition*, the backbone of all modern competition research in phytoplankton ecology. Section 3.6 is a summary of the taxonomic trends found so far in steady-state competition experiments. Section 3.7, on competition under fluctuating conditions, and Section 3.8, on the time demand for competitive exclusion, cover some "nonequilibrium" aspects. In Section 3.9, a discussion of three different concepts for the change of competitive interactions during succession brings us back to the central question—to what extent does resource competition select seasonal dominants within a successional sequence? Finally, a more speculative section addresses the question of how the species pool of a lake might have been assembled by past competition (Section 3.10).

At the onset, it must be stated that Tilman's (1982) equilibrium theories of competition are not universally accepted. Most frequently, they have been criticized because plankton is rarely, if ever, in steady state (e.g., Harris, 1986). However, "disequilibrium" does not necessarily mean that change cannot tend toward some equilibrium condition. Indeed, there is evidence that internal changes in community structure are strongly directional (Reynolds, 1988). Moreover, starting competition research with equilibrium models does not imply that disequilibrium is ignored. It simply means doing the first step first, according to the principle "seek simplicity, but distrust it." The next step ("distrust simplicity") is to explore which of the findings of steady state competition experiments may be retained and which have to be modified. Section 3.7 will show that such research is currently under way.

3.2 Nutrient Limitation Models

Nutrient limitation of phytoplankton has been the focus of an enormous variety of experimental and descriptive studies. The limitation models applied in those studies fall into four categories: limitation of uptake rates by external (i.e., dissolved and biologically available) concentrations of the limiting nutrient; limitation of reproductive rates by external concentrations of the limiting nutrient; limitation of reproductive rates by intracellular concentrations of the limiting nutrient ("cell quota"); limitation of total biomass by total (i.e., dissolved plus particulate) concentrations of the limiting nutrient. Models in this last class usually take the form of log-log regressions; the former three are based on saturation curves with an asymptotic approach to maximal uptake or reproductive rates. Usually only one nutrient is considered to be limiting for a given species at a given time (Liebig's "law of the minimum").

Regression models of total biomass vs. total nutrient concentrations ("yield limitation") have played an important role in the eutrophication discussion (e.g., Dillon and Rigler, 1974) but they do not significantly affect a discussion of mechanisms of species succession. The question of yield limitation should therefore be put aside. Limitation of uptake rates and of growth rates, however, are central to the concept of nutrient competition. Since some excellent reviews have recently appeared (Droop, 1983; Hecky and Kilham, 1987; Kilham and Hecky, 1987; Turpin, 1987) it is unnecessary to discuss the mathematics of nutrient limitation in detail. Only three models shall be presented here, which completely suffice for our purpose: The Monod equation (Monod, 1950) describes the dependence of the per-capita reproductive rate (μ) on the external concentration of the limiting nutrient (S):

$$\mu = \mu_{max} \cdot S/(S + k_s) \tag{3.1}$$

where μ_{max} is the "maximal" (i.e., nutrient- and light-saturated) reproductive rate and k_s is the half-saturation constant for growth. Maximal reproductive rates vary between about 0.3 d^{-1} (large dinoflagellates) and about 2.1 d^{-1} (small Chlorococcales); half-saturation constants for phosphorus vary between 0.003 μM (*Synedra*) and 1.9 μM (*Volvox*; for further data see Table 1 in Tilman, Kilham, and Kilham (1982) and Tables A40 to A58 in Jørgensen, 1979). In the case of silicate-limited growth of diatoms, sometimes a threshold concentration has to be introduced into the model (Paasche, 1980).

Dugdale's (1967) model of uptake limitation is mathematically analogous to the Monod equation:

$$v = v_{max} \cdot S/(S + k_m) \qquad (3.2)$$

where v is the specific uptake rate (uptake per cell or unit biomass). If conventional biomass parameters are replaced by the concentration of the limiting nutrient already incorporated into biomass, v becomes a growth rate of the cellular nutrient and eqs. 3.1 and 3.2 can directly be compared.

Droop's (1973) model describes the dependence of the reproductive rate on the intracellular concentration of the limiting nutrient ("cell quota," q):

$$\mu = \mu_{max}' \, (1 - q_o/q) \qquad (3.3)$$

where q_o is the minimal cell quota. The μ_{max}' of the Droop equation is higher than the μ_{max} of the Monod equation, because it is a theoretical value that could only be attained at an infinite cell quota.

For the field ecologist, the Monod equation is the most attractive one, because it directly relates an environmental variable (S) to a population-level response (μ). However, caution is needed in the application to field data. Under steady-state conditions, when nutrient consumption equals nutrient suply, and external concentrations remain constant, eq. 3.1, 3.2, and 3.3 are mathematically equivalent (Burmaster, 1979). Temporal variation of the nutrient regime, however, will lead to a breakdown of this equivalence. Short-term maximal uptake rates of some nutrients (e.g., P and N) are higher than maximal reproductive rates. This means that under transient conditions of saturating nutrient concentrations algae multiply their nutrient content faster than their biomass or cell number, i.e., the cell quota will increase ("luxury consumption"). If nutrient-saturated conditions are followed by nutrient limitation, the storage pool accumulated under rich condition will temporarily permit a higher growth rate than would be predicted from the Monod equation. This does not occur with silicate limitation of diatom growth, because the intracellular, metabolically accessible pool of silicate is negligibly small in comparison with the amount of silicate needed for the construction of new theacae (Paasche, 1980).

Algal species have three options to cope with a temporally variable nutrient regime (Sommer, 1985a).

1. They may be efficient users of low concentrations, i.e., the initial slope of their Monod and Dugdale curves (μ_{max}/k_s and v_{max}/k_m) should be high.

2. They may utilize transient nutrient enrichment for a rapid growth pulse that compensates for decline under nutrient impoverished conditions, i.e., they should have a high μ_{max}.

3. They utilize transient nutrient enrichment for the build-up of an intracellular storage pool; i.e., v_{max} should be much higher than μ_{max}.

However, storage can only be a successful strategy if the time between nutrient pulses is not too long, because without new uptake, each cell division leads to a halving of the cell quota. For the sake of brevity, these three strategies shall be called 1) affinity strategy, 2) growth strategy, and 3) storage strategy.

3.3   Does Nutrient Limitation Occur In Situ?

For several decades nutrient limitation of phytoplankton growth has been taken for granted, when one or several nutrients had been depleted down to indetectable or nearly indetectable levels. The Monod model has provided the rationale for that assumption. It should be noted, however, that half-saturation constants of the most efficient species are below the analytical limit of detectability for some nutrients. Tilman, Kilham, and Kilham (1982) report a k_s of 0.003 μM for P-limited growth of *Synedra filiformis*; the limit of detectability for soluble reactive phosphorus (SRP) is three times that concentration! It follows that at apparently zero concentrations of SRP the Monod model would predict anything between 0 and 75 percent of μ_{max}. Similar problems have been encountered with N-limited growth of marine algae. Moreover, it may be questioned whether the chemically measured nutrient species coincide with the biologically available ones. The conventionally measured soluble molybdenum-blue reactive phosphorus, for instance, contains much more than the orthophosphate ion (Rigler, 1968). Although Løvstad and Wold (1984) found that the Monod kinetics of *Synedra acus* and *Diatoma* sp. do not differ significantly for orthophosphate and SRP, the discussion is still far from being settled. Comparable questions may arise with respect to the biological availability of different species of inorganic nitrogen and of urea.

   The most serious criticism against the line of evidence from low nutrient concentration to nutrient limitation arises from nonsteady-state conditions in natural waters. Variability of nutrient concentrations occurs at many different spatial and temporal scales. Sources of such variability include the generation of micropatches by zooplankton excretion (Goldman et al., 1979), diel patterns of algal uptake and zooplankton excretion, abundance fluctuations of nutrient-consuming and nutrient-regenerating organisms, and episodic mixing events. As discussed above, under these circumstances the Monod relationship breaks down for those nutrients which can be stored within the cell.

Use of the Droop model instead of the Monod model is impeded by the difficulties in measuring cell quotas of individual populations: natural seston is a mixture of many algal species with diverse heterotrophs and detritus. However, if absolute per-capita reproductive rates (μ) are replaced by relative ones ($\mu_{rel} = \mu/\mu_{max}$) species-specific differences in μ_{max} disappear. Only some uniformity in the biomass-specific minimal cell quota is required in order to use the chemical composition of a plankton mixture as an indicator of the average nutritional state of all populations. eq. 3.3 can be rewritten as:

$$\mu_{rel} = 1 - q_o/q \text{ or } 1 - \mu_{rel} = q_o/q \tag{3.4}$$

$1 - \mu_{rel}$ may be defined as *intensity of nutrient limitation* (IL) and is inversely proportional to the cell quota. If the cell quota is expressed on a per carbon basis, IL is proportional to the C:N, C:P, etc. ratio. Based on culture experiments with several marine algae, Goldman et al. (1979) have recommended the use of an atomic ratio of C:N:P of 106:16:1 (the "Redfield ratio") as indicator of near maximal reproductive rates, i.e., absence of N- or P-limitation.

Biomass-based minimal cell quotas for N are very scarce in literature: minimal P-quotas are more common. Forty values (for both freshwater and marine algae) were found. Using Redfield stoichiometry they predict an average value of 0.86 for μ_{rel}, but the scatter is very wide (see frequency distribution plot in Figure 3.21; Section 3.10). Even if there were less interspecific scatter, several problems remain:

1. A mixture of a P-limited species with a surplus of N in the cells with an N-limited species with a surplus of P in the cells may give Redfield stoichiometry.

2. Phytoplankton limited by some other factor (light, trace elements) may have Redfield stoichiometry for C, N, and P (Tett et al., 1985).

3. Detritus or detrital-bacterial aggregates may leak the limiting nutrient (the general assumption) or enrich it (Gächter and Mares, 1985).

Despite these objections, comparison of seston stoichiometry with other indicators for nutrient limitation (Sommer, 1988) and comparison of seston fractions with different detritus content (Healey and Hendzel, 1980) suggest that conclusions based on seston ratios are only slightly, if at all, biased by detritus.

For future work involving particle stoichiometry, it will be best to appropriate fractionation techniques (e.g., size fractionation, density gradient separation) to obtain monospecific plankton fractions. Not only does this overcome some of the methodological problems, it also provides species-specific information, and this is just the kind of information needed for the purpose of understanding the influence of nutrient limitation in phytoplankton succession. Whenever possible, chemical analyses should be supplemented by an independent method of assessing nutrient limitation. Simple enrichment assays may be used, if care is taken not to impose a stronger nutrient limitation in the control bottle than in situ. This can easily happen, since in the bottles algae are protected from several loss processes (such as grazing and sinking) and thus may accumulate a higher biomass than in situ. This problem can be circumvented if the original plankton suspension is sufficiently diluted by filtered lake water, and data analysis is restricted to the period of the first few cell divisions (Løvstad, 1986; Sommer, 1988).

Goldman et al. (1979) were the first to use the stoichiometry approach to assess the intensity of nutrient limitation in situ. In their influential article, they presented a compilation of oceanic seston stoichiometry data which clustered around the Redfield ratio. They concluded that oceanic phytoplankton might be growing at μ_{max}. Despite the caution with which they expressed their conclusion, despite the fact that their compilation includes data from deeper water layers that are certainly light limited (Tett et al., 1985), despite the other methodological problems, and despite the restriction of their data set to the tropical oceans, Harris (1983) cited their paper as follows: "If growth rates are indeed maximal in most, if not all environments, because of nutrient recycling (Goldman et al., 1979), then changes in the observed rates of growth in natural populations will be more a function of the scales of environmental perturbations than nutrient concentrations." In his recent book, Harris (1986) again insists on the absence of nutrient limitation of phytoplankton growth.

Meanwhile, data have been published which, according to stoichiometry, indicate nutrient limitations. Sakshaug et al. (1983) found maximal C:P ratios of 357 in the Trondheimfjord and 400 in eutrophic Lake Haugatjern. Maximal C:N ratios were 14 in the Trondheimfjord, 14.7 in the North Sea off Bergen, and 12.9 in Haugatjern. Heaney et al. (1987) found C:P ratios of up to 717 during the *Ceratium* bloom in eutrophic Esthwaite Water. In Lake Constance, maximal C:P ratios in the light-saturated layer are around 400 (Sommer, 1987a). According to Goldman (1979) a C:P ratio of 400 indicates a μ_{rel} of 0.6 to 0.7; a C:N ratio of 14 suggests a μ_{rel} of about 0.25. Except for Esthwaite Water, these data have in common that elevated levels of the C:P or C:N ratios are restricted to short periods and are interrupted by periods when the seston stoichiometry is again near the Redfield ratio. This suggests that nutrient limitation occurs more as a bottleneck for only a few generations, rather than as a constant steady state.

In the moderately eutrophic Schöhsee in northern Germany, the stoichiometry of total seston and of size fractions have been compared with the results of enrichment bioassays (Sommer, 1988c). There was good agreement between the C:P and C:N ratios and an averaged index of P- and N-limitation in the bioassays (average of IL weighed by biomass for each size fraction). Again P- and N-limitation was interrupted in time and relatively modest in most cases (Figure 3.2). Individual species, however, were—for short periods—strongly P-or N-limited (IL > 0.5). Furthermore, the temporal pattern in the intensity of limitation was different between different species.

The case for Si-limitation of diatoms has always been more readily accepted than P- and N-limitation, partly because of the absence of significant intracellular storage, and partly because of Lund's (1950) influential study on the termination of an *Asterionella* bloom by silicate shortage. Here, *Asterionella* responded to extreme Si-shortage not only by cessation of reproduction, but even by physiological death. Later studies

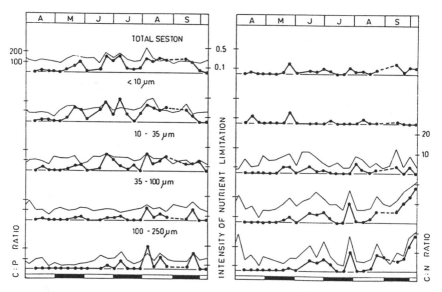

Figure 3.2 Intensity of P- and N-limitation of Schöhsee phytoplankton reproduction as estimated by enrichment bioassays (thick lines, weighted averages for total phytoplankton and for size fractions) compared with the C:P and C:N ratios of total seston and seston size fractions (thin line).

have confirmed this behavior of *Asterionella* both experimentally (Moed, 1973) and in the field (Sommer and Stabel, 1983). Since diatoms leave a visible skeleton after cell death, it is possible to calculate their reproductive rates from counts of live cells, dead cells, and cells collected in sediment traps (Reynolds, 1984; Sommer, 1984a). Thus the μ in situ can be compared with ambient Si-concentrations. Such a comparison provided convincing evidence for Si-limitation of diatoms at the end of their summer bloom in Lake Constance (Sommer, 1987a, 1987b). Also, in Schöhsee *Asterionella* was found to be strongly Si-limited throughout the entire stratified period.

The differences between Si-limitation and P- or N-limitation in the Schöhsee were remarkable. Si-limitation often was strong and persistent through time. P- and N-limitation, however, were often weak and were frequently interrupted by periods of nutrient saturation. Those differences may be explained by the completely different dynamics of those nutrients. Nitrogen and phosphorus are excreted by zooplankton in dissolved and biologically available form and are cycled rapidly between the different biological compartments and the dissolved phase. In contrast, sedimentation is the main process of loss for diatoms (Reynolds et al., 1982) and even if they are grazed, the frustules are converted into particulate debris from which dissolved silicate leaks only very slowly (Sommer, 1988a).

3.4   Taxonomic Trends in Nutrient Limitation

Monod-type experiments have revealed some gross taxonomic trends (see Table 1 in Tilman, Kilham, and Kilham, 1983; Tables A40 to A58 in Jørgensen, 1979). Diatoms of the family Fragilariaceae have very low demands for P (k_s = 0.003 to 0.02 μM) but high ones for Si (k_s = 1.5 to 19.7 μM), whereas some centric diatoms have higher P-demands (k_s = 0.13 to 0.25 μM) and lower Si-demands (k_s = 0.12 to 1.44 μM). Blue-green and green algae have high P-demands (k_s = 0.03 to 1.89 μM) and no significant demand for silicate. Unfortunately Chrysophyceae, Dinophyceae, and Cryptophyceae—all very important in freshwater phytoplankton—are rarely, if ever, used in culture experiments. In one of the few exceptions, Lehman (1976) found half-saturation constants around 0.5 μM for P-uptake by several *Dinobryon* species (note that k_m is not necessarily identical with k_s).

The enrichment bioassays using Schöhsee phytoplankton (Sommer, 1988c) were of special interest, because in this lake all three nutrients (P, N, Si) have the potential to become limiting. Remarkable differences among various higher taxa appeared both with respect to quality and to intensity of nutrient limitation (Figure 3.3).

Bacillariphyceae (represented in the data set by *Asterionella formosa, Synedra acus, Diatoma elongatum, Stephanodiscus rotula,* and *Stephanodiscus min-*

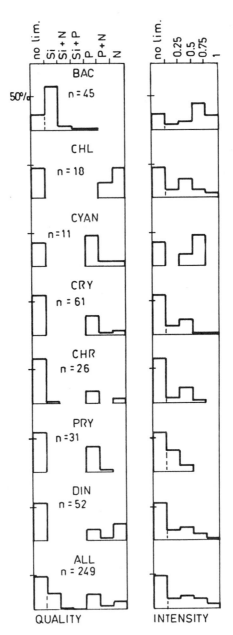

Figure 3.3   Frequency distribution of the quality (identity of the limiting nutrient) and the intensity of nutrient limitation (IL) of the reproductive rates for the different divisions of Schöhsee phytoplankton. Each species on each sampling date is an individual case; bioassays were made weekly. Abbreviations: BAC, Bacillariophyceae; CHL, Chlorophyceae; CYAN, Cyanophyceae; CRY, Cryptophyceae; CHR, Chrysophyceae; PRY, Prymnesiophyceae; DIN, Dinophyceae; ALL, total phytoplankton.

utus) were in most cases Si-limited, and often nutrient limitation was strong (IL $>$ 0.5 in 58 percent of cases).

Chlorophyceae were represented by *Sphaerocystis schroeteri, Ankyra judayi,* and *Closterium acutum.* If nutrient limited, these were either N-limited (44 percent) or they responded only to combined addition of P and N (22 percent). Usually nutrient limitation was not strong.

Cyanophyta (*Anabaena flos-aquae* and *Chroococcus limneticus*) were more often P-limited (45 percent) than N-limited (9 percent); if limited, nutrient limitation was relatively strong (IL $>$ 0.5 in 45 percent).

Cryptophyceae (*Rhodomonas minuta, Rhodomonas lens,* and *Cryptomonas ovata*) were not nutrient limited at all in 57 percent of cases, in 93 percent IL was $<$ 0.5. In case of nutrient limitation, P-limitation was most common.

Chrysophyceae were not nutrient limited at all in 65 percent of cases, in 96 percent IL was $<$ 0.5. Most cases of nutrient limitation were P-limitation of *Rhizochrysis* sp. *Dinobryon sociale, D. divergens,* and *D. bavaricum, D. crenulatum* were not nutrient limited in most cases. *Mallomonas caudata* was Si-limited once.

Prymnesiophyceae (only *Chrysochromulina parva*) were not nutrient limited in 58 percent, and IL was always $<$ 0.5. If limited, P-limitation was most common.

Dinophyceae (*Ceratium hirundinella, Peridinium bipes, P. cinctum, P. umbonatum,* and *P. inconspicuum*) were not limited in 54 percent; in 88 percent IL was $<$ 0.5. *Ceratium* was more often N-limited; *Peridinium* spp. were more often P-limited.

There was no size-related trend apparent in the data (Figure 3.4), but a clear difference appeared between coccoid algae and flagellates. Coccoid algae are more prone to nutrient limitation than flagellates. This tendency is preserved, though less pronounced, even if the cases of Si-limitation of diatoms are not considered. A lower frequency of limitation of flagellates may result from their motility. Diel vertical migrations may permit access to water that is richer in nutrients at or below the thermocline (Salonen et al., 1984; Sommer and Gliwicz, 1986). Chemotactic flagellates (Sjoblad et al., 1978) may exploit nutrient micropatchiness, even if their migrational amplitudes are insufficient for profitable vertical migration. Elevated short-term uptake velocities of limiting nutrients make even random encounters with nutrient micropatches profitable (Goldman, 1984).

3.5   The Theory of Steady-state Competition

The question has to be asked, should a competition model start from uptake kinetics (e.g., following Dugdale) or from growth kinetics (e.g., following Monod)? Uptake kinetics express the ability to exert competition by consumption of resources, whereas growth kinetics express the ability to with-

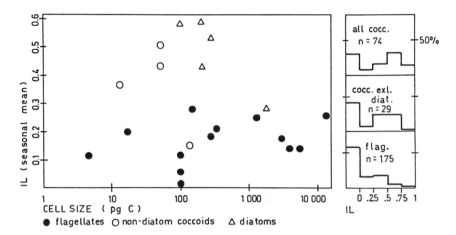

Figure 3.4 Left: Seasonal averages of the intensity of nutrient limitation of Schöhsee-phytoplankton species as related to cell size (pg C) and "life-form" (Flagellate, non-siliceous coccoid, diatom). Right: Comparison of the intensity of nutrient limitation (IL) for the different life forms based on frequency distributions of individual cases.

stand competition by leaving sufficient offspring despite depleted levels of resources. If uptake kinetics and growth kinetics are closely correlated, the difference does not matter very much. This is usually the case when cellular demands for a given nutrient (expressed by q_o) are similar, but if cellular demands are very dissimilar (as is quite common for Si), the effect of high demands may outweigh the effect of uptake affinity. In such a case an individual with high specific uptake rates may leave less offspring than an individual with lower specific uptake rates. Sooner or later the offspring of the individual with lower specific uptake rate will become so numerous, that their total resource consumption will be higher than the total resource consumption by those with higher specific uptake rates. It follows that growth kinetics are a better starting point for a competition model than uptake kinetics.

Tilman's (1977, 1982) steady-state model of competition is built on the theory of chemostat culture (Monod, 1950). A chemostat is a flow-through culture, in which fresh medium is supplied at the same, and temporally invariable rate, as that at which organisms are removed by the overflow. In such a culture a steady state is attained, where the per-capita reproductive rate equals the "loss" rate (λ) and where the input of fresh nutrients equals the consumption of nutrients. The residual concentration (R^*) of the limiting nutrient can be calculated by rearrangement of the Monod equation:

$$R^* = \mu \cdot k_s / (\mu_{max} - \mu); \mu = \lambda \tag{3.5}$$

Note that in chemostat culture, the ambient nutrient concentration has become the dependent variable. The steady-state biomass can be calculated from the incorporated amount of the limiting nutrient ($S_1 - R^*$, where S_1 is the nutrient concentration of the inflow medium) and the yield coefficient (Y), which is the reciprocal of the cell quota. If several species compete for the same limiting nutrient, the one with the lowest value for R^* under the given conditions will outcompete all the others ("competitive exclusion"; Hardin, 1960). At this concentration the other species are unable to attain a reproduction rate in balance with the losses. Note that the model includes the possibility of interspecific differences in loss rates (see Figure 11 in Tilman, 1982), although it is technically difficult to impose different loss rates in a chemostat experiment. Three cases of comparative kinetics can be imagined:

1. If the Monod curves do not intersect and if losses are equal between competitors, at all loss rates the same species will win.

2. If the Monod curves intersect and losses are equal, the species with the higher maximal reproductive rate will win at high loss rates and the species with the higher initial slope will win at low loss rates.

3. If losses are different between competitors, advantages in loss resistance may offset disadvantages in growth kinetics. From the point of view of competition, loss resistance becomes a "resource conservation strategy." This consideration shows that the concepts of resource controlled (bottom-up) and of predation-controlled community dynamics (top-down) are not necessarily mutually exclusive.

Coexistence under steady-state conditions becomes possible if the competing species are limited by different resources. The superior competitor (species 1) for resource 1 can be prevented from outcompeting the subordinate competitor for resource 1 (species 2), if limitation by some other resource (2) prevents species 1 from depressing resource 1 below the R^* value of species 2. Tilman (1982) has developed a simple graphical technique by which the possibilities of exclusion and coexistence in a system with two potentially limiting resources can be analyzed (Figure 3.5).

First, establish the R^*-values for the species and resources in question. Draw a "zero-net-growth isocline" (ZNGI) which connects all combined concentrations of the two resources that permit a reproductive rate in balance with the loss rate (see Figure 3.5A). According to the law of the minimum, the ZNGI must be rectangular. The corner of the ZNGI defines the "optimal ratio" of the two resources, i.e., the transition point from limitation by resource 1 to limitation by resource 2. Optimal foraging considerations suggest that at steady state the two resources will be consumed in optimal ratios. The combined concentrations of the two re-

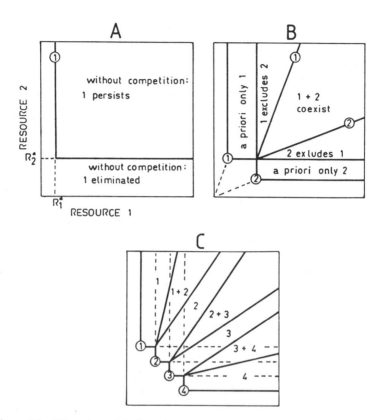

Figure 3.5   Tilman's model of steady-state competition for two resources. A: Definition of the "zero-net-growth isocline" (ZNGI) of species 1 by the R^*-values (for definition see eq. 3.5) for two limiting resources. If prior to consumption by organisms the concentration of both nutrients is higher than the R^*-values, species 1 can persist in the absence of competition. B: Conditions for coexistence and exclusion of species 1 (superior competitor for resource 1) and species 2 (superior competitor for resource 2). Both species are characterized by their ZNGI's and their consumption vectors (diagonal lines, slope corresponds to optimal ratio of the two resources). C: Four species share a resource-ratio gradient. 1 is the best competitor for resource 1, 2 is the second best, 3 third, and 4 fourth; the competitive rank order for resource 2 is vice versa. Single numbers indicate regions of 1-species dominance, double numbers indicate regions of 2-species coexistence.

sources before uptake by organisms (e.g., concentrations in the inflow medium of a chemostat) can be characterized by a "supply point." If the supply point falls into the region between the ZNGI and the two axes, the species in question will be eliminated from the system even in the absence of competitors. If the supply point lies in the region outside of the ZNGI, the species can establish itself in the absence of competitors.

If the ZNGI's of competing species do not intersect, the species with the ZNGI closest to the axes will always outcompete the others.

If the dominant competitor for resource 1 is the subordinate competitor for resource 2 and vice versa, the ZNGI's of the competing species intersect (Figure 3.5B). The competing species can coexist in steady state when the supply point lies in the region confined by the intersection point of the ZNGI's and two vectors, the slope of which is defined by the optimal ratios for the two species (consumption vectors). In this case, the residual concentrations of the two limiting resources are focused to the intersection point ("two-species equilibrium point") by the combined action of resource consumption and resource resupply. If the supply point lies in the region closer to the y-axis (low resource 1:resource 2 ratios) species 1 excludes species 2; if the supply point lies in the region closer to the x-axis (high resource 1:resource 2 ratios) species 2 excludes species 1.

Numerous species can share a gradient of resource ratios, provided that the rank order of competitive dominance for resource 1 is the reverse of the rank order of competitive dominance for resource 2 (Figure 3.5C). Note, however, that at each single-resource ratio no more than two species can coexist; i.e., more generally, the number of coexisting competitors cannot exceed the number of limiting resources.

The steady-state competition model has now been tested in a number of chemostat experiments (see Table 3.1). Whenever the Monod kinetics of the competing species were established, the outcome of competition followed the predictions derived from the Monod kinetics (e.g., Tilman, 1977, 1981; Holm and Armstrong, 1981; Kilham, 1984; Sommer, 1986b). Taxonomic trends have been consistent between experiments carried out at different laboratories and the predictions about the conditions of coexistence and exclusion have been verified. At present, Tilman's competition model is one of the experimentally best-verified concepts in physiological ecology.

3.6 Taxonomic and Size-related Trends

Chemostat competition experiments have either used pairs of species from clonal cultures or multispecies assemblages taken directly from lakes (see Table 3.1). The former approach has the advantage that the outcome of competition can be compared with the prediction derived from single-species Monod kinetics; the latter approach has the advantage that the competitive dominants within a large species pool can be identified. The experimental variables employed so far have included nutrient-supply ratios, dilution rate (i.e., at equilibrium reproductive rates), and temperatures. Among resource ratios, the Si:P ratios has most often been used. Nutrient:light ratios have not

Table 3.1   Steady-state competition experiments with planktonic algae

Competing species	Experimental variable	Reference
Asterionella formosa *Cyclotella meneghiniana*	Si:P ratio Dilution rate	Tilman, 1977
Asterionella formosa *Fragilaria crotonensis* *Synedra filiformis* *Tabellaria flocculosa*	Si:P ratio	Tilman, 1981
Asterionella formosa *Synedra ulna*	Si:P ratio Temperature	Tilman, Matson, and Langer, 1981
Asterionella formosa *Microcystis aeruginosa*	Si:P ratio	Holm and Armstrong, 1981
Natural plankton from Lake Mephremagog	Dilution rate	Smith and Kalff, 1983
Natural plankton from Lake Constance	Si:P ratio	Sommer, 1983
Stephanodiscus minutus *Synedra* sp.	Si:P ratio	Kilham, 1984
Natural plankton from Lake Superior	Si:P ratio P:N ratio Temperature	Tilman and Kiesling, 1984
Fragilaria crotonensis *Tabellaria flocculosa*	Si:P ratio Inoculum size of competitors	Tilman and Sterner, 1984
Natural plankton from Lake Michigan	Si:P ratio	Kilham, 1986
Natural plankton from Lake Constance	Dilution rate	Sommer, 1986a
Natural plankton from the Antarctic Sea	Si:N ratio Dilution rate	Sommer, 1986b
Natural plankton from Lake Superior and Eau Galle Reservoir	Si:P ratio N:P ratio Temperature	Tilman et al., 1986

yet been explored; the Si:N gradient has only been tested for Antarctic marine phytoplankton.

In all multispecies experiments published so far, coccoid or filamentous algae have become dominant. Flagellates were usually excluded relatively early during the course of experiments. In 2-species experiments they have never been used. The poor performance of flagellates in competition experiments contrasts with the observation that in situ flagellates tend to be less nutrient limited than coccoid algae. Two hypothetical explanations may be given. They might be more sensitive to the mechanical stress of stirring, which is necessary to minimize wall growth and to assure a homogeneous distribution of cells. Alternatively, if—as hypothesized in Section 3.4—flagellates are better adapted to exploit spatial heterogeneity (be it micropatchiness or vertical gradients) they lose their main advantage in an experimental system where everything is done to prevent spatial heterogeneity.

Owing to the poor performance of flagellates, available information on taxonomic trends in nutrient competition comprises only three divisions: diatoms, green algae, and cyanobacteria. Among them, several diatoms dominate at high Si:P ratios, i.e., they are the superior P competitors (Sommer, 1983; Kilham, 1986; Tilman et al., 1986); as N competitors they are subordinate to cyanobacteria (Tilman and Kiesling, 1984). Cyanobacteria dominate at low N:P ratios. Green algae are the poorest N-competitors, in P-competition they are dominant over cyanobacteria but subordinate to diatoms. As a consequence, they dominate at low Si:P, but high N:P ratios. These basic patterns are modulated by temperature, because μ_{max} always and k_s sometimes is a function of temperature (Tilman, Matson, and Langer, 1981). At higher temperatures diatoms require higher Si:P ratios in order to become dominant over green algae in multispecies experiments (Figure 3.6). Green algae are most competitive at medium temperatures (15 to 20°C), cyanobacteria at high temperatures (> 20°C; Tilman and Kiesling, 1984).

The above classification of higher taxa relies only on the few species which have been successful competitors in the experiments. Other species of

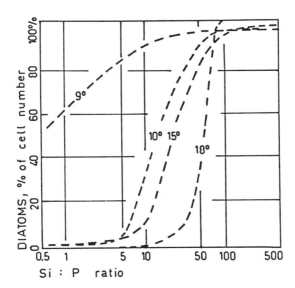

Figure 3.6   Percent contribution of diatoms to total cell number at termination of competition experiments on dependence of Si:P ratios on temperatures. (Sources: 9°C and 15°C: Tilman et al., 1986; 10°C: Kilham, 1986; 18°C: Sommer, 1983).

the same divisions may well behave differently. It cannot be predicted that at high Si:P ratios any diatom will win against any other alga. It can be predicted, however, that in a species-rich assemblage with a sufficient number of diatom species, one diatom species will be the dominant competitor at high Si:P ratios. Whenever natural inocula were used the following members of the genus *Synedra* became dominant at the highest Si:P ratios and at dilution rates < 1.6 d⁻¹: *Synedra acus* in Lake Constance (Sommer, 1983), *S. acus* and *S. ulna* in Lake Mephremagog (Smith and Kalff, 1983), and *Synedra filiformis* in Lake Michigan (Kilham, 1986).

Comparison of Monod kinetics and competition experiments revealed that several diatom species are able to share the gradient of Si:P ratios, i.e., that their competitive abilities for these two nutrients are inversely ranked (see Figure 3.5C). According to Table 1 in Tilman, Kilham, and Kilham (1982) *Synedra filiformis* is the best competitor for P, *Asterionella formosa* the second best, *Fragilaria crotonensis* third, *Diatoma elongatum* fourth, and *Stephanodiscus minutus* fifth. The rank order in Si-competition is the reverse of that for P. Comparison with kinetic data of several nondiatomaceous algae shows, however, that *Stephanodiscus minutus* would lose in a P-competition against *Oscillatoria agardhii*, *Mougeotia thylespora*, *Scenedesmus acutus*, and *Chlorella minutissima* (Figure 3.7). Since these species have higher μ_{max} values, they would also win at higher dilution rates than shown. The diatoms predicted to win P-competition against nondiatoms belong to a single family, Fragilariaceae.

The influence of dilution rates has been studied less often than the influence of resource ratios. A change in the taxonomic outcome should occur, if the Monod curves of the competing species intersect. Smith and Kalff (1983) used dilution rates from 0.055 d⁻¹ to 0.93 d⁻¹ at a high Si:P ratio (they gave no exact figure) for competition experiments with natural plankton and found no change in the dominant species (*Synedra acus*). Although they did not mention it in the text, their tables suggested that several other species were able to coexist with *Synedra*. Among them, shifts along the gradient of dilution rates could be found (Sommer and Kilham, 1985). A wider gradient of dilution rates (up to 2.0 d⁻¹) and two different nutrient compositions (Si:P = 800:1 and 0:1) for similar experiments with phytoplankton from Lake Constance were used in Sommer (1986a). At the high Si:P ratio *Synedra acus* was the only persisting species at all dilution rates up to 1.6 d⁻¹; only at 2.0 d⁻¹ was *Achnanthes minutissima* the winner. In the silicate-free treatment (complete exclusion of diatoms) *Mougeotia thylespora* was the winning species at 0.3 and 0.5 d⁻¹; *Scenedesmus acutus* at 0.7 and 0.9 d⁻¹, and *Chlorella minutissima* at the highest dilution rates.

The assumption that smaller algae are better competitors irrespective of taxonomic position is a spin-off from the general inverse size-metabolic rate law (Peters, 1983). Smith and Kalff (1983) used a regression "exclusion rate" vs. cell/colony size to prove that assumption (larger algae were more rapidly excluded (Figure 3.8). However, many divisions are represented only by one

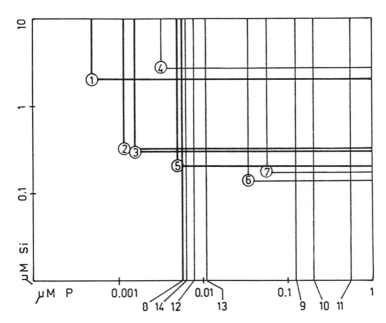

Figure 3.7   ZNGI's of 14 phytoplankton species for silicate and phosphate at a dilution rate of 0.1 d^{-1} at a temperature of 20°C (species 1 to 11; data from Tilman, Kilham, and Kilham, 1982) or 18°C (species 12 to 14; data from Sommer, 1986a). 1: *Synedra filiformis*; 2: *Asterionella formosa*; 3: *Fragilaria crotonensis*; 4: *Tabellaria flocculosa*; 5: *Diatoma elongatum*: 6: *Stephanodiscus minuta*; 7: *Cyclotella meneghiniana*; 8: *Oscillatoria agardhii*; 9: *Microcystis aeruginosa*; 10: *Volvox aureus*; 11: *Volvox globator*: 12: *Scenedesmus acutus*; 13: *Chlorella minutissima*: 14: *Mougeotia thylespora*.

species in their data set and the fate of the smallest algae (picoplankton), although certainly present in their natural inoculum, was not studied. If the size of successful competitors in steady-state experiments is compared to the possible range of freshwater phytoplankton cell volumes (1 to 10^5 μm^3) no support for a size-related trend can be found, except that none of the winners belongs to the largest order of magnitude and only two winners at very high dilution rates belong to the smallest order of magnitude. Thus, size alone does not explain competitive success.

3.7   Competition Under Fluctuating Conditions

The existence of spatial and temporal variability in nutrient concentrations and in external parameters influencing nutrient utilization (e.g., temperature) is beyond doubt. The question of interest is rather whether biological processes average over these fluctuations or not, i.e., whether the biological re-

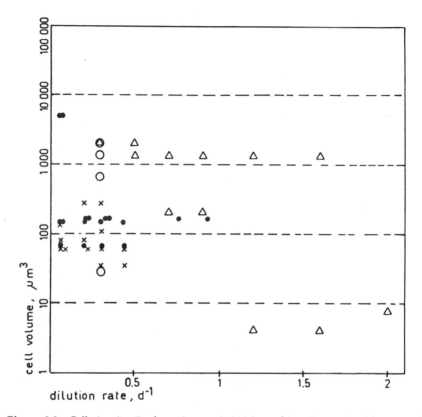

Figure 3.8   Cell size (μm³) of species persisting in multispecies competition experiments at different dilution rates. Sources: •: Smith and Kalff, 1983, acknowledged in the text as persisting; x: Smith and Kalff, 1983, extracted from the tables; O: Sommer, 1983; Δ: Sommer, 1986a.

sponse to fluctuating environmental conditions is similar to the response to homogeneous conditions with the same long-term averages of relevant environmental parameters or not. Obviously the response depends on the intensity and the duration of deviations from average. Responsiveness of biological systems to different scales of patchiness may be different at different levels of aggregation. The algal physiologist will be primarily interested in the response at the level of uptake rates; the ecophysiologist will be interested in the response at the level of reproductive rates and taxonomic-outcome competition; the community ecologist will be interested in the response at the level of species richness, diversity, and similar highly aggregated variables.

Goldman et al. (1979) explained phytoplankton growth near μ_{max} in the oligotrophic ocean by micropatchiness of nutrients caused by zooplankton excretion. This hypothesis was criticized by Jackson (1980) on hydrophysical grounds. He calculated that diffusion would destroy such micropatches before phytoplankton could take substantial profit from elevated concentrations. However, subsequent experimental work (Goldman and Glibert, 1982; Lehman and Sandgren, 1982) confirmed highly elevated short-term maximal uptake rates of nutrient-limited algae if exposed to a transient pulse of ammonium or phosphate. According to Lehman and Sandgren (1982) small cyanobacteria in Third Sister Lake were able to double their cell quota within two minutes. The short-term increase of v_{max} over its steady-state level means that algae acquire more nutrients in a patchy environment than in a homogeneous one of equal average concentrations. This contradicts Currie's (1984) expectation that due to the geometry of saturation curves gains from above-average nutrient concentrations should be smaller than losses from equivalently below-average concentrations. Interspecific competition under microscale patchiness has not yet been studied experimentally. Influence of micropatchiness on the taxonomic outcome of competition should be expected, though, if interspecific differences in the relative advantage of microscale patchiness are large enough to translate into marked differences in the response of reproductive rates.

Until now competition experiments under nonsteady-state conditions have used macroscale disturbances at time intervals from 1 to 28 days (d) (Table 3.2). The nature of disturbances consisted either of nutrient injection into a continuous flow culture or of discrete dilution events. The first technique imposes only a temporal pattern of nutrient availability and hence of reproductive rates; the latter technique also imposes a temporal pattern of loss rates. Usually a regular regime of disturbances was used; Robinson and Sandgren (1983) found only minor differences between regular dilution intervals and randomly variable ones with similar long-term averages. Like steady-state experiments, pulsed-state competition experiments also reach a final "equilibrium" phase where further competitive exclusion no longer occurs. In contrast to steady-state experiments, at least some of the persisting species show regular oscillations instead of constant population densities. Species showing rapid increase after enrichment/disturbance and population decline under subsequently impoverished conditions may be assigned to the type "growth strategist," species showing a weak or even insignificant growth rate response to enrichment are either "affinity strategists" or "storage strategists."

All nonsteady-state competition experiments published so far have shown a taxonomic response to the temporal pattern of disturbance. Within the marine plankton assemblage used by Turpin and Harrison (1980) *Chaetoceros* sp. became dominant under steady-state conditions and a 1-d dilution interval; *Skeletonema costatum* became dominant at 3-d intervals; *Thalassiosira nordenskioldii* at 7-d intervals. A repetition with clonal cultures of these species confirmed the results. In the experiments of Robinson and Sandgren (1983) *Dictyosphaerium planktonicum* was the numerical dominant under all disturbance regimes (1-, 7-, 28-d intervals), but major changes occurred at the level

Table 3.2   Nonsteady-state competition experiments with planktonic algae; time interval of disturbances in days

Competing species	Intervals (days)	Nature of disturbance	Reference
Natural marine plankton *Chaetoceros* sp. *Skeletonema costatum* *Thalassiosira nordenskioldii*	0*, 1, 3, 7	Dilution	Turpin and Harrison, 1980
Chlamydomonas sp. *Cryptomonas obovata* *Dictyosphaerium planctonicum* *Dinobryon cylindricum* *Euglena gracilis* *Mallomonas pappilosa* *Ochromonas danica* *Peridinium* sp. *Staurastrum pingue* *Stephanodiscus* sp. *Synedra delicatissima* *Synura spinosa* *Trachelomonas hispida*	1, 7, 28; regular and randomly variable	Dilution	Robinson and Sandgren, 1983
Natural plankton from Lake Constance	0, 7	P-injection	Sommer, 1984b
Natural plankton from Lake Constance	0, 7	P-injection P- and Si-injection	Sommer, 1985a
Natural plankton from Lake Constance	1, 2, 3, 5, 7, 10, 14	Dilution	Gaedeke and Sommer, 1986
Synedra sp. *Fragilaria crotonensis*	0, 8	P-injection	Grover, 1989

* 0 indicates a steady-state experiment.

of subdominant species. In two-species competition experiments with *Synedra* sp. and *Fragilaria crotonensis* as competitors and P as limiting nutrient (Grover, in press), *Synedra* excluded *Fragilaria* under steady-state conditions; under pulsed supply (8-d intervals) of P, it was not clear whether *Fragilaria* could coexist with *Synedra* or whether it was only excluded at a much lower rate.

Among Lake Constance phytoplankton *Mougeotia thylespora* is the dominant steady-state competitor for P, if lack of silicate excludes diatoms (Sommer, 1983); at 1-d dilution intervals it is replaced by *Koliella spiculiformis*, at 2-d intervals by *Pseudanabaena catenata*, and at intervals of 3-d or longer, diverse assemblages persist with either *Chlamydomonas* sp. or *Scenedesmus* sp. dominant by biomass (Gaedeke and Sommer, 1986).

Pulsed-state competition experiments with Lake Constance phytoplankton showed a marked displacement of dominance regions along the resource ratio gradient relative to steady-state experiments. In these experiments (Sommer, 1985a) either one (P) or two (P and Si) potentially limiting nutrients were injected at 7-d intervals into continuous flow cultures. Under pulsed conditions, diatoms needed higher Si:P ratios to become dominant over green algae (Figure 3.9). The P-pulses were used by green algae both for rapid growth response (*Scenedesmus* sp., *Chlorella minutissima*, and *Pandorina morum*) and for P-storage (presumably *Mougeotia thylespora*, *Pediastrum duplex*, and *Staurastrum cingulum*). The only cyanobacteria of some importance, *Aphanizomenon flos-aquae*, probably also may be considered a storage strategist. Excellent storage capacities of this species are known from the physiological literature (Uehlinger, 1980). Diatoms did not take a comparable advantage from Si pulses, since appreciable intracellular storage of silicate is impossible and the best steady-state competitors at high Si:P ratios (*Synedra*, *Asterionella*) have only moderate maximal reproductive rates. The diatom best conforming to the type "growth strategist" was *Nitzschia acicularis* (see Figure 3.9).

Tilman (1982) suggests two different mechanisms by which variability in nutrient supply can increase species richness beyond the number of limiting resources (graphically demonstrated in Figure 3.10, for a rigorous proof see Tilman, 1982; Levins, 1979). If many species share a ratio gradient of two resources (Figure 3.10A), the resource supply point may be replaced by an area representing the range of variation of both resources. If this area overlaps the regions of persistence of more than two species, more species can coexist than there are limiting resources.

Figure 3.10B demonstrates the possibility of coexistence of two species competing for only one limiting resource. If, as discussed above, nutrient patchiness is advantageous in terms of long term average reproductive rates, the R^* value for the resource average decreases with increasing variance of the resource. This is visualized by the curved ZNGI's in Figure 3.10B. Species 2 takes a relatively greater advantage from patchiness than species 1. Consequently, species 2 consumes the resource variance (var. R) and the resource average (av. R) at a higher ratio, as indicated by the slope of the consumption vectors. If the supply point lies in the region between the intersection point of the ZNGI's and the consumption vectors, both species can coexist. According to the terminology of Levins (1979) var. R becomes an "additional resource." As with two chemically distinct resources, numerous species can share a gradient of var. R:av. R ratios.

The relative importance of the two mechanisms (they are not mutually exclusive) can be distinguished experimentally. If variable resource ratios alone mattered, the dominant species can be predicted from the average ratio and only a number of subdominant species would be added. If all species were simultaneously limited by the same resource, no increase of the species number could be brought about by resource variability. Parallel variation in the absolute levels of two limiting resources (white vector in Figure 3.10A) would

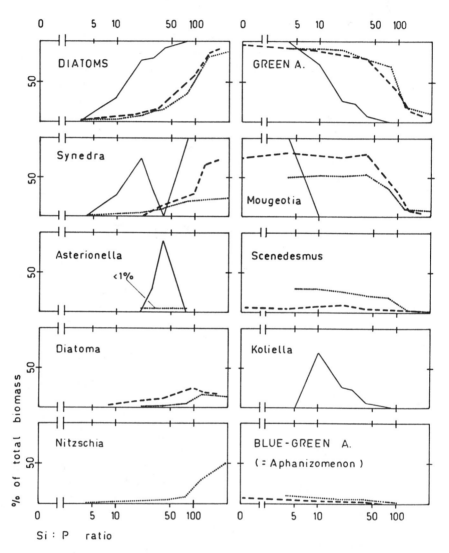

Figure 3.9   Percent contribution of phytoplankton species and divisions to total bio-volume in the terminal phase of steady state- (solid line), pulsed P-(dashed line), and pulsed P- and Si- (dotted line) competition experiments with natural phytoplankton from Lake Constance (after Sommer, 1985a). Note that the values for the pulsed-state experiments are averages for the cycles of oscillations.

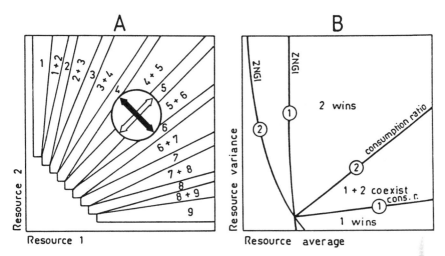

Figure 3.10   Coexistence of species under variable nutrient supply. A: 9 species share a gradient of 2 resources If variance of the resource supply is constrained by the circle, species 4, 5, and 6 can coexist. B: Species 1 and 2 compete for one variable resource. Species 1 takes a greater advantage from var. R and consumes a higher var. R:av. R ratio.

be inert in terms of diversity. Only change in the direction of the black vector would increase diversity. In contrast, the model in Figure 3.10B permits a change of dominant species by resource patchiness, coexistence even if only one resource is limiting, and increase of species number over the number of resources even if resource ratios do not change. Nonsteady-state competition experiments where only one of two potentially limiting nutrients is added discontinuously offer a strong variability of the resource ratio, whereas experiments with simultaneous addition of both limiting nutrients or discontinuous-dilution experiments show only weak variation in the resource ratio.

Both in discontinuous-dilution experiments (Gaedeke and Sommer, 1986) and in nutrient-injection experiments (Sommer, 1984b, 1985a) where—in total absence of Si—only P was limiting, coexistence of several species was found, when the time interval between disturbances was at least 3 d. Nutrient-injection experiments in which Si and P were injected simultaneously (little variability of resource ratios) yielded a higher species number and more diversity than experiments with injection of P alone (strong variability of resource ratios). This effect was found over the entire range of tested Si:P ratios (Figure 3.11). It follows that model B in Figure 3.10 is a more realistic representation of the mechanism that enables coexistence under nonsteady-state conditions.

The results of discontinuous-dilution experiments so far published suggest that the interval between dilution has to exceed one generation time to increase diversity beyond the level that could be attained at steady state under

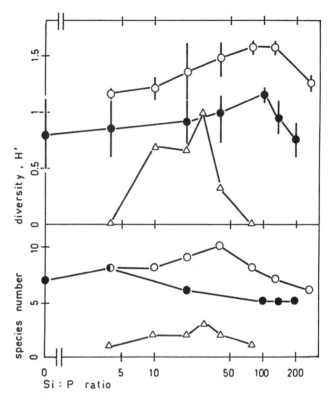

Figure 3.11   Species number and diversity index (Shannon-Weaver, H') in steady state (Δ), pulsed P (●), and pulsed P and Si (○) competition experiments (Sommer, 1985a). The vertical bars indicate the fluctuations of H' within the 7-d intervals between nutrient pulses.

double nutrient limitation (Figure 3.12). It is not yet clear whether the diversity peak at an interval of about three generation times in the experiments of Gaedeke and Sommer (1986) is valid or only an artifact caused by the small size of the culture vessels. Such an artifact could arise from the fact that in order to maintain the same long-term mean dilution rate over the entire range of tested intervals, dilutions have to be very drastic at the longer intervals which implies the danger of random extinction of rare species. It should be expected, however, that at very long intervals between disturbances, diversity should decrease again (see Connell's "intermediate disturbance hypothesis," 1978; for field data in support of this hypothesis see Reynolds, 1988). At very long intervals, the period of nutrient depletion should become too long to permit persistence of species that have to rely on enrichment pulses. It is noteworthy that the diversity peak at about three generation times is relatively close to the diversity maximum in the lake itself (H' about 2.5).

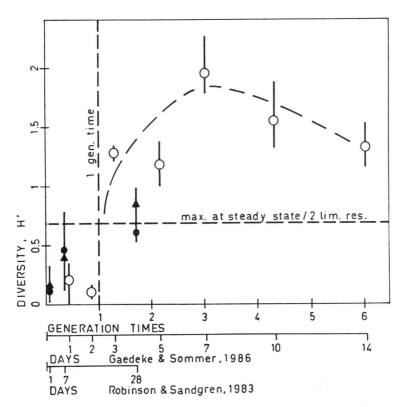

Figure 3.12   Diversity index (Shannon-Weaver, H') in versus the intervals between dilutions in discontinuous dilution experiments; time of intervals expressed as multiples of the average generation time (calculated as steady state doubling time from dilution rates). O: Gaedeke and Sommer (1983); •: Robinson and Sandgren (1983), regular intervals; ▲: Robinson and Sandgren (1983), randomly variable intervals; broken horizontal line; maximal theoretically possible diversity at steady state with 2 limiting resources.

3.8   Temporal Dynamics of Competition

Theoretically, competitive exclusion is an asymptotic process which demands infinite time until completion. The rate at which the losing species (L) is excluded (exclusion rate, ϵ) can be calculated from the R^* of the winning species (W), the dilution rate of the chemostat ($=$ loss rate; at steady state, identical with μ_w), and the Monod parameters of the loser.

$$\epsilon = \mu_{max,L} \cdot R^*_w / (R^*_w + k_{s,L}) - \mu_w \qquad (3.6)$$

Before the limiting resource is depleted down to the level of R^*_w, ϵ should be smaller. Figure 3.13 shows the dependence of ϵ on the loss rate for three combinations of Monod kinetics of winner and loser:

1. μ_{max} different, k_s
2. μ_{max} similar, k_s different;
3. intersecting Monod curves.

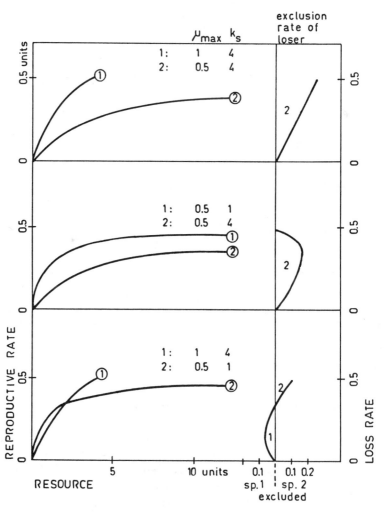

Figure 3.13   Exclusion rate of loser in steady-state competition as a function of the Monod kinetics of the competing species and of the loss rates (identical loss rates for both species assumed). Top: μ_{max} different, k_s similar. Center: μ_{max} similar, k_s different. Bottom: intersecting Monod curves.

In all cases ϵ is minimal at minimal loss rates. In case no. 1, ϵ increases linearly with the loss rate; in case no. 2, ϵ is maximal at an intermediate, but relatively high loss rate; in case no. 3, ϵ of the growth strategist is maximal at an intermediate loss rate and ϵ of the affinity strategist increases with the loss rate. For all cases a slightly paradoxical result may be stated: Although resource competition is necessarily mediated by resource limitation, there is no positive correlation between the intensity of resource limitation and the intensity of competition, if defined as the rapidity of competitive exclusion. Thus, the observation that algae grow at high rates (Goldman et al., 1979) neither supports nor rejects the importance of competition. Competition can occur, and may be quite strong, even at high turnover rates.

Kalff and Knoechel (1978) have argued that exclusion of the loser in Tilman's (1977) experiments was quicker than calculated from the kinetic data. They took this as an indication that something beyond exploitative competition was happening. However, after recalculating the value of ϵ from Tilman's graphs for those experiments in which *Cyclotella meneghiniana* was excluded, one obtains the results shown in Figure 3.14. The values for ϵ were widely scattered, but only for the lowest dilution rates

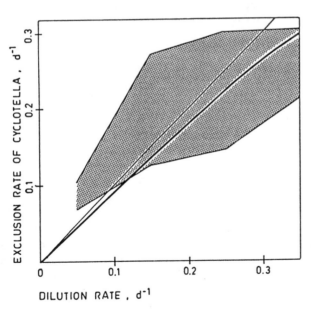

Figure 3.14   Exclusion rates of *Cyclotella meneghiniana* in Tilman's (1977) competition experiments (*Asterionella* versus *Cyclotella*) under P-limitation for both species as a function of the dilution rate. Shaded area: range of observed values; thick line: Maximal estimate of ϵ calculated by eq. 3.6; thin line: ϵ if there were no reproduction at all ($\mu = 0$, but no mortality).

were all values higher than anticipated theoretically. With increasing dilution rates contradiction between anticipated and observed data decreased. Deviations of observed growth kinetics from the Monod model at very low relative growth rates ($< 0.2\ \mu_{max}$) are quite common (e.g., Droop, 1983) and may be explained by the fact that under such conditions algae are not simply limited but are also physiologically "ill," which leads to cell mortality (Müller, 1972).

Usually competition experiments are terminated when some arbitrary criterion is fulfilled: 99 percent or 95 percent of total biomass/cell number for the winning species or winning couple of species and/or continuous decline of the losing species over some minimal time period, usually expressed as a multiple of the chemostat's water renewal time (Sommer, 1983). Typically, the duration of chemostat competition experiments is 25 to 50 days. Opponents of competition theory argue that such long periods of steady state rarely occur in natural lakes and, therefore, competition theory has no "real-world" application (Harris, 1986). However, the duration of experiments is dictated by the demand of securing confidence in the results. If experiments were terminated as soon as the first symptoms of competitive pressure were found, their duration would be much shorter. When two-species competition experiments are started with a total biomass relatively similar to the steady-state biomass the subordinate competitor starts to decline from the very beginning (Tilman, 1977); if inoculum biomass is much smaller than steady state-biomass, the loser starts to decline as soon as steady state biomass is approached (Tilman, 1977, 1981). In experiments with natural inocula where none of the species is physiologically adapted to the situation in the chemostat it takes about two weeks until the last loser starts to decline continuously. Despite differences in inoculum composition, dilution rate, and chemical composition, this time was remarkably uniform in all experiments published so far (Smith and Kalff, 1983; Sommer, 1983, 1986a): 14.4 ± 3.1 d (S.D.). The time needed for the winning species or winning couple of species to reach 95 percent of total biomass was 23.3 ± 5.0 d in those experiments.

Tilman and Sterner (1984) performed experiments where one species was permitted to reach steady state and another species was permitted to invade after 35 d. If the invading species was the dominant competitor (in all cases *Fragilaria crotonensis*), the subordinate competitor (always *Tabellaria fenestrata*) started to decline within the first few (< 5 days) after the invasion of *Fragilaria*. The time needed for the abundance curves to intersect depended on the *Tabellaria:Fragillaria* ratio at invasion date (Figure 3.15). If *Tabellaria* was used as invading species, it started to decline immediately.

In summary, a closer look at the time course of competition experiments shows that inferior competitors usually respond very rapidly to the deleterious effects of the appearance of a superior competitor. There-

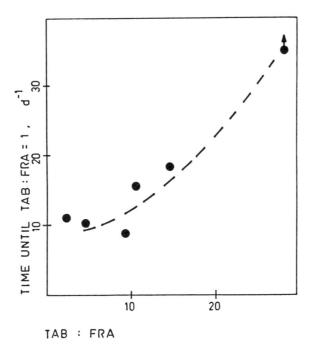

Figure 3.15   Time needed for invading superior competitor *Fragilaria* to equal the abundance of *Tabellaria* as a function of the *Tabellaria:Fragilaria* ratio at the start of invasion (data from Tilman and Sterner, 1984).

fore, the duration of competition experiments cannot be used as an argument against the occurrence of competition in nature.

3.9   Three Models for Competitive Interactions

Phytoplankton seasonal succession differs from primary terrestrial succession in one respect. In phytoplankton succession the availability of *all* resources tends to decrease, whereas in terrestrial successions some resources (e.g., light) decrease in availability, whereas some other (e.g., nitrate) increase in availability, owing to the process of soil formation at least during the initial stages of succession. There are three different models that can be used to explain the differences in competitive abilities between early and late successional species. These conceptual models are shown in Figure 3.16.

In model A succession is considered to be a directional change in limiting resource ratios ("resource-ratio hypothesis of plant succession," Tilman, 1982): During succession the silicate supply rates decline more rapidly than supply

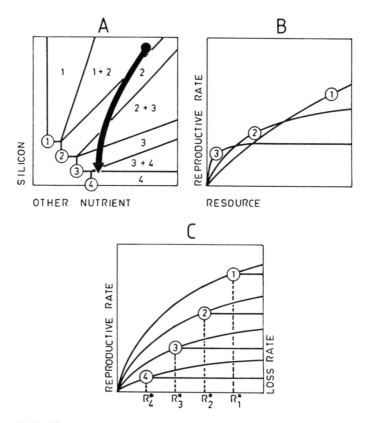

Figure 3.16   Three conceptual models for the change of competitive relationships in a successional sequence. Low numbers characterize early successional species, higher numbers later ones. A: Resource-ratio hypothesis. Species are characterized by their ZNGI's. B: From resource saturation to resource limitation: species are characterized by reproductive rate vs. resource curves. C: Increasing loss rates.

rates of other resources. The change of resource ratios leads to a change in competitive advantage. Species are characterized by their ZNGI's.

Model B conceives succession as a progression from resource sufficiency to resource deficiency and competition, but only the final stage of succession is considered to be in competitive equilibrium. Species are characterized by reproductive rates vs. resource curves. Initially species with high maximal reproductive rates dominate. With increasing colonization resources become more limiting, reproductive rates under limiting conditions become increasingly important. Early vs. late successional species conform to "r- vs. K-strategists" following Kilham and Kilham (1980).

Model C conceives succession as a progression from selection for rapid reproduction to selection for loss resistance ("r- and K-selection" following

Summer, 1981). Species with higher maximal reproductive rates are less loss resistant. Initially such species colonize the lake. With increasing colonization grazer populations build up (increasing loss rates). Loss-resistant algae become superior competitors, even if their reproductive rates are less at all resource levels. Again, only the final stage of succession is considered to be in competitive equilibrium.

The models are not mutually exclusive. Note, for example, that by co-determining the R^* value (eq. 3.5) loss resistance also is a constitutive component of the competitive power of a species. In many cases it may be very difficult to decide which model is the best representation of reality. If late successional stages are characterized by low Si:P ratios, low absolute P availability, and high potential impact of loss factors, it seems wise for a species to have a low optimal Si:P ratio, to be a good P-competitor, and to be resistant against the prevailing loss factors. Each of the three models would place this species into the category "late successional." However, the different models have quite different implications for early successional stages. The different predictions of the three models are summarized in Table 3.3.

Changing resource ratios   The empirical basis for the resource-ratio hypothesis is the variation in recycling efficiency of grazing zooplankton for different nutrients. Silicate is only poorly recycled within the epilimnion. Even if diatoms are grazed (the colonial species are usually relatively resistant against grazing) zooplankton excretes only a particulate silicate debris, most of which is lost by sedimentation before it dissolves (Sommer, 1988a). Re-

Table 3.3   Predictions of the three models of change in competitive relationships

	Model		
	A	B	C
Early successional species	Good competitors at high Si:P and Si:N ratios	High μ_{max}	High μ_{max} Edible for zooplankton
Late successional species	Good competitors for P or N with no Si-demand	Good competitors for limiting nutrient	Loss resistant Inedible
Successional trend in community average turnover rates	No prediction	Decreasing	Decreasing
Successional trend in organism size	No prediction	No prediction*	Increasing

* Some researchers suggest that there may be selection for smaller size with increasing competition, but that is not supported by the results of competition experiments (see Section 3.6 and Figure 3.9).

plenishment of Si in the epilimnion occurs only by vertical transport (eddy diffusion, episodic mixing events) or by allochthonous input from the catchment. Nitrogen and phosphorus, however, are released by zooplankton in their best available forms (ammonium and orthophosphate). P-recycling may greatly diminish when the C:P ratio of the food is higher than the C:P ratio of zooplankton biomass (Olsen and Østgaard, 1985; Goldman et al., 1987). As a consequence, Si:P and/or Si:N ratios should decline as long as stratification persists. The existence of a consistent successional trend in N:P ratios is still an open question.

The resource-ratio hypothesis predicts that diatoms, particularly Fragilariaceae, are early successional species. Input of hypolimnetic (Si-rich) water into the productive zone by episodic mixing events may permit nonsuccessional mid-season "reversions" (Reynolds, 1980) to a diatom-dominated plankton. The taxonomic identity of the late successional ("climax") algae depends, first, on whether N or P is the limiting nutrient and, second, on the mechanism by which the late successional algae achieve a low R^*. It has to be remembered that a low R^* value can be obtained through several strategies, such as high affinity Monod kinetics, loss resistance, and exploitation of spatial/temporal heterogeneity in nutrient supply. If N rather than P is in short supply, heterocystous blue-green algae (Nostocaceae) should be the dominant "climax" species. If P is more limiting and steady-state affinity holds true, *Mougeotia thylespora* or *Oscillatoria agardhii* (see Figure 3.7) could be potential candidates. If P is limiting and loss resistance holds true, large dinoflagellates (*Ceratium hirundinella*, *Peridinium* spp.) and colonial cyanobacteria should be the climax species. Both are too large to be efficiently grazed (Knisely and Geller, 1986) and both are resistant to sinking losses. If low R^*-values for P are to be achieved by exploitation of environmental gradients, vertically migrating species (flagellates, gas-vacuolate cyanobacteria) should be favored. Since the velocity and amplitude of migration increase with size (Sommer, 1987c), large rather than small flagellates should be typical for late successional stages.

The interpretation of field data needs some caution, because different nutrients have different cycling velocities (Si slow, N and P rapid). In this case the easily measurable ratios of nutrient concentrations are not identical with the ratios of nutrient availability ("supply rates"). To obtain the latter, concentrations have to be multiplied by the cycling velocities. However, if supply rates are monotonously increasing functions of concentrations, qualitative patterns of dominance along concentration ratio gradients should be similar to dominance patterns along supply-rate ratio gradients.

Strictly speaking, the resource-ratio hypothesis is self-contradictory, because it applies a steady-state solution to a transient-state phenomenon like succession. Nevertheless, it may work as an approximation if the change of resource supply-rate ratios is slow relative to the time needed for superior competitors to obtain numerical or biomass dominance. Even in such a case, time lags may occur between the occurrence of optimal resource ratios and the numerical dominance peaks of the winning competitors, as is shown in Figures 3.17 and 3.18.

From saturation to limitation   In their attempt to apply the concept of "r- vs. K-selection" to phytoplankton succession, Kilham and Kilham (1980) characterized succession as a directional change in suppply:demand ratios of nutrients. Initially supply exceeds demand for growth (nonlimited reproduction, r-selection). Later supply:demand ratios approach unity, and algae begin to compete for nutrients (nutrient limitation, K-selection). The species composition of the initial successional stages is not determined by competitive interactions but by a "race" where each species grows independently at μ_{max}. The PEG-model of plankton succession (Sommer et al., 1986 and Section 1.2) makes this assumption for the spring bloom of phytoplankton (statement 1)

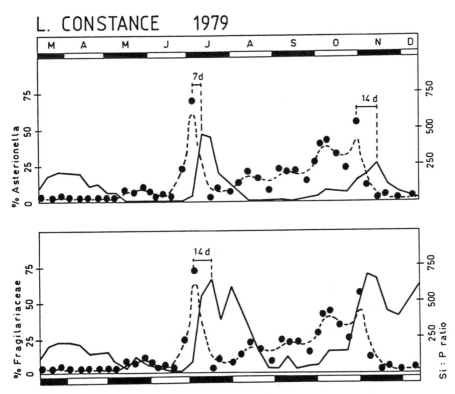

Figure 3.17   Percent contribution of *Asterionella formosa* and of all Fragilariaceae to total phytoplankton biomass as a delayed response to Si:P concentration ratios in the euphotic zone of Lake Constance.

SCHÖHSEE   1986

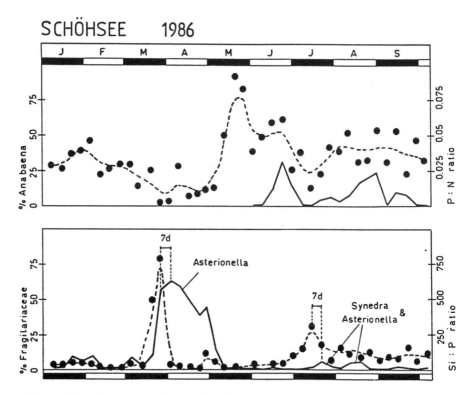

Figure 3.18   Percent contribution of *Anabaena flos-aquae* as a delayed response to P:N concentration ratios in the euphotic zone of the Schöhsee, and percent contribution of Fragilariaceae (mainly *Asterionella formosa* and *Synedra acus*) to total biomass as a delayed response to Si:P concentration ratios.

and for the first biomass increase after the grazing-induced clear-water phase in mid-summer (statement 8).

Contrary to the resource-ratio hypothesis, model B does not classify Fragilariaceae as early successional species because their maximal reproductive rates are only intermediate. In this model, initial colonization is considered to be a function mainly of species with high maximal reproductive rates (at spring temperatures). Using the Lake Constance data (Sommer, 1981, 1987a), algae like *Rhodomonas* spp. and *Cryptomonas* spp. can be hypothetically assigned to this type, owing to their rapid increase at the start of the spring bloom. However, due to the technical difficulties of maintaining Cryptophyceae in culture, true μ_{max} data are still lacking. At least some of the minor species typical for the Lake Constance spring bloom show maximal reproductive rates far in excess of the mid-successional large diatoms (data from Sommer, 1983): *Chlorella minutissima*: 2.19 d⁻¹, *Monoraphidium contortum*: 2.0

d^{-1}, *Stephanodiscus "hantzschii"*: 1.91 d^{-1}. The larger diatoms which are usually mid-successional in Lake Constance have μ_{max} values between 1.11 and 1.54 d^{-1}. Late successional species, such as *Ceratium hirundinella* and colonial cyanobacteria have still lower μ_{max} values.

In general, model B predicts a decline of reproductive rates in the course of succession. Unfortunately, the measurement of the reproductive rates of natural populations is extremely laborious, be it direct estimates via the fractions of cells undergoing mitosis (Braunwarth and Sommer, 1985), be it indirect estimates via analysis of the loss rates (Reynolds et al., 1982; Sommer, 1984a). A community average of μ, however, can easily be calculated from photosynthesis/biomass quotients ($\mu^* = \ln(1 + P/B)$, Tilzer, 1984). As predicted and as shown in Figure 3.19, μ^* of Lake Constance phytoplankton show a marked depression during the summer period of sequential (first P, then Si) nutrient depletion, after very high values during the spring bloom and the clear water phase. These are again followed by higher values in the autumnal period after deepening of the mixed layer relieves nutrient stress.

Increasing importance of loss factors   Model C makes similar predictions for the initial stages of succession as does model B (see Table 3.3). Pioneer

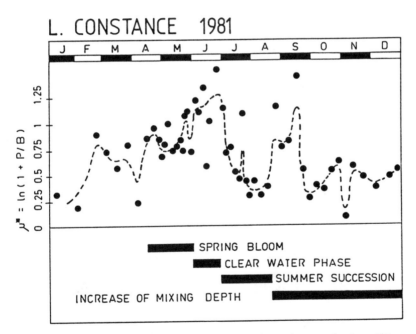

Figure 3.19 Community average of μ^* (recalculated from photosynthesis and biomass data; Tilzer, 1984) of Lake Constance phytoplankton. Horizontal bars mark the extent of the spring bloom, the clear water phase, the summer succession period with increasing nutrient stress, and the autumnal period of increasing mixing depths.

species with high maximal reproductive rates are usually small (Sommer, 1981) and—by virtue of their small size—well edible for filter-feeding zooplankton. In particular, Cryptophyceae are generally considered to be optimal food both for herbivorous Cladocera and calanoid Copepoda (Knisely and Geller, 1986). Large diatoms are intermediate in maximal reproductive rates and are moderately (e.g., *Asterionella*) to very (e.g., *Fragilaria*) resistant against grazing. They are, however, more prone to sedimentary losses than other algae. Large dinoflagellates, large colonial Volvocales, and colonial cyanobacteria are nearly totally resistant against grazing by general herbivores (but not resistant against specialists like *Ascomorpha*) and against sedimentary losses. This transition from small (well-edible) to large (inedible) algae is quite common phenomenon in many meso- and eutrophic lakes and described in statement 8 of the PEG-model. Figure 3.20 illustrates this phenomenon in Lake Constance over a two-year period.

Like model B, model C also predicts generally declining reproductive rates with the progress of succession. The main difference between the two models lies in the fact that model B anticipates late successional species are good competitors even in the absence of interspecific differences in loss rates, whereas model C predicts that late successional species have low R^* values only by virtue of their loss resistance. The former is certainly true when

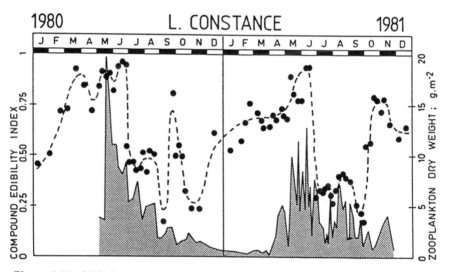

Figure 3.20 Shift from well-edible to poorly-edible phytoplankton in response to zooplankton pressure. Edibility of phytoplankton is expressed by the compound edibility index (full circles and smoothed, broken line): a weighted average of the edibility of phytoplankton biomass. Edibility indices of individual species (from 0 to 1) have taken from the experimental work of Knisely and Geller (1986). Shaded area: zooplankton dry weight.

shortage of reduced N selects for N-fixing cyanobacteria; the latter is probably true for large dinoflagellates (Sherr et al., 1982), unless migration to nutrient-rich deeper water layers plays a major role.

Synthesis   One of the major, easily tested differences between the three models shown in Figure 3.16 is their prediction about the seasonal occurrence of diatoms, especially Fragilariaceae. Model A predicts that they will begin seasonal succession, while models B and C predict they will be mid-successional. Plenty of field observations can be found to support both predictions. In relatively shallow lakes, where seasonal cycles often start during spring circulation, diatoms tend to be dominant during the spring bloom quite independently of the trophic state of the lake (e.g., most of the lakes in the vicinity of Plön, FRG, including Schöhsee, see Figure 3.18). In deep lakes, where the beginning of the growth season may be delayed until the onset of stratification, large diatoms dominate the spring bloom in oligotrophic lakes only and are replaced by Cryptophyceae together with other nanoplankton algae in more eutrophic lakes (Sommer, 1986c). In those lakes, large diatoms are quite often mid-successional.

Before we can try to reconcile these patterns with the three different succession models we have to recall an additional property of the large diatoms, which has not yet been discussed. At least some of them (*Asterionella formosa* according to Reynolds and Reynolds, 1985; *Asterionella formosa* and *Fragilaria crotonensis* according to Sommer, 1987a) have been shown to have low-light requirements for growth and to do well under the highly variable light conditions typical for periods of deep mixing. If light is treated in the same way as a chemical resource, this means that they are not only good competitors at high Si:P ratios, but also at high Si:light ratios.

Let us first consider the start of seasonal succession in "relatively shallow" lakes where spring mass growth of phytoplankton starts during circulation, usually immediately after icebreak. In spite of usually high surface irradiance, algae will experience a low average supply of light because mixing transports them up and down along the vertical light gradient (Riley, 1957; "effective light climate" following Ramberg, 1979). If such a lake is fairly rich in nutrients, light will be the primary limiting resource. High Si:light ratios will favor dominance of diatoms. If a shallow lake is so poor in P that after a few cell divisions P is more limiting than light, high Si:P ratios will also select for diatoms. For an N-limited lake with initially high Si:N ratios a prediction is not yet possible due to the lack of experimental data for competition at high Si:N ratios. With the progress of succession, Si-consumption by diatoms and poor recycling of Si (especially after the onset of stratification) will depress the amount of Si relative to other factors and so shift the competitive advantage first to less Si-demanding diatoms and then to nonsiliceous algae. It follows, that for relatively shallow lakes the resource-ratio hypothesis describes the successional trends in competitive interactions from the beginning

on. Note, however, that competition may be interrupted by physical disturbance (flushing, episodic mixing events) and by periods of overexploitation
by herbivorous zooplankton (grazing rate $> \mu_{max}$).

In deep lakes, the detrimental effect of mixing on the effective light
climate is so strong that not even diatoms can start their spring bloom during
curculation (red species of *Oscillatoria* may be more tolerant; see Chapter 2).
The start of the growth season is delayed until the onset of stratification.
Unless there are strong storms during this period, mixing depths are typically
small at the beginning of the stratified period. It follows that the algae will
experience a relatively high supply of light at the start of the growth season.
In phosphorus-poor lakes where P may become limiting after a few cell divisions, initially high Si:P ratios will select for diatoms. Subsequently, Si:P
ratios will decline and the competitive advantage will be shifted along the
Si:P gradient. Therefore, model A applies from the very beginning. If lakes
are rich in P, it takes a longer time until external P concentrations are depleted,
and cell quotas are diluted by subsequent divisions to a level limiting for
reproduction. Under such circumstances, the build-up of dense zooplankton
populations and high grazing pressure may be more rapid than the appproach
to nutrient limitation.

Thus, the species composition of the spring peak is a mixed function of
the "inoculum size" (i.e., overwintering conditions) and the maximal reproductive rates in cold water. If grazing rates exceed maximal reproductive rates
of spring phytoplankton, a clear-water phase develops (statements 4 and 5
of the PEG-model). During this period there is no further depletion of nutrients, which means that after the clear-water phase, phytoplankton growth
is again nutrient saturated (statements 8 and 9 of the PEG-model). In Lake
Constance, for example, this is reflected by near Redfield-proportions (see
Section 3.3) or an even P-richer stoichiometry of the particulate matter until
the middle of July (Sommer, 1987). Then Si:P ratios rise, because P-depletion
by the rapidly reproducing early summer algae (*Cryptomonas, Pandorina*) is
quicker than Si-depletion of the diatoms. Nutrient competition under high
Si:P ratios then shifts the competitive advantage to diatoms (statement 11 of
the PEG-model), which then reach their maximal dominance about one to
two weeks after the peak of Si:P ratios (see Figure 3.17). The following decline
of the Si:P ratios again shifts the comppetitive advantage away from the
diatoms (statement 12 of the PEG-model). It follows that the resource-ratio
hypothesis applies only to a relatively brief period during summer (although
it is the period of the annual biomass maximum) whereas the entire period
from the onset of stratification until the transition of the mid-successional
diatom stage has to be conceived of as primarily noncompetitive.

The low resource requirements of late successional species can be achieved
either by steady-state affinity, by efficient use of transient nutrient pulses or
spatial heterogeneity, or by loss resistance. The relative importance of each
of these mechanisms probably varies strongly among lakes, and regular pat-

terns in this variation have still to be found. Grazing pressure is probably further controlled by fish predation on zooplankton which, in the more densely populated regions, is a consequence of fisheries management rather than an intrinsic ecosystem property.

Due to the restrictions of available experimental evidence the discussion has so far concentrated on Si:P ratios and differential loss resistance. For further interpretation of phytoplankton succession in terms of competition we still need to know more about the following: competitive dominants along the Si:N gradient and along any nutrient:light gradient; successional trends in N:P ratios; the competitive importance of nitrogen speciation (NO^{2-}_3 vs. NH_4); the competitive performance of flagellate taxa; and estimates of species-specific grazing losses.

3.10   Hutchinson's Paradox Reversed and the "Ghost of Competition Past"

Over the most recent quarter of a century it has become commonplace to introduce articles about phytoplankton competition with a reference to Hutchinson's (1961) "paradox of the plankton," i.e., the apparent contradiction between the competitive exclusion principle and the diversity of phytoplankton. Meanwhile, both theoretical and experimental analyses of competition have produced a number of mechanisms that permit coexistence despite competition. Therefore, the time has come to turn Hutchinson's paradox upside down and to ask the reverse question, "Why are there so few species of phytoplankton in a given lake?"—several hundred vs. several tens of thousands of potentially planktonic algal species. In other words, "By what mechanism has the species pool of a lake been assembled?". This book puts its emphasis on the mechanisms that select seasonal dominants out of a lake's species pool, but we should also ask whether the same mechanisms are responsible for the determination of the species pool.

The worldwide distribution of most phytoplankton taxa and the rareness of "island biogeographical" phenomena, such as endemism, indicate that dispersal is not a major problem for freshwater phytoplankton. It will be recalled, for example, that most tree species in climatically similar regions of America and Europe are different, whereas most phytoplankton species in similar lakes are the same. Therefore, immigration history seems to be important only in recently filled reservoirs or in lakes that have recently undergone major environmental change. The phytoplankton-species pool of most lakes is mainly determined by the factors that influence reproductive rates, mortality rates, and the perennation of temporarily adverse conditions. With only slight exaggeration it can be stated that the worldwide total of potentially planktonic algal species is the source pool out of which the algal flora of every lake is selected.

Even if resource competition occurs only during parts of the year, it could be imagined that the adverse effects of competitive pressure on some species accumulate over the years until their complete exclusion occurs. The generally good performance of natural phytoplankton under nutrient depletion (high μ_{rel}; see Section 3.3) could be explained by the assumption that species with a poorer performance have already been excluded or never could establish successful populations. Admittedly, such an argument means invoking the "ghost of competition past" (Connell, 1980). However, past competition is not as unverifiable as Connell believes. Direct observation is possible by long-term studies on lakes undergoing change, such as recovery from eutrophication. If a species decreasing in number in such a lake showed a trend of decreasing cell quota of the limiting nutrient over the years, this would be a strong indication of competitive exclusion.

Even if long-term observations are impossible the "ghost of competition past" should leave some tracks which are measurable today. If competition for a particular resource is important in the determination of a lake's species pool, then the lake's species should be better competitors for that resource than a randomly drawn subsample of all species or the species in a lake, where that particular resource is not limiting. The Schöhsee data in Figure 3.2 (Section 3.3) and a comparison with a literature survey of P-limited Droop kinetics provide a basis for a first step in that direction. In order to test the suitability of the Redfield-ratio as an indicator of an approach to nutrient saturation (see Section 3.3) linear regressions of μ_{rel} on the C:P ratio have been made and the intensities of P-limitation (IL $= 1 - \mu_{rel}$) for Redfield-proportions have been calculated from the regressions (Sommer, 1988c). This has been done for the total seston, for the different size fractions, and for those groups of samples separately, where an individual species dominated a fraction. The predicted IL values at Redfield-proportions were between 0.02 and 0.14 (Figure 3.21). For comparison, all the possible data on biomass-based Droop kinetics (eq. 3.3) from the literature have been used to calculate the IL values at the Redfield C:P ratio for them (with all the uncertainties of dry weight-carbon and cell volume-carbon conversions). The calculated IL data show approximately a log-normal distribution (n $= 40$, mean $= 0.142$, mean $+$ SD $= 0.31$, mean $-$ SD $= 0.065$, total range: 0.02 to 1). It follows that Schöhsee phytoplankton on average do better and behave more uniformly than the algae tested in culture so far. Of course such a comparison is only a first step: it may be questioned whether the algae used in laboratory cultures are a random subsample of the total; the lake data refer mainly to mixed fractions instead of individual species; the Droop equation alone is not yet a sufficient measure of competitive power because higher cellular requirements may be offset by higher uptake affinity. Still this example shows that it is possible to find the footprints of the "ghost of competition past."

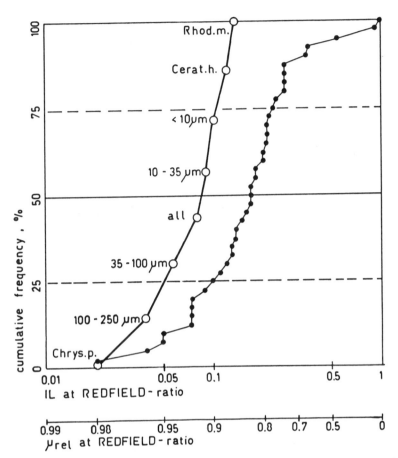

Figure 3.21 Cumulative frequency plot of the intensity of P-limitation at Redfield stoichiometry (log-scale) for Schöhsee seston fractions (○) and for laboratory cultures (literature; •).

3.11 Concluding Remarks

The concept of interspecific competition has a long tradition in theoretical ecology, but it was only during the last decade (Tilman, 1977) that a design for the direct experimental test of competition has been developed. Before that, competition was inferred from quite remote pieces of evidence, such as species-area curves, abundance distribution patterns, character displacement, niche breadths, and size ratios between species within the same guild. Sometimes inferences have been extremely indirect, for instance when niche parameters have been characterized by morphological traits.

Application of the Lotka-Volterra competition model to empirical data mainly has been performed by fitting the interaction coefficients a posteriori to population dynamics of supposed competitors without prior, independent estimate of their competitive power. Being a superior competitor was defined as being superior in competition. Quite correctly, this approach was denounced as tautological by Peters (1976). By independently predicting a species' competitive power from resource-dependent growth kinetics, Tilman was the first to break the tradition of circular reasoning. Unfortunately, his competition theory has received little attention from the new anti-competition literature of the Tallahassee school (Connor and Simberloff, 1979; Simberloff, 1983; Strong et al., 1983).

One of the standard criticisms against Tilman's competition theory is the equilibrium-disequilibrium controversy. It is frequently argued that external conditions are not stable for sufficiently long periods to permit the completion of competitive exclusion. However, such arguments confuse the process (competition) with its end product (competitive exclusion). Even if a process is often interrupted before its equilibrium outcome reached, it does not imply the absence of this process. For example, in a competition experiment where *Asterionella* finally excludes *Cyclotella*, competition between them occurs only before and not after the exclusion of *Cyclotella*. Moreover, the large discrepancy between the actual number of phytoplankton species in a lake and the maximal possible one shows that it is at least a plausible hypothesis, that quite a lot of competitive exclusion has occurred already. (Section 3.10 discusses how it could be tested.) Admittedly, modern experimental competition research has started with extremely simplified situations, such as chemostat-type steady states. But this is because experiments always have to be "simpler than nature," and in a new area of research experiments have to be very simple. Meanwhile, we have accepted the challenge of more complicated situations and started to study the influence of different patterns of nonsteady state both on the maintenance of diversity and on the taxonomic outcome of competition.

We will probably never be able to predict the outcome of all possible competitive situations. The number of species involved and the number of possible combinations of resources and external factors are simply too high. But phytoplankton competition research is one of the few examples, where a classic problem in theoretical ecology has successfully been tested in controlled experimental research. The results of this research have rewarded the efforts of the authors listed in Tables 3.1 and 3.2 and is still worth further efforts on the part of phytoplankton researchers.

References

Braunwarth, C. and Sommer, U. 1985. Analyses of the in situ growth rates of Cryptophyceae by use of the mitotic index technique. *Limnology and Oceanography* 30: 893–897.

Burmaster, D.E. 1979. The continuous culture of phytoplankton: Mathematical equivalence among three steady state models. *American Naturalist* 113: 123–124.

Connell, J.H. 1978. Diversity in tropical rain forests and coral reefs. *Science* 199: 1304–1310.

Connell, J.H. 1980. Diversity and coevolution of competitors, or the ghost of competition past. *Oikos* 35: 131–138.

Connor, E.F. and Simberloff, D. 1979. The assembly of species communities: chance or competition? *Ecology* 60: 1132–1140.

Currie, D.J. 1984. Microscale nutrient patches: Do they matter to phytoplankton. *Limnology and Oceanography* 29: 211–214.

Dillon, P.J. and Rigler, F.H. 1974. The phosphorus:chlorophyll relationships in lakes. *Limnology and Oceanography* 23: 767–773.

Droop, M.R. 1973. Some thoughts on nutrient limitation in algae. *Journal of Phycology* 9: 264–272.

Droop, M.R. 1983. Twenty-five years of algal growth kinetics. *Botanica Marina* 26: 99–112.

Dugdale, R.C. 1967. Nutrient limitation in the sea: dynamics, identification and significance. *Limnology and Oceanography* 12: 685–695.

Gächter, R. and Mares, A. 1985. Does settling seston release soluble reactive phosphorus in the hypolimnion of lakes? *Limnology and Oceanography* 30: 364–372.

Gaedeke, A. and Sommer, U. 1986. The influence of the frequency of periodic disturbances on the maintenance of phytoplankton diversity. *Oecologia* 71: 98–102.

Goldman, J.C. 1979. Physiological processes, nutrient availability, and the concept of relative growth rate in marine phytoplankton ecology, pp. 179–194, in Falkowski, P.G. (editor), *Primary Productivity in the Sea*. Plenum, New York.

Goldman, J.C. 1984. Conceptual role for microaggregates in pelagic waters. *Bulletin of Marine Science* 35: 462–476.

Goldman, J.C., Caron, A.D., and Dennett, M.R. 1987. Nutrient cycling in a microflagellate food chain: IV. Phytoplankton-microflagellate interactions. *Marine Ecology Progress Series* 38: 75–87.

Goldman, J.C. and Glibert, P.M. 1982. Comparative rapid ammonium uptake by four species of marine phytoplankton. *Limnology and Oceanography* 27: 814–827.

Goldman, J.C., McCarthy, J.J., and Peavey, D.G. 1979. Growth rate influence on the chemical composition of phytoplankton in oceanic waters. *Nature* 279: 210–215.

Grover, J.P. 1988. The dynamics of competition in a variable environment: Experiments with two diatom species. *Ecology* 69: 408–417.

Hardin, G. 1961. The competitive exclusion principle. *Science* 131: 1292–1298.

Harris, G.P. 1983. Mixed layer physics and phytoplankton populations: Studies in equilibrium and non-equilibrium ecology. *Progress in Phycological Research* 2: 1–51.

Harris, G.P. 1986. *Phytoplankton Ecology*, Chapman and Hall, London, 384 pp.

Healey, F.P. and Hendzel, L.L. 1980. Physiological indicators of nutrient deficiency in lake phytoplankton. *Canadian Journal of Fisheries and Aquatic Sciences* 37: 442–453.

Heaney, S.I., Smyly, W.J.P., and Talling, J.F. 1987. Interactions of physical, chemical and biological processes in depth and time within a productive English lake during summer stratification. *Internationale Revue der gesamten Hydrobiologie* 71: 441–494.

Hecky, R.E. and Kilham, P. 1988. Nutrient limitation of phytoplankton in freshwater and marine environments: a review of recent evidence on the effects of enrichment. *Limnology and Oceanography* 33: 796–822.

Holm, N.P. and Armstrong, D.E. 1981. Role of nutrient limitation and competition in controlling the populations of *Asterionella formosa* and *Microcystis aeruginosa* in semicontinuous culture. *Limnology and Oceanography* 26: 622–634.

Hutchinson, G.E. 1961. The paradox of the plankton. *American Naturalist* 95: 137–147.

Jackson, G.A. 1980. Phytoplankton growth and zooplankton grazing in oligotrophic oceans. *Nature* 294: 439–441.

Jørgensen, S.E. 1979. *Handbook of Environmental Data and Ecological Parameters*. International Society for Ecological Modelling, Copenhagen, 1162 pp.

Kalff, J. and Knoechel, R. 1978. Phytoplankton and their dynamics in oligotrophic and eutrophic lakes. *Annual Review of Ecology and Systematics* 9: 475–495.

Kilham, P. and Hecky, R.E. 1988. Comparative ecology of marine and freshwater phtytoplankton. *Limnology and Oceanography* 33: 776–795.

Kilham, P. and Kilham, S.S. 1980. The evolutionary ecology of phytoplankton, pp. 571–592, in Morris, I. (editor), *The Physiological Ecology of Phytoplankton*. Blackwell, Oxford.

Kilham, S.S. 1984. Silicon and phosphorus growth kinetics and competitive interactions between *Stephanodiscus minutus* and *Synedra* sp. *Verhandlungen des internationalen Vereins für theoretische und angewandte Limnologie* 22: 435–439.

Kilham, S.S. 1986. Dynamics of Lake Michigan natural phytoplankton communities in continuous cultures along a Si:P loading gradient. *Canadian Journal of Fisheries and Aquatic Sciences* 43: 351–360.

Knisely, K. and Geller, W. 1986. Selective feeding of four zooplankton species on natural lake zooplankton. *Oecologia* 96: 86–94.

Lehman, J.T. 1976. Ecological and nutritional studies on *Dinobryon* Ehrenb.: Seasonal periodicity and the phosphate toxicity problem. *Limnology and Oceanography* 21: 646–658.

Lehman, J.T. and Sandgren, C.D. 1982. Phosphorus dynamics of the procaryotic nannoplankton in a Michigan lake. *Limnology and Oceanography* 27: 828–838.

Levins, R. 1979. Coexistence in a variable environment. *American Naturalist* 114: 765–783.

Løvstad, O. 1986. Biotests with phytoplankton assemblages: Growth limitation along temporal and spatial gradients. *Hydrobiologia* 134: 141–149.

Løvstad, O. and Wold, T. 1984. Determination of external concentrations of available phosphorus for phytoplankton populations. *Verhandlungen des internationalen Vereins für theoretische und angewandte Limnologie* 22: 205–210.

Lund, J.W.G. 1950. Studies on Asterionella. 2. Nutrient depletion and the spring maximum. *Journal of Ecology* 38: 1–35.

Moed, J.R. 1973. Effect of combined action of light and silicon depletion on *Asterionella formosa* Hass. *Verhandlungen des internationalen Vereins für theoretische und angewandie Limnologie* 18: 1367–1374.

Monod, J. 1950. La technique de la culture continue: Theorie et applications. *Annales d'Institut Pasteur, Lille* 79: 390–410.

Müller, H. 1972. Wachstum und Phosphatbedarf von *Nitzschia actinastroides* (Lemm.) von Goor in statischer und homokontinuierlicher Kultur unter Phosphatlimitierung. *Archiv für Hydrobiologie, Supplement* 33: 206–236.

Olsen, Y. and Østgaard, K. 1985. Estimating release rates of phosphorus from zooplankton: Model and experimental verification. *Limnology and Oceanography* 30: 844–853.

Paasche, E. 1980. Silicon, pp. 258–284, in Morris, I. (editor), *The Physiological Ecology of Phytoplankton*. Blackwell, Oxford.

Peters, R.H. 1976. Tautology in evolution and ecology. *American Naturalist* 110: 1–12.

Peters, R.H. 1983. *The Ecological Implications of Body Size*, Cambridge University Press, 329 pp.

Porter, K.G. 1977. The plant-animal interface in freshwater ecosystems. *American Scientist* 65: 159–170.

Ramberg, L. 1979. Relations between phytoplankton and light climate in two Swedish forrest lakes. *Internationale Revue der gesamten Hydrobiologie* 64: 749–782.

Reynolds, C.S. 1980. Phytoplankton assemblages and their periodicity in stratifying lake systems. *Holarctic Ecology* 3:141–159.

Reynolds, C.S. 1984. *The Ecology of Freshwater Phytoplankton*, Cambridge University Press, 384 pp.

Reynolds, C.S. 1988. The concept of biological succession applied to seasonal periodicity of freshwater phytoplankton. *Verhandlungen der internationalen Vereiniguna für theoretische und angewandte Limnologie* 23: 683–691.

Reynolds, C.S. and Reynolds, J.B. 1985. The atypical seasonality of phytoplankton in Crose Mere, 1972: an independent test of the hypothesis that variability in the physical environment regulates community dynamics and structure. *British Phycological Journal* 20: 227–242.

Reynolds, C.S., Thompson, J.M., Ferguson, A.J.D., and Wiseman, S.W. 1982. Loss processes in the population dynamics of phytoplankton maintained in closed systems. *Journal of Plankton Research* 4: 561–600.

Riley, G.A. 1957. Phytoplankton of the north central Sargasso Sea 1950–1952. *Limnology and Oceanography* 2: 252–270.

Rigler, F. 1968. Further observations inconsistent with the hypothesis that the molybdenum blue method measures inorganic phosphorus in lake water. *Limnology and Oceanography* 13: 7–13.

Robinson, J.V. and Sandgren, C.D. 1983. The effect of temporal environmental heterogeneity on community structure: a replicated experimental study. *Oecologia* 57: 98–102.

Sakshaug, E., Andresen, K., Myklestad, S., and Olsen, Y. 1983. Nutrient status of phytoplankton communities in Norwegian waters (marine, brackish, and fresh) as revealed by their chemical composition. *Journal of Plankton Research* 5: 175–196.

Salonen, K., Jones, R.I., and Arvola, L. 1984. Hypolimnetic phosphorus retrieval by diel vertical migrations of lake phytoplankton. *Freshwater Biology* 14: 431–438.

Sherr, E.B., Sherr, B.F., Berman, T., and McCarthy, J.J. 1982. Differences in nitrate and ammonia uptake among components of a phytoplankton population. *Journal of Plankton Research* 4: 961–965.

Simberloff, D. 1983. Competition theory, hypothesis testing, and other community ecological buzzwords. *American Naturalist* 122: 626–635.

Sjoblad, R.D., Chet, I., and Mitchell, R. 1978. Quantitative assay for algal chemotaxis. *Applied and Environmental Microbiology* 36: 847–850.

Smith, R.E. and Kalff, J. 1983. Competition for phosphorus among co-occurring freshwater phytoplankton. *Limnology and Oceanography* 28: 448–464.

Sommer, U. 1981. The role of r- and K-selection in the succession of phytoplankton in Lake Constance. *Acta Oecologica/Oecologia Generalis* 2: 327–342.

Sommer, U. 1983. Nutrient competition between phytoplankton in multispecies chemostat experiments. *Archiv für Hydrobiologie* 96: 399–416.

Sommer, U. 1984a. Sedimentation of principal phytoplankton species in Lake Constance. *Journal of Plankton Research* 6: 1–15.

Sommer, U. 1984b. The paradox of the plankton: Fluctuations of the phosphorus availability maintain diversity of phytoplankton in flow-through cultures. *Limnology and Oceanography* 29: 633–636.

Sommer, U. 1985a. Comparison between steady state and nonsteady state competition: Experiments with natural phytoplankton. *Limnology and Oceanography* 30: 335–346.

Sommer, U. 1986a. Phytoplankton competition along a gradient of dilution rates. *Oecologia* 68: 503–506.

Sommer, U. 1986b. Nitrate- and silicate competition among antarctic phytoplankton. *Marine Biology* 91: 345–351.

Sommer, U. 1986c. The periodicity of phytoplankton in Lake Constance (Bodensee) in comparison to other deep lakes of central Europe. *Hydrobiologia* 138: 1–7.

Sommer, U. 1987a. Factors controlling the seasonal variation in phytoplankton species composition—A case study for a deep, nutrient rich lake. *Progress in Phycological Research* 5: 123–178.

Sommer, U. 1988a. Growth and survival strategies of plankton diatoms, pp. 227–260, in Sandgren, C.D. (editor), *Growth and Survival Strategies of Freshwater Phytoplankton*. Cambridge University Press, Cambridge.

Sommer, U. 1988b. Some size relationships in phytoflagellate motility. *Hydrobiologia* 161: 125–131.

Sommer, U. 1988c. Does nutrient competition among phytoplankton occur in situ? *Verhandlungen der internationalen Vereiniguna für theoretische und angewandie Limnologie* 23: 707–712.

Sommer, U. and Gliwicz, Z.M. 1986. Long range vertical migration of *Volvox* in tropical Lake Cahora Bassa (Mozambique). *Limnology and Oceanography* 31: 650–653.

Sommer, U., Gliwicz, Z.M., Lampert, W., and Duncan, A. 1986. The PEG-model of seasonal succession of planktonic events in fresh waters. *Archiv für Hydrobiologie* 106: 433–471.

Sommer, U. and Kilham, S.S. 1985. Phytoplankton natural community competition experiments: A reinterpretation. *Limnology and Oceanography* 30: 436–440.

Sommer, U. and Stabel, H.H. 1983. Silicon consumption and population density changes of dominant planktonic diatoms in Lake Constance. *Journal of Ecology* 73: 119–130.

Strong, D., Simberloff, D., and Abele, L. (editors). 1983. *Ecological Communities: Conceptual Issues and Evidence*. Princeton University Press.

Tett, P., Heaney, S.I., and Droop, M.R. 1985. The redfield ratio and phytoplankton growth rate. *Journals of the Marine Biological Association of the United Kingdom* 65: 487–504.

Tilman, D. 1977. Resource competition between planktonic algae: an experimental and theoretical approach. *Ecology* 58: 338–348.

Tilman, D. 1981. Test of resource competition theory using four species of Lake Michigan algae. *Ecology* 62: 802–815.

Tilman, D. 1982. *Resource Competition and Community Structure*. Princeton University Press, 269 pp.

Tilman, D. and Kiesling, R.L. 1984. Freshwater algal taxonomy: taxonomic tradeoffs in the temperature dependence of nutrient competitive abilities, pp. 314–319, in Klug, M.J. and Reddy, C.A. (editors), *Current Perspectives in Microbial Ecology*. American Society of Microbiology, Washington.

Tilman, D., Kiesling, R., Sterner, R., Kilham, S.S., and Johnson, F.A. 1986. Green, bluegreen and diatom algae: Taxonomic differences in competitive ability for phosphorus, silicon and nitrogen. *Archiv für Hydrobiologie* 106: 473–485.

Tilman, D., Kilham, S.S., and Kilham, P. 1982. Phytoplankton community ecology: the role of limiting nutrients. *Annual Review of Ecology and Systematics* 13: 349–372.

Tilman, D., Matston, M., and Langer, S. 1981. Competition and nutrient kinetics along a temperature gradient: an experimental test of mechanistic approach to niche theory. *Limnology and Oceanography* 26: 1020–1033.

Tilman, D. and Sterner, R.W. 1984. Invasions of equilibria: tests of resource competition using two species of algae. *Oecologia* 61: 197–200.

Tilzer, M.M. 1984. Estimation of phytoplankton loss rates from daily photosynthetic rates and observed biomass changes in Lake Constance. *Journal of Plankton Research* 6: 309–324.

Turpin, D.H. 1987. The physiological basis of phytoplankton resource competition. In Sandgren, C.C. (editor), *Growth and Reproductive Strategies of Freshwater Phytoplankton*. Cambridge University Press, Cambridge.

Turpin, D.H. and Harrison, P.J. 1980. Cell size manipulation in natural marine, planktonic, diatom communities. *Canadian Journal of Fisheries and Aquatic Sciences* 37: 1193–1195.

Uehlinger, U. 1980. Experimentelle Untersuchungen zur Autökologie von *Aphanizomenon flos-aquae*. *Archiv für Hydrobiologie, Supplement* 60: 260–288.

4

The Role of Grazers in Phytoplankton Succession

Robert W. Sterner

Department of Biology
University of Texas, Arlington
Arlington, Texas, U.S.A.

4.1 Introduction

At times, freshwater zooplankton consume phytoplankton populations at rates similar to or faster than that at which they are growing (Hargrave and Geen, 1970; Gulati, 1975; Horn, 1981; Persson, 1985; Børsheim and Anderson, 1987). Such high losses certainly must help direct seasonal succession as they force a subset of algal species to suffer high mortality rates. Some studies have concluded that losses in general (Kalff and Knoechel, 1978; Reynolds, et al., 1982) and grazing losses in particular (Porter, 1973, 1976, 1977; Lynch and Shapiro, 1981; Crumpton and Wetzel, 1982; Kerfoot, 1987) are important in seasonal succession. In addition, the influence of zooplankton on algal succession is not limited to their selective effect on algal numbers. Zooplankton also interact indirectly with phytoplankton by making some nutrients more available to them (Gliwicz, 1975; Lehman, 1980a, b; Redfield, 1980; Lehman and Scavia, 1982; Sterner, 1986a). Zooplankton thus act not only as predators in the classic sense, but they also have an effect on the competition among algae (Elser et al., 1988).

The role of this chapter is to discuss these interactions between zoo- and phytoplankton and to hypothesize when and where they may be important in succession. Organismal biology constrains the structure of communities (e.g., Werner, 1986); thus, many details of grazing and nutrient regeneration must be considered. However, a relatively low number of dominant interactions may explain much of what occurs in natural communities. The focus will be on how differential mortality and ratios of nutrient supply affect algal communities.

Many approaches have been used to study the importance of grazers.

One has been to separately estimate the various components of the net growth of algal populations, i.e., reproduction minus all categories of losses (sinking, grazing, parasitism, and any others). Then an effort is made to determine which of these "causes" populations to increase and decrease. This is an obvious and potentially meritorious approach. Of course, it is only as reliable as are the individual estimates of growth and loss and these are notoriously difficult to quantify. For example, an early attempt by Cushing (1976) was rebutted by Lewis (1977) and Knoechel (1977) who questioned the assumptions and mathematical constructs. More recent studies (Crumpton and Wetzel, 1982; Reynolds et al., 1982) have taken advantage of significantly improved methods with fewer assumptions, but wide confidence intervals still make definitive conclusions difficult. In addition, many unavoidable methodological simplifications bias the estimates. Furthermore, this approach considers only the effects of grazers on algal mortality rates and not their other effects, such as on nutrient levels.

In this chapter we will concentrate on four other types of studies:

1. The mechanistic bases of feeding and nutrient regeneration; these predict from first principles how grazers influence algal populations.

2. Seasonal trajectories of in situ grazing loss rates.

3. Experimental manipulations of grazer density.

4. Modeling grazing and nutrient regeneration using several simple equations.

The first part of the chapter covers mostly grazer effects on algal mortality, and the second part concerns effects on reproduction.

4.2   Some Mathematics of Clearance and Selectivity

Consider edible algal species to be ones for which grazing zooplankton remove some nonnegligible fraction of the population. Edible algae thus are species that are encountered, ingested, and killed. Grazing will be said to be selective when the per capita mortality experienced by one algal species as a result of grazing differs from that experienced by other species. It will soon become clear that this definition conforms to electivity indices already in use. Selectivity could direct a successional trajectory from dominance by edible to dominance by inedible species, as has often been suggested (Porter, 1973, 1976; McCauley and Briand, 1979; Lynch, 1980; Shapiro, 1980; Lynch and Shapiro, 1981; Crumpton and Wetzel, 1982; Kerfoot, 1987). The idea that large algae can dominate in the summer at least partly because they are not as heavily grazed as smaller species is incorporated in the PEG-model in step 8 (Sommer et al., 1986).

From the standpoint of algal population dynamics, consider the clearance rate (also called the "filtering rate" or the "grazing rate") on algal species i, C_i, to be the amount of fluid that contains the quantity of algae removed from the population by grazing animals in a unit time. This definition is purposely crafted to avoid the complexities of digestion resistance and nutritional inadequacy. Clearance is back-calculated from feeding; it is not a direct measurement of a distinct volume of fluid. Fluid does pass through the feeding chamber (in species possessing one), and fluid does pass through the intersetular spaces of the feeding appendages of at least some grazer species, the magnitude of this quantity being a matter of controversy. Some reports suggest the fluid volume passing between the setules is small to zero for copepods (Koehl and Strickler, 1981; Vanderploeg and Ondricek-Fallscheer, 1982; Koehl, 1984; Strickler, 1984) as well as for *Daphnia* (Gerritsen and Porter, 1982; Ganf and Shiel, 1985a, b) because of the viscosity at low Reynolds number (Zaret, 1980; Vogel, 1981). These reports contradict the analogy of particle clearance to a leaky sieve (Boyd, 1976). Other reports find inefficient clearance by *Daphnia* of particles smaller than the intersetular distance and suggest that a leaky sieve is an appropriate analogy (see also picoplankton clearance by cladocerans, below). Fryer (1987) strongly criticizes the anatomical aspects of the *Daphnia* evidence against the leaky sieve, and Brendelberger et al. (1987) argue that the closed filtering chamber of *Daphnia* is qualitatively different from the open system of copepods, explaining why a sieve may be appropriate for some taxa but not others. Regardless of this debate, the clearance rate need not and usually will not exactly equal any fluid volume easily defined by the mechanisms of feeding (see e.g., Friedman, 1980).

Clearance is a combined result of encounter, capture (which together constitute collection), mastication, and digestion. Differential rates of *any* of these would cause a grazing animal to be selective. If encounter rates were directly proportional to the proportional representation of different algal types, encounter could not cause selectivity. Glasser (1984), however, points out that larger cells may be encountered more frequently than proportions would predict. Also, microcinematography has found that particle movement in the vicinity of feeding copepods (Koehl and Strickler, 1981; Koehl, 1984; Strickler, 1984; Vanderploeg and Paffenhöffer, 1985) and *Daphnia* (Scavia et al., 1984) make it difficult—if not impossible—to judge when a particle has been encountered. Long-distance detection of particles by chemoreception or mechanotactile cues in copepods (Friedman and Strickler, 1975; Friedman, 1980; Poulet and Marsot, 1980) divorces the probability of encounter still further from proportional representation (Price and Paffenhöfer, 1985). Particle encounter rates probably underlie some selectivity in grazing. Differential particle capture is usually what is thought of as causing selectivity. Sorting of collected particles during passage to the mouth opening (including postcapture rejection of particles) may also be selective. Some selectivity results from viable gut passage of digestion-resistant algal taxa (Porter, 1973).

Though some details are not yet completely resolved, many studies have shown that clearance rates are high and relatively independent of algal density below the algal density where feeding becomes satiated (the incipient limiting concentration, or ILC), whereas clearance rates decrease with increasing algal density above the ILC (e.g., McMahon and Rigler, 1965; Porter et al., 1982; Ganf and Shiel, 1985a) (see Figure 4.9 for illustration). Some studies argue that clearance rates also are low at extremely low algal density (Frost, 1975; Lehman, 1976, and others), but such a lower threshold for feeding seems to be taxonomically restricted as it has not been observed in studies of *Daphnia magna* (Porter et al., 1982), or the *Daphnia longispina* group or *Eudiaptomus gracilis* (Muck and Lampert, 1980). In fact, marine calanoids may be the only zooplankton taxa to show a lower threshold for feeding.

Peters (1984) reviewed the methodology of measuring clearance rates; other useful comments can be found in *Hydrobiological Bulletin* 19(1) (1985). Two basic methods have been employed most often: radiotracer measurements of particle collection (durations of minutes), and measurements of particle disappearance (durations of hours). As not every cell collected is removed from the population, these two methods need not always measure the same rates. A third basic method combines measurements of the quantity of chlorophyll in the animals' guts with gut residence time (Mackas and Bohrer, 1976) and is being critically evaluated for marine copepods (compare Conover et al., 1986, with Kiørboe and Tiselius, 1987) as well as for freshwater species (Christoffersen and Jespersen, 1986). This method is, however, unsuitable for studying clearance on individual algal species and so is not useful in the context of algal succession.

The particle disappearance method is most illustrative of the relationship between clearance rates and algal population dynamics. Below the ILC, clearance rates can be calculated directly from the instantaneous rates of loss of food particles (the slope of log density vs. time) (Coughlan, 1969; Frost, 1972). Above the ILC, because clearance rates are not independent of particle density and therefore change during the measurement, the instantaneous rate at the average food concentration is calculated (Peters, 1984, gives the formulae). At all algal densities, clearance rate is directly proportional to the instantaneous per capita mortality from grazing:

$$K = \text{Algal per capita mortality} = \text{Clearance rate} \times \text{Grazer density} \quad (4.1)$$

Here, the quantity K is defined as the "grazing rate." In spite of its simplicity, eq. 4.1 seems not to be widely appreciated. Studies of in situ grazing (see Section 4.4) have almost invariably multiplied K (day^{-1}) by 100 and considered this to be "percent cleared per day." This practice should be discontinued as it is misleading. Considering the following numerical example: In a hypothetical population of 0.1 unselective "animals" (referring either to number of individuals or biomass) per ml where each animal clears five ml per day, the grazing rate equals 0.5 day^{-1}. This does not, however, mean that 50 percent of the environmental volume (or of an algal population) is cleared in one day, which would require a grazing rate of:

$$-\ln(0.50) = 0.693 \text{ per day} \qquad (4.2)$$

or a clearance rate of almost seven ml per animal per day, not five. A grazing rate is an instantaneous rate of exponential decline, not a percent removed in a finite length of time. Dividing grazing rate values reported as percent per day by 100 converts them to instantaneous mortality rates, which is an informative dimension.

Because clearance rates measure the volume of environment cleared per unit time, comparing clearance rates also gives information on selectivity. When $C_i > C_j$, algal species i is being removed at higher rates than predicted by proportional representation, and species i has a higher per capita mortality from grazing than species j. Comparison of clearance rates is therefore the basis for many indices of selectivity. Jacobs (1974) pointed out that ratios of mortality rates (which are equal to ratios of clearance rates) have the desirable property that they are independent of the relative proportions of food items and thus should not vary just because food composition changes. Ivlev's index does not have this property. Paloheimo (1979) reaches similar conclusions. Many studies have utilized clearance rate ratios to evaluate selection by zooplankton (e.g., Bartram, 1980; Persson, 1985; Bogdan and Gilbert, 1984, 1987; Okamoto, 1984; Knisely and Geller, 1986), and this has led to a plethora of names and symbols all with common meanings. In a series of studies (Vanderploeg and Scavia, 1979a, 1979b, Vanderploeg, 1981; Vanderploeg and Scavia, 1983; Vanderploeg et al., 1984), several such indices were developed and applied. One of them, W_i' normalizes the clearance rate on food type i to the clearance rate on the food type with the highest clearance rate:

$$W_i' = C_i/C_{pref} \qquad (4.3)$$

High values of W_i' will be said here to indicate a high "efficiency" of particle removal, with the most preferred species said to be selected with 100 percent efficiency. When efficiency of clearance is low, that algal species will be said to have a refuge from grazing.

Selectivity is said to be "invariant" if ratios of clearance rates are constant even when the environmental context changes, for example when the algal community composition changes. Invariant selectivity implies invariant ratios of grazer-induced mortality, which, if true, would greatly simplify the study of how grazing affects algal succession. But, perhaps more importantly, invariant selectivity implies an absence of frequency-dependent switching behavior. Switching is well-known to stabilize predator-prey interactions in other systems when the most common prey type is most heavily grazed. If switching was important in planktonic grazing, it could help account for the maintenance of high algal species diversity. Optimal foraging predicts yet a different type of variant selectivity: hungrier animals should be less selective than well-fed animals. Such patterns have been reported for copepods (see Chapter 6).

Though ratios of clearance rates have many useful properties, successions in percent dominance of different algal species are more closely related to numerical differences in clearance rates. If two algal populations have identical reproductive rates and at $t = 0$ have identical population sizes (relaxing these assumptions does not materially alter the conclusions), the ratio of their population sizes obeys the following equation:

$$N_{1t}/N_{2t} = \exp\left[(C_2 - C_1){\cdot}G{\cdot}t\right] \qquad (4.4)$$

where N_{it} = density of algal species i at time t (individuals per ml), C_i = clearance rate on i (ml per animal per hour), and G = grazer density (individuals or biomass per ml).

The rate at which the relative abundances of the two algal populations change depends upon the term $(C_2 - C_1){\cdot}G$ (Figure 4.1). Eq. 4.4 shows how small differences in W_i' (eq. 4.3) may cause rapid succession when grazing rates (K) are high. For example, at 100 individuals per liter, the bottom curve results from an absolute difference in clearance rates of only 0.08 ml per animal per hour. With maximum clearance rates (C_{pref}) of 5 ml per animal per hour, as with large *Daphnia*, succession from equal numbers of two algal species to nearly complete dominance by one species occurs in only ten days with a difference of only 1.7 percent between the clearance rates on these

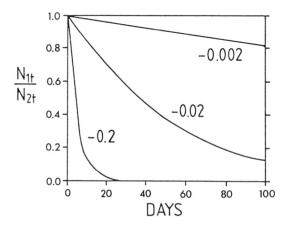

Figure 4.1 Succession of two algal species with identical reproductive rates but different susceptibility to grazing. N_{it} = density of algal species i on day t. The two species begin at equal density on day 0. Algal species 1 is selected over algal species 2 ($C_1/C_2 > 1$). The three curves show successional rates for three values of the term $(C_2 - C_1){\cdot}G$ (eq. 4.4).

two food types. This is a W_i' of 0.983, well within the experimental error of most measurements of selectivity. It is risky to treat any grazers as being truly unselective when grazing rates are high. Large-bodied *Daphnia* have the highest C_{pref} rates of all the freshwater zooplankton, so although they are also thought to be the most unselective planktonic herbivores (see below), their small degree of selectivity may still be quite important to phytoplankton successional dynamics. Of course the converse is also true: a high degree of selectivity (by W_i') for species with lower C_{pref}, or a small population size of grazers with high C_{pref} will have only a weak impact on succession.

4.3   Taxonomic Trends in Selectivity

From the standpoint of understanding seasonal succession, one would like to be able to rank algal taxa as to their relative susceptibilities to zooplankton grazing. In Chapter 6 similar questions will be considered from the perspective of the grazer populations. Upon first examining the extensive literature on zooplankton feeding this does not appear to be a subject that can be generalized. There is tremendous variety in modes of feeding within the various groups of freshwater zooplankton, and there is similar variety in the characteristics of phytoplankton that might determine their edibility. It would seem hopeless at first to describe a factorial combination of two such enormous varieties. Nevertheless, some useful generalities exist.

Cladocera feeding   The anatomical basis of feeding is well described for some cladocerans (e.g., Watts and Young, 1980; Geller and Müller, 1981; Fryer, 1987). The genus *Daphnia*, the most thoroughly studied of the freshwater zooplankton, and also the genus which either explicitly or implicitly is the model herbivore in many discussions of planktonic grazing, is the best understood in this regard. However, there is more than one mechanism of feeding within this order. For example, differences between the Daphnidae and the Bosminidae, which cause important differences in the types of phytoplankton most selected, have been well characterized (DeMott, 1982; DeMott and Kerfoot, 1982; Bleiwas and Stokes, 1985). DeMott and Kerfott (1982) have revitalized the distinctions drawn by Naumann (1921, cited in DeMott and Kerfott, 1982) among the *Daphnia* type of feeding (*Daphnia* and *Cerio-daphnia*), the *Sida* type (*Diaphanasoma, Holopedium,* and *Sida*) and the *Bosmina* type (*Bosmina* and *Eubosmina*). Though these differences must be acknowledged, the distribution of available literature and the importance often associated with *Daphnia* grazing necessitates that the following discussion will concentrate on the *Daphnia* type of feeding.

Morphology, especially size, is the characteristic most associated with cladoceran feeding selectivity, though it is thought that taste (Porter and Orcutt, 1980; DeMott, 1986; Chapter 7), presence of flagella (Knisely and

Geller, 1986), and surface properties (e.g., electrostatic charge) (Gerritsen and Porter, 1982) play some role in some cladocerans. Among more arcane choices, volume, longest linear dimension, and longest dimension at optimal orientation have all been used to characterize algal size in studies of selectivity. It is not easy to summarize the shape of the complex geometric shapes that some algal species have. Particularly vexing are pennate diatoms, many of which are ingested (Reynolds et al., 1982; Infante and Edmondson, 1985; Lehman and Sandgren, 1985; Sterner, 1989; Knisely and Geller, 1986; Sommer, 1988), even though length might classify them as inedible (e.g., McCauley and Briand, 1979). Also cyanobacterial colonies are highly variable in gross morphology and sometimes are consumed (often with low efficiency), but sometimes are not (see discussion in Knisely and Geller, 1986). Further permutations come from the manner of treatment of colonial species; some studies choose to measure the colony, while others focus on the cell, perhaps because the morphology of colonies can be extremely variable and is often subject to alteration during preservation. Colonies may also be broken apart before they are eaten. Colonies with uncompact morphology are large when considered as a colony, but small when the volumes of the cells are summed. Two independent studies found that volume and surface area predicted response to grazing better than longest linear dimension, as well as several other morphometric parameters (Bergquist et al., 1985; Horn, 1985a). However, particle selection is not based upon any single morphological criterion. Upper and lower ends of the size spectrum are inefficiently cleared for very different reasons. In spite of these complexities of shape, rough indications of size are often the only parameter reported in studies of selectivity, and must therefore often be relied upon.

Picoplankton-sized algae (<2 μm), including *Synechococcus* and *Synechocystis*, are inefficiently cleared by many cladocerans. In comparisons across zooplankton taxa, the distances between the setules on the feeding appendages predict the size of the smallest algae efficiently collected (Geller and Müller, 1981; Gophen and Geller, 1984; Brendelberger, 1985; DeMott, 1985; Hessen, 1985; Brendelberger and Geller, 1985). Adults of some *Daphnia* species (*D. galeata, D. pulicaria*, and *D. hyalina*) have mesh sizes between 1.3 and 2.3 μm (Geller and Müller, 1981) and collect *Synechococcus* with low efficiency in field grazing trials (Lampert and Taylor, 1984, 1985). *Bosmina* is similar (Bogdan and Gilbert, 1987). Adults of other *Daphnia* species, as well as *Diaphanasoma* and *Ceriodaphnia*, and juveniles of many cladoceran species have smaller mesh sizes and do not discriminate so highly (Brendelberger, 1985), clearing picoplankton with high efficiency. Some picoplankton may survive gut passage (Infante and Edmondson, 1985; Stockner and Antia, 1986). Picoplankton are somewhat inedible to some cladocerans.

Momentarily setting aside the complexities of shape, spherical plastic beads characterize the efficiency of particle clearance across the size spectrum. That cladocerans are inefficient at clearing large beads was shown early on

(Burns, 1968; Gliwicz, 1969, 1977; McCauley and Downing, 1985 refine the statistical analysis of Gliwicz's data). This refuge at the upper end of the size spectrum is important to more algal species than is the refuge at the lower end of the size spectrum. The maximal size of ingestible particles varies with animal body size (Burns, 1968) so that the larger the algae, the fewer the herbivores that can endanger it. Also, Gliwicz (1980) gives some anatomical reasons as to why the probability of capture by a single zooplankton individual should gradually decrease as particle size increases. The analysis by McCauley and Downing (1985, their Table 1) shows statistically significant underrepresentation of beads in animal guts beginning at 12.5–17.5 μm for *Daphnia cucullata* and at 12.5–22.5 μm for *Daphnia longispina* (two cases at 2.5 μm ignored). Burns's (1968) data shows the largest sizes of beads that were ingested by five species of *Daphnia* (not including *D. magna*) ranged between 12 and 42 μm. The size structures of both the algal and the grazer population must sometimes be taken into account. Holm et al. (1983) report that single trichomes of *Aphanizomenon flos-aquae* are cleared by *Daphnia*, but trichomes aggregated into flakes are not. Fulton and Paerl (1987) found that *Microcystis* colony size affected selectivity by several zooplankton species. Lehman and Sandgren (1985) suggested that small *Daphnia* do not efficiently clear *Asterionella* whereas larger individuals do. In its extreme, the presence of many large particles may interfere with food collection (Infante and Riehl, 1984) and can lower clearance rates on *all* algal species (Gliwicz, 1977, 1980; Gliwicz and Siedlar, 1980; Richman and Dodson, 1983; Porter and Mc-Donough, 1984; Fulton and Paerl, 1987).

Within a broad intermediate size range, *Daphnia* are often considered relatively unselective. Studies of the diets of natural populations of *Daphnia* tend to bear this out, but also uncover some subtleties. Gut-content analyses show that coexisting *Daphnia* species consume a wide spectrum of algal species and exhibit very high overlap. However, precise measurements of selectivity are impossible with this approach because of the inestimable effects of breakage and digestion (Infante and Edmondson, 1985; Kerfoot et al., 1985). In a study of particle clearance, Knisely and Geller (1986) found that *Daphnia hyalina* and *D. galeata* in Lake Constance consumed both flagellates and coccoid species but preferred the former, even though the sizes were similar. Sommer (1988) did not find similar preference for flagellates in long-term laboratory culture and found that *Daphnia longispina* and *D. magna* were very similar in their preferences for 11 intermediate-size algal species.

There is very broad consensus that particles about 3 to 20 μm in length and without spines or protective coverings are highly susceptible to cladoceran grazing. This would include green algae, such as *Chlamydomonas* and some *Scenedesmus* species, cryptophytes such as *Cryptomonas* and *Rhodomonas*, chrysophytes such as *Chrysochromulina*, and diatoms such as *Cyclotella*. That these same genera are often the dominant genera early in succession, before grazing pressure becomes high, is consistent with the assertion that their populations are depressed later in the season by grazing (Reynolds et al., 1982; Chapter 3). On the other hand, some researchers characterize crypto-monads as indicators of high grazing pressure later in the season (Shapiro and Wright, 1984), which must be caused by some mechanism other than selectivity.

Because of gut passage, cells possessing gelatinous sheaths or hard cell walls may not be cleared, even though they are ingested (Porter, 1973, 1976). These species are sometimes found to be uninhibited or stimulated when grazer density is experimentally increased (Lynch and Shapiro, 1981; Vanni, 1987). This has led some to treat all such cells as completely inedible. However, reports of clearance of some gelatinous greens (Lehman and Sandgren, 1985; Sterner, 1989; Sommer, 1988) show that this refuge is not always absolute.

Variant selectivity has been reported for *Sida* (Downing, 1981), but whether *Daphnia* select particles by variant or invariant means is open to conjecture. An early report using Coulter-counter techniques (Berman and Richman, 1974) characterized *Daphnia* selectivity as highly plastic. More recently, Knisely and Geller (1986) reported selection to be invariant in their seasonal survey of *Daphnia* diets, and some laboratory studies (Lampert and Taylor, 1985; DeMott, 1982) but not all (Meise et al., 1985) conclude *Daphnia* are invariant. Strict and complete invariance does seem unlikely. The behavior described by Gliwicz (1980) and by Gliwicz and Siedlar (1980) would make selectivity variant: cladocerans confronted with high proportions of large, interfering algae narrow their gape opening to avoid entanglement and the resulting energy expenditure. When interfering algae are rare, the gape is open, so the occasional large particle is taken readily into the feeding chamber, where it may be ingested or otherwise killed. Studies of *Daphnia longispina* grazing on size-fractionated seston (Okamoto, 1984) found lowest selectivities on large algal species when large species (e.g., *Staurastrum dorsidentiferum* and *Pediastrum biwae*) dominanted. Note that this would be destabilizing to the phytoplankton community: the greater the dominance by large algae, the greater these species are spared the risk of mortality in the feeding chamber, which should lead to enhanced dominance by larger species. For this reason, Dawidowicz and Gliwicz (1987) suggest that small populations of filamentous algae can be controlled by grazers, but larger populations slip out from grazing pressure. Further potential mechanisms for variance in feeding by cladocerans are:

1. As food density increases, rejections with the postabdominal claw increase (Porter et al., 1982; Scavia et al., 1984) which may dislodge some algal species more than others.

2. Viable gut passage may be enhanced by high food density, which limited evidence suggests decreases gut passage time (Murtaugh, 1985).

Copepoda feeding   Three major contrasts between feeding by adult, calanoid copepods and *Daphnia* are that:

1. Copepods have more distinct size preferences (Bogdan and McNaught, 1975; Muck and Lampert, 1980, 1984; Peters and Downing, 1984; Horn, 1985b).
2. Copepods have the ability to use chemical cues to select among individual particles that differ in chemical makeup (Chapter 6).
3. Copepods have lower clearance rates than *Daphnia* even when animals of equivalent body size are compared (Bogdan and McNaught, 1975; Peters and Downing, 1984; Porter et al., 1982).

Studies of calanoid copepod feeding, most of which have concentrated on marine species, have uncovered sophisticated mechanisms for selectivity of individual particles involving chemoreception (Friedman and Strickler, 1975; Friedman, 1980; Poulet and Marsot, 1980; DeMott, 1986) and mechanoreception (Friedman and Strickler, 1975; Friedman, 1980) with raptorial, active removal of large particles (Paffenhöfer et al., 1982). Koehl (1984) provides a thorough review of particle capture by marine copepods. Selectivity may change when the particle spectrum changes (Paffenhöfer, 1984) in a "peak tracking" manner (e.g., Poulet, 1973; Wilson, 1973), though the methodology which finds peak tracking (Coulter counters) has been forcefully criticized (Vanderploeg, 1981; Vanderploeg et al., 1984). Peak tracking would be functionally equivalent to switching. An alternative mode of variant selectivity is optimal foraging (Chapter 6), where the degree of selectivity is dependent on the animal's nutrition, and where discrimination is largely based on biochemistry, not on size. Studies on freshwater species are few and profoundly disagree as to the importance of variant selectivity. Richman et al. (1980) and Chow-Fraser (1986) characterize *Diaptomus* feeding as variant, and Okamoto (1984) found seasonal changes in *Eudiaptomus* food selection, but other studies using different methods strongly disagree (Vanderploeg, 1981; Vanderploeg et al., 1984) and characterize *Diaptomus* as a nearly rigidly invariant feeder preferring particles of about 15 µm equivalent spherical diameter (e.s.d.). Cinematographic study suggested that the marine calanoids *Eucalanus* and *Paracalanus* differ from the freshwater *Diaptomus* in the movement of their feeding appendages (Vanderploeg and Paffenhöfer, 1985) in a way that may explain a difference between these taxa. Recent work (Chapter 6) supports the optimal foraging mode of particle selection.

Cinematography shows that particle capture involves two different processes. The first is a "passive," relatively unselective capture of small algal particles, while the second, more "active" process is raptorial collection. The active process is the primary reason why copepod feeding is more selective than *Daphnia* feeding. The passive mode is similar in some respects to cladoceran feeding, except that the feeding appendages in copepods are not enclosed by a bivalved carapace and copepod appendages have greater independence of movement, characteristics which Porter et al. (1982) argue may account for greater clearance rates but lower selectivity by *Daphnia*.

Picoplankton and even somewhat larger algal cells have a distinct refuge from grazing by adult (McQueen, 1970; Vanderploeg, 1981; Vanderploeg and Ondricek-Fallscher, 1982; Vanderploeg et al., 1984; Bogdan and Gilbert, 1987) as well as naupliar (Bogdan and Gilbert, 1987) stages of *Diaptomus* and adult *Eudiaptomus* (Horn, 1985b; Persson, 1985; Riemann et al., 1986). Other nanoplanktonic, unprotected algae that are highly susceptible to *Daphnia* grazing, such as *Chlamydomonas* and *Cryptomonas* are also often efficiently cleared by adult (Muck and Lampert, 1980, 1984; Knisely and Geller, 1986; Bogdan and Gilbert, 1987), and naupliar stages (Bogdan and Gilbert, 1987) of both freshwater calanoids and cyclopoid nauplii (Havens and DeCosta, 1985).

Though raptorial feeding of freshwater calanoids on large particles is well-known, there is very little information on selectivities on individual algal taxa. Even the efficiency of clearance of certain major algal groups, such as colonial cyanobacteria or species with protective coverings, are very sparse. In one study that concentrated on algal taxa, Knisely and Geller (1986) found that *Eudiaptomus* had low selectivities on 14 taxa of colonial algae compared to unicellular flagellates. Kibby (1971) found high clearance rates for *Diaptomus gracilis* feeding on unialgal suspensions of *Ankistrodesmus, Carteria, Chlorella, Diplosphaeria, Haematococcus, Nitzchia*, and *Scenedesmus*, but low clearance rates on bacteria and *Pediastrum*. Fulton and Paerl (1987) found *Eurytemora affinis* consumed very little of either unicellular or colonial *Microcystis aeruginosa*. Except for the taste factor, the preferences so far shown for copepods are not radically different from those shown by *Daphnia*.

Descriptions of the algal community solely by size help to describe copepod feeding preferences. McQueen (1970) found maximal efficiency of *Diaptomus oregonensis* on cells between 100 and 800 μm^3 (4.6–9.2 e.s.d.), with lower efficiency on larger as well as smaller cells. The experiments of Richman et al. (1980) and Vanderploeg (1981) show distinct peaks of selectivity for particles 10–25 μm e.s.d. Studies using seston size fractionations are less specific, but have found higher efficiency of clearance on nanoplankton than on net plankton (22-μm cutoff) for *Diaptomus minutus* (Bogdan and McNaught, 1975), but varying selectivity of *Eudiaptomus japonicus*, with larger-size fractions ($> 25 \mu m$) being preferred when large algae were common and smaller fractions preferred when smaller algal species dominated (Okamoto, 1984). These studies are consistent with the premise that diaptomids show clear selectivity peaks for algal species that are neither <10 μm nor >50 μm, but again, shape and taste of the algae as well as the nutritional status of the animal may greatly change patterns of selectivity.

Reports of studies of cyclopoid selectivity are rare, perhaps because they are often simplistically cast in the role of being exclusively predacious. Knisely and Geller (1986) found *Cyclops* sp. had high efficiency on flagellates as well as on colonial pennate diatoms (*Fragilaria crotonensis* and *Asterionella formosa*) but low efficiency on other colonial algal species.

Rotifera feeding   Feeding by rotifers is usually concentrated on small cells, those about 20 μm in diameter or less (Pourriot, 1977). Reviews of this topic can be found in Dumont (1977), Pourriot (1977), and Starkweather (1980). The following discussion will concentrate on more recent literature. After first noting many idiosyncrasies in rotifer selectivities, Gilbert and Bogdan (1984) described the functional morphology and feeding modes of diverse species of rotifers and offered the following generalization. A generalist mode of feeding, exhibited by some species of *Keratella* and *Kellicottia*, is distinguished by a corona with a broad, funnel-shaped buccal field, and by selection with preference for particles larger than bacteria but smaller than large *Cryptomonas*. In contrast, a specialist mode, as seen in *Polyarthra* and *Syncheata*, is characterized by a sparsely ciliated corona and by efficient clearance of larger (*Cryptomonas*-sized) particles. Later, Bogdan and Gilbert (1987) distinguished two rotifer feeding guilds by principal component analysis. One guild showed high efficiency on picoplankton-sized cells whereas the other guild had low efficiency for these cells. It appears that picoplankton have a refuge from grazing by some species of rotifers, but not from others.

Many of the same algal taxa that suffer most from *Daphnia* and copepod grazing (unprotected, nanoplanktonic unicells) are also often highly susceptible to rotifer grazing. Bogdan and Gilbert (1982) found that *Keratella cochlearis*, *Polyarthra vulgaris*, and *P. dolichoptera* had high efficiencies of clearance on *Chlamydomonas*, and Bogdan and Gilbert (1987) found that selectivities by *Conochilus unicornis*, *Keratella crassa*, *K. cochlearis*, and *Polyarthra vulgaris* showed high efficiency for *Chlamydomonas* and for three strains of *Cryptomonas*. Algal cells with this type of gross morphology apparently are highly selected by virtually all members of the macrozooplankton community, which justifies treating them as "highly edible" regardless of the zooplankton community composition.

Fulton and Paerl (1987) found that *Brachionus calyciflorus* cleared colonial *Microcystis* at 1.9 times the rate of *Chlamydomonas* when *Microcystis* colony size was very small: 15.3–47.2 cells. Clearance rates on *Chlamydomonas* and *Microcystis* did not statistically differ when colony size was larger. It appears that very large algal species usually do not suffer greatly from rotifer grazing.

To summarize, unprotected nanoplanktonic unicells are vulnerable to nearly all grazing zooplankton. Large, colonial algae, especially cyanobacteria, have a refuge from many herbivores. Picoplankton have a partial refuge, being inedible to many herbivore species but not to others. Other major functional groups of phytoplankton, such as colonial diatoms, large dinoflagellates, and some gelatinous greens, are inedible to many herbivore species, but not to all. There is some latitude to speak of "edible" and "inedible" algal species without specifying the dominant herbivore species.

4.4 Community Grazing Pressure

The magnitude and the temporal pattern of grazing rates is clearly an integral aspect of seasonal succession. Grazing is important to phytoplankton primarily when grazing rates are high compared to reproductive rates and other losses. When grazing rates vary through the season, they could favor one suite of algal species at one time of the year and another suite of species later. This could occur, for example, if there is a tradeoff between being inedible and being competitive. Low grazing rates would favor edible species; high grazing rates would favor inedible species (Figure 4.2) (Sterner, 1989; Kilham, 1988 presents another example). This tradeoff is suggested on a course scale by the dominance patterns in whole community chemostats (see Figure 3.9) where neither the smallest (picoplankton) nor the largest (large colonies) algal species have competitively dominated. Relatively inedible species seem to be inferior nutrient competitors to species in the middle, more edible size classes. Though two attempts to explain seasonal succession by such a tradeoff have not supported this mechanism (Sommer, 1985; Sterner, 1989), further investigation seems warranted. Even in nonequilibrium communities, the ratio of densities of two algal species that are differently susceptible to grazing changes most quickly when grazing rates are high (see Figure 4.1). For these reasons, seasonal changes in grazing pressure should be important aspects of the seasonal succession of phytoplankton.

As is clear from the preceding section, the grazing loss rate on any one algal species at any one time will depend upon many factors, the most important including the taxonomic composition, community structure, and density of herbivores; the total density of edible algae and nonalgal seston; and the density of inedible phytoplankton (which can interfere with and slow food collection). However, empirical study has discovered that grazing rates in the field are largely dependent on herbivore biomass (Chow-Fraser and Knoechel, 1985; Knoechel and Holtby, 1986; Lampert, 1987). Applying allometric equations of clearance rates vs. body size to zooplankton census data thus can give an approximation to the seasonal patterns in grazing rate.

A more rigorous approach is to measure grazing rates directly. Burns and Rigler (1967) and Haney (1971, 1973) pioneered the measurement of community grazing by offering radioactively labeled, edible cells to natural grazer communities under more or less natural conditions. The flux of radiotracer

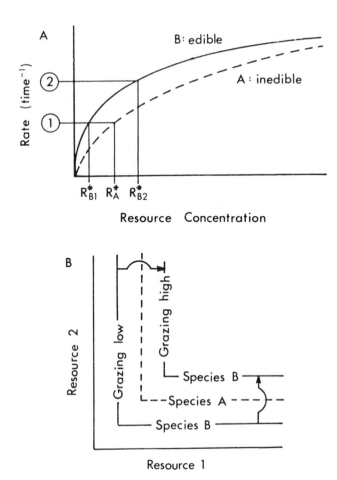

Figure 4.2   Seasonal variations in grazing rate may drive seasonal succession in equlibrium communities. A. Monod curves of growth rate as a function of resource concentration for two different algal species. Species *A* and *B* have similar resource-saturated growth rates, but different growth rates at low resource concentration. Competitive ability is determined by R*, the concentration of resource necessary for growth to exactly offset a given mortality rate. The lower R*, the better the competitive ability. In this example, species *B* is a superior competitor whenever mortality rates are equal (e.g., point no. 1) because then its R* is always lower than that of *A*. If there is a tradeoff between edibility and growth at low resource concentration such that species *B* is more susceptible to grazing than species *A*, the mortality experienced by the edible species will be high when grazing pressure is high (point no. 2). Species *B*'s R* may then exceed species *A*'s R*, and competitive dominance switches to the inedible species. Note that, in general, rankings of competitive abilities of various species are altered by changes in mortality rates. B. An extension to two resources (see Section 3.5 for methods of constructing resource isocline diagrams like these). When isoclines do not cross, the species whose isocline is nearest the axes will be the competitive dominant. At low grazing rates, the edible species is dominant; at high grazing rates, the inedible species wins.

from the food suspension to the entire herbivore guild is the basis for calculating community grazing rate.

This same general methodology has since been used to examine a variety of different lakes around the world, and the results can be compared and contrasted with each other. Since no two investigations have followed exactly the same protocol, caution is called for. Table 4.1 shows the methodology used for research in ten different lakes. To make these comparisons, one must presume that in spite of the differences in methods, grazing rates on the fraction of the algal community that is most edible to all grazers was measured in each study. There seems to be no better way to begin to summarize seasonal patterns of grazing pressure than this. For more information on the pros and cons of this methodology see these papers themselves (see also Gulati, 1985).

Absolute magnitude of grazing rates   Figures 4.3A, B, and C show changes in the grazing rate over a one-year period for the ten lakes listed in Table 4.1. Table 4.2 (page 128) lists the dominant grazers found in these ten lakes. To compile these data, values were taken from data in the published reports, and relabeled as instantaneous rates per day. Depth averages were taken when possible and necessary. Figures 4.3A, B, and C show that for some of these lakes, the succession from dominance by edible species to dominance by inedible species occurs during periods of high grazing rates.

Grazing rates for these ten sites differed dramatically. Maximum community grazing pressure ranged from only about 0.1 per day in Lake le Roux (Figure 4.3C), a 14,000-ha, silt-laden reservoir, to more than 2.5 per day in both Heart Lake (Figure 4.3A), a 17.6-ha seepage lake, and Hartbeespoort Dam (Figure 4.3B), a 2,000-ha reservoir. Given these differences, it is important to search for any patterns. First, there is no evidence that grazing rates systematically vary along an oligotrophic—eutrophic gradient. The highest grazing rates were measured in sites considered meso-oligotrophic (Schöhsee, Figure 4.3C), eutrophic (Heart), and hypertrophic (Hartbeespoort), whereas intermediate grazing rates were found in sites classified as oligotrophic (Lawrence, Figure 4.3B) and eutrophic (Blelham, Figure 4.3A, and Star, Figure 4.3B), and lowest maximal grazing rates were measured in oligotrophic (le Roux) and eutrophic (Vechten, Balaton, and Tjeukemeer) sites. Generalizations about the magnitude of grazing mortality in lakes of different trophy seem to be difficult to make. Gliwicz (1969) and Gulati (1984) have also examined grazing in relation to trophic gradients and do not reach any firm conclusions as to a dependable association of grazing levels and lake trophy. A well-documented tendency for zooplankton biomass to increase with trophic status (McCauley and Kalff, 1981; Canfield and Watkins, 1984; Hanson and Peters, 1984; Pace, 1986) is probably offset by a lowering of clearance rates because of increased algal biomass (both edible and inedible).

Zooplankton size structure, which is strongly influenced by fish predation (Chapter 7), does seem to be associated with the magnitude of grazing rates.

Table 4.1 Some aspects of methodology used to measure community grazing rates in ten lakes

Size	Measurements performed Day	Night	Depth integral (m)	Food organism	Chamber vol. (L)	Minimum grazer size (μm)	Reference[1]
Heart Lake	+	−	Aerobic stratum	*Rhodoturula* (yeast)	2	75	1
Blelham Tarn	+	+[2]	0–5 m	*Chlorella emersonii*	1	70[3]	2
Hartbeespoort Dam	+	−	0–20 m	*Chlorella* sp.	3	60	3
Lawrence Lake	+	+[2]	Epilimnion	*Chlorella pyrenoidosa*	20	153	4
Star Lake	+	−	0–2.5 m	*Chlamydomonas*	7.5	48	5
Lake Vechten	+	−	0–5 m	<15 μm seston	1	105	6
Lake le Roux	+	+	0–10 m	*Scenedesmus*	5.7	75	7
Schöhsee	+	+	0–12 m	*Scenedesmus*	3	106	8
Lake Balaton			Entire water column	<60 μm seston	flasks	nauplii	9
Tjeukemeer	+	−	0–2 m	<15 μm seston	1	105	10

[1] 1, Haney, 1973; 2, Thompson et al., 1982; 3, Jarvis, 1986; 4, Crumpton and Wetzel, 1982; 5, Bogdan and Gilbert, 1982; 6, Gulati et al., 1982; 7, Hart, 1986; 8, Lampert and Taylor, 1984; 9, Zánkai and Ponyi, 1986; 10, Gulati, 1975.
[2] Darkened daytime.
[3] >240 μm concentrated.

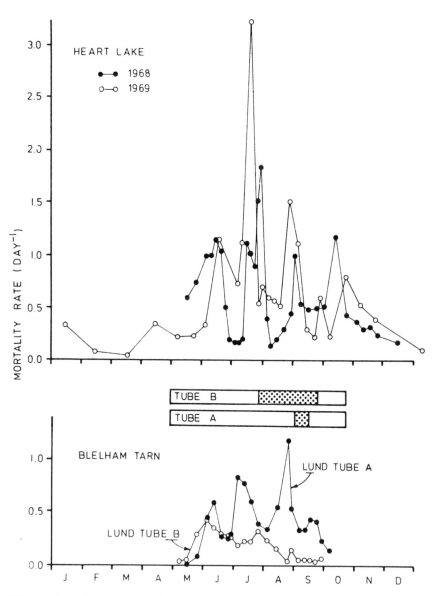

Figure 4.3A   Community grazing pressure in Heart Lake and Blelham Tarn. Data for two years are shown for Heart Lake. Horizontal bars for Blelham Tarn: unshaded portion, edible algae constitute greater than 50 percent of biovolume; shaded portion, inedible algae dominate.

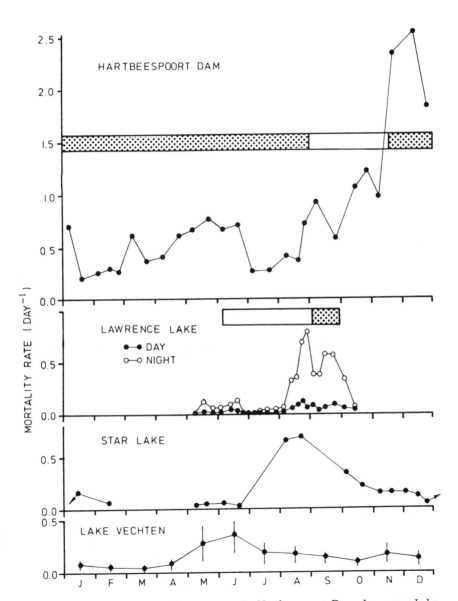

Figure 4.3B Community grazing pressure in Hartbeespoort Dam, Lawrence Lake, Star Lake, and Lake Vechten. For the first two, the horizontal bars indicate: unshaded portion, edible algae dominate; shaded portion, inedible algae dominate. The error bars for Lake Vechten represent 95 percent confidence intervals for seven years of monthly estimates (n=7).

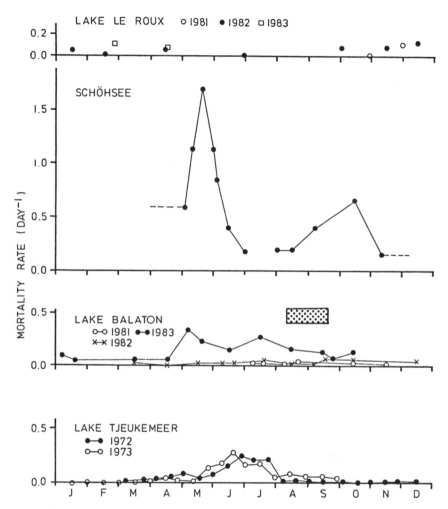

Figure 4.3C   Community grazing pressure in Lake le Roux, Schöhsee, Lake Balaton, and Tjeukemeer.

The highest grazing rates all occurred during dominance by species of *Daphnia* (Schöhsee, *D. galeata*; Heart, *D. galeata* and *D. rosea*; and Hartbeespoort, *D. pulex*). Other sites with dominance by *Daphnia* species had intermediate grazing rates (Blelham, *D. hyalina*; Lawrence, *D. pulex*; among others, July to September). The four lakes with herbivore guilds dominated by smaller species (primarily rotifers and *Bosmina*) all had lower community grazing rates (Star, Vechten, Tjeukemeer, and Balaton). There are several studies that support this trend of herbivore body size and grazing loss rates. The eutrophic Loosdrecht Lakes are dominated by *Bosmina*, *Chydorus*, and rotifers, and monthly estimates performed for 14 consecutive months revealed the following low annual means for community grazing at five locations: 0.20, 0.16, 0.13, 0.22, and 0.24 day^{-1} (Gulati, 1984). Shapiro and Wright (1984) reported grazing rates (determined by allometric relations) in Round Lake, Minnesota, were considerably higher in two years of *Daphnia* dominance than in the year of *Bosmina* dominance. Effects of zooplankton size structure on algal abundance are numerous. Pace (1986) concludes from an observational study of 13 lakes in the state of New York that higher concentrations of macrozooplankton relative to microzooplankton are associated with lower Chl *a* per unit total phosphorus and that large cladocerans are particularly important in this trend. Data in McQueen et al. (1986) concur with Pace's conclusions. Though Haney (1973) reports very high community grazing rates within *Bosmina* "swarms," these probably are not representative for loss rates within the whole lake basin. The evidence appears very strong that zooplankton communities dominated by large *Daphnia* show higher grazing rates than other communities.

This in turn suggests that the reduction in phytoplankton biomass that follows fish-induced shifts toward larger zooplankton (Hrbáček et al., 1961; Lynch and Shapiro, 1981; Shapiro and Wright, 1984; Carpenter and Kitchell, 1987a; see also Chapter 8) is caused by increased community grazing rates.

At first this may seem completely intuitive, as larger animals have higher clearance rates. But this trend is surprising in some ways. As with many other size-specific physiological rates (Peters, 1983), weight-specific clearance rate decreases with increasing body size, at least in cross-taxa comparisons spanning several orders of magnitude (Peters and Downing, 1984; Knoechel and Holtby, 1986). Thus, any given biomass of large zooplankton should have lower grazing rates than the same biomass of smaller animals. To illustrate, Figure 4.4 shows that when the allometric relations between body and clearance rate given by Peters and Downing (1984) are taken, and total grazer biomass is held constant at 1000 μg per 1, animals of 1 μg body mass should produce algal mortality rates several times higher than animals of 100 μg body mass. Pace (1986) and Bogdan and Gilbert (1982) have argued on these grounds that microzooplankton are more significant grazers is than usually assumed. It thus seems paradoxical that zooplankton assemblages dominated by small species have lower community grazing rates than ones dominated by large-bodied species. Apparently, total herbivore biomass is not independent of zooplankton size structure; herbivore guilds dominated by larger species have higher biomass than guilds dominated by smaller species. Simply put, zooplankton assemblages dominated by *Daphnia* have high grazing rates not because the large individual animals have high clearance rates, but because of some factor that allows *Daphnia* population biomass density to become high. Perhaps *Daphnia* possess some autoecological energetic advantage compared to other taxa (see Chapter 7), or alternatively, *Daphnia* may suffer lower losses when they dominate as other taxa do when they dominate.

Table 4.2 Some characteristics of the ten lakes

Site	Location	Trophic status	Maximum chl a conc. (μg/L)	Dominant grazers
Heart Lake	Ontario, Canada	Eutrophic	N.A.	Daphnia rosea D. galeata mendotae Ceriodaphnia reticulta
Blelham Tarn	Cumbria, U.K.	Eutrophic	98.6[1]	Daphnia hyalina
Hartbeespoort Dam	South Africa	Hypertrophic	3000	Daphnia pulex D. longispina Ceriodaphnia reticulata
Lawrence Lake May to July:	Michigan, USA	Oligotrophic	12[2]	Copepods Bosmina
July to August:				Daphnia retrocurva D. galeata mendotae D. pulex
Star Lake	Vermont, USA	Eutrophic	N.A.	Bosmina longirostris Keratella cochlearis Polyarthra
Lake Vechten	Holland	Mesotrophic	N.A.	Bosmina longirostris Daphnia cucullata Eudiaptomus gracilis

Lake le Roux	South Africa	Oligotrophic	5	*Metadiaptomus meridianus* *Daphnia gibba* *D. barbata* *Moina brachiata*
Schöhsee	Holstein, FRG	Meso-eutrophic[3]	6	*Eudiaptomus gracilis* *Eudiaptomus graciloides* *Daphnia galeata*
Lake Balaton	Hungary	Eutrophic	20.4	Nauplii *Cyclops* *Diaphanasoma* *Eudiaptomus* *Daphnia hyalina* *D. galeata* *D. cucullata*
Tjeukemeer	Holland	Nutrient rich, turbid	N.A.	*Bosmina longirostris* *B. coregoni* *Daphnia cucullata* *D. hyalina* *Ceriodaphnia pulchella* *Diaphanosoma brachyurum* *Chydorus sphaericus* *Eurytemora affinis*

N.A., not available.
[1] Lack and Lund, 1974.
[2] Wetzel, 1983.
[3] Lampert et al., 1986.

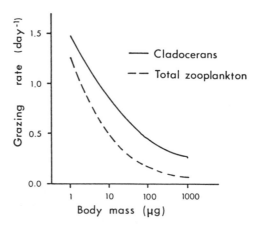

Figure 4.4 Community grazing rate versus animal body size given allometric relationships between body size and clearance rate (Peters and Downing, 1984) and holding total grazer biomass constant at 1000 μg per l. This relationship predicts that phytoplankton co-occurring with herbivore communities of small body size should have higher mortality rates than phytoplankton co-occurring with larger herbivores, which is opposite to the associations actually observed (see Figure 4.3).

Taylor (1984) advances an argument to explain higher total biomass when larger herbivores dominate that seems to be based on the premise that zooplankton populations are adjusted so that phytoplankton gains are exactly balanced by losses. This is not evolutionarily tenable. Duarte et al. (1987) report that maximum biomass attained under ideal laboratory conditions is independent of body size, arguing against an energetic advantage to large body size. Of course, any advantageous factor possessed by *Daphnia* may not be related to their body size at all; some other trait may be involved (large diaptomids are very different from large *Daphnia*). The success of large *Daphnia* in monopolizing resources (Kerfoot, 1987) and achieving high biomass creates a successional arena of high grazing-loss rates.

Fluctuations in grazing rates   Grazing rates fluctuate on many temporal scales. Diurnal rhythms in grazing rates were reported for Lawrence Lake and the Schöhsee (the latter not shown, see Lampert and Taylor, 1985, and Lampert, 1987). Much smaller diel differences were found in Lake le Roux and in the darkened daytime treatments in Blelham Tarn. Diurnal differences in grazing in general may be due partly to diel rhythms in grazing rates of the individual animals (Haney, 1985), but are more likely to be linked to vertical migration. Diel rhythms in grazing losses have been suggested to affect algal dynamics by a temporal disjunction between growth in the light and death in the dark (Lampert, 1987).

Data for several of the sites where *Daphnia* dominated (Heart, Hartbeespoort, and at least one of the Lund Tubes in Blelham Tarn) suggest periodic oscillations in grazing rate similar to those that have been found for *Daphnia* population densities (Murdoch and McCauley, 1985; McCauley and Murdoch, 1987). These density cycles were said to provide evidence for a predator-prey coregulation of herbivore and algal populations. Algal succession in *Daphnia*-dominated sites appears to be subject not only to a season-wide waxing and waning of grazing pressure, and sometimes to a diurnal rhythm in grazing pressure, but to intermediate frequency cycles as well.

It is clear from Figure 4.3 that grazing loss rates fluctuate considerably on a seasonal time scale. The PEG-model (see Section 1.2) emphasizes heaviest grazing losses during spring and early summer, especially in eutrophic lakes: PEG steps 2 to 5 describe a buildup of herbivore biomass and an accompanying increase in grazing pressure, leading to a clear water phase. In step 6, the herbivore biomass is said to decrease. In step 8, at the start of phytoplankton summer crops, grazing pressure is said to be reduced. Figure 4.3 allows an examination of this aspect of the PEG-model.

In the northern latitudes, Schöhsee, Vechten, Balaton (one of three years), and one of the Lund Tubes did indeed have grazing maxima early in the season. In contrast, Heart Lake and the other three northern-latitude lakes had maximum grazing pressure later in the season, closer to the time of maximal epilimnetic temperature, and Tjeukemeer was intermediate in the timing of its grazing maximum. In the southern latitudes, Lake le Roux had no clear grazing maximum, and Hartbeespoort Dam had a grazing maximum which preceded the maximum of water temperature by two months (temperature data in Robarts and Zohary, 1984). Though distinct peaks in grazing loss rates are the rule, the timing of the peak appears to be variable from lake to lake. There is no strong association of the timing of grazing maximum with trophic status. Thus, the PEG description of early season grazing maxima occurred only in some sites. In this sample there were no maxima in grazing rates after temperature maxima, but even this sometimes occurs (Sterner, 1989).

Periods when grazing rates are about 1.0 or greater, which is similar to the maximal growth rates of many species of algae, sometimes last several months (several peaks in Heart Lake and a single peak in Hartbeespoort Dam) or may be shorter, and in some lakes grazing rates approach zero for long periods of time, even during the growing season. In this way grazing pressure resembles the timing of phosphorus and nitrogen competitive pressure already discussed by Sommer (see Chapter 3). Rather than being constant and inexorable, grazing pressure actually produces a strong focusing of the phytoplankton species assemblage during distinct parts of the seasonal succession. In this way, grazing resembles other ecological interactions which have been studied for long time courses, such as competition among birds (Wiens, 1977; Grant, 1986). Conclusions as to the importance of these biotic interactions require close interval sampling of the entire seasonal cycle. However, the evidence suggests that site-to-site variation in grazing pressure is greater than year-to-year variation within sites (Lake Balaton being a possible exception). Within obvious limits, sites can be characterized as to their level of grazing pressure, and characterizations of algal interactions under high loss rates vs. low loss rates (e.g., Sommer, 1986) have natural analogs.

Effect on algal dominance   When grazing rates become high, mortality of edible phytoplankton species is high, favoring dominance by inedible species. The timing of succession from edible species to inedible ones provides a test of the importance to succession of grazing. For example, in the Schöhsee, the May and June peak in grazing was shown to be responsible for a decrease in population sizes of small, edible algal species (*Rhodomonas, Stephanodiscus*) so that larger, less edible species (*Ceratium, Closterium, Staurastrum, Sphaerocystis*, and *Fragilaria*) became dominant (Lampert et al., 1986; Lampert, 1987; algal data not shown in Figure 4.3). In Hartbeespoort Dam, dominance by edible chlorophytes and chrysophytes gave way to dominance by *Microcystis* during the grazing maximum (Robarts and Zohary, 1984, give another interpretation). In Lawrence Lake, dominance by two different species of the edible diatom *Cyclotella* yielded to the less edible *Sphaerocystis* precisely at the grazing maximum. In each of the two Lund Tubes in Blelham Tarn, *Microcystis* achieved dominance during or somewhat after the respective grazing maxima, but, curiously, the tube with higher overall grazing rates had a later and shorter period of *Microcystis* dominance. These examples are consistent with the view that the mortality experienced by edible species during grazing peaks enables inedible species to dominate.

Dominance by inedible species is not, however, always associated with high grazing rates. A bloom of *Anabaenopsis raciborskii* in Lake Balaton was reported for 1982, a year with exceptionally low grazing loss rates, and it failed to reappear the following year when grazing rates were much higher (algal counts not reported so dominance patterns not shown). *Anabaenopsis* and other nostocalean algae, such as *Anabaena* and *Aphanizomenon*, are superior nitrogen competitors owing to their ability to fix nitrogen (Elder and Parker, 1984; Tilman et al., 1986). High mortality on potential competitors is not necessary for nitrogen-fixers to dominate an algal community when nitrogen is the most limiting element (Sterner, 1989). In general, grazing should be most important to algal dominance patterns when inedible species are the inferior competitors in the absence of grazing.

4.5   Indirect Effects

A single grazer population consuming a single algal population constitutes a direct effect—there are no intermediates and no other populations play a role. A complete analysis of the ways that grazers affect algal succession must also

consider any indirect effects. Miller and Kerfoot (1987) describe three qualitatively different types of indirect effects. One type, trophic linkage, occurs
because of a propagation of the effects of one species' density on the rates
of change of other species through a food chain. When A affects B and B
affects C, A has an indirect effect upon C. A second type, behavioral effects,
occurs when species A alters the behavior of species B and therefore affects
B's interactions with a third species, C. The third type, a chemical response,
resembles the second type except that A alters B's interactions with C because
of a chemical released by A that affects B.

   Much current research in aquatic population interactions revolves around
indirect effects (Kerfoot and DeMott, 1984; many papers in Kerfoot and Sih,
1987). Indirect effects of grazers on phytoplankton have received attention
(Schoenberg and Carlson, 1984; Sterner, 1986a; Lampert, 1986; Vanni, 1987;
Sommer, 1988). There are no reports of chemical response, the third type of
indirect effect, on the interaction between grazers and phytoplankton.

   Indirect behavioral effects have not been so identified, but they have
been described. In one suite of indirect behavioral effects, large, interfering
algae affect cladoceran food gathering, lowering the clearance rate (and thus
the mortality) on all algal species (see Section 4.3). When the density of the
interfering algal species increases, mortality rates should decrease for all species, making this an indirect mutualism. Another indirect mutualism can occur
between edible algal species. Above the incipient limiting concentration, clearance rates decrease with increasing total edible algal density. Thus, clearance
rates on all algal species decline when the density of one edible algal species
increases (Sterner, 1986b; Sterner, 1989; Sterner, in prep.). These indirect
mutualisms act over the short term; long-term effects may or may not be
qualitatively—much less quantitatively—equal (Abrams, 1987).

   The evidence for indirect trophic-linkage effects of zooplankton on algae,
mediated via the inorganic nutrient pool, is extensive. As will be discussed
below, field manipulations of grazers often produce responses from the phy-
toplankton populations that are difficult to explain solely by selective grazing.
Other studies have shown that experimental alterations in grazing pressure
result not only in changes in algal mortality but also in algal productivity
(Gliwicz, 1975; Porter, 1976; Redfield, 1980; Bergquist and Carpenter, 1986;
Sterner, 1986a), which must be caused by indirect, rather than direct effects.
Increases in productivity may be about as large as increases in mortality
(Sterner, 1986a). Related evidence comes from studies showing that net growth
rates of algal populations sometimes assume either a positively or a uni-
modally shaped response to increasing grazer density (Lehman and Sandgren,
1985; Bergquist and Carpenter, 1986; Sterner, 1986a). Also, factorial manipulations of nutrient and grazer levels show statistically significant interaction
terms, suggesting an interdependence among nutrient limitation patterns and
levels of grazing pressure (Sterner, 1989). When these lines of evidence are
added to studies concluding that a major portion of nutrient supply to phy-

toplankton is from zooplankton release (Barlow and Bishop, 1965; Hargrave and Geen, 1968; Lehman, 1980a; Axler et al., 1981; Ejsmont-Karabin et al., 1983; Mitamura and Saijo, 1986), the evidence is weighty that indirect effects through the trophic linkage grazer to nutrients to algae, are relatively common and are also relatively "strong" (i.e., are important relative to other factors in determining algal population rates of change). In this section, after first reviewing available experimental results, a framework will be provided for considering these indirect effects of zooplankton on phytoplankton, and then that framework will be the basis for some theoretical results.

One aspect of these indirect effects, microscale patchiness of nutrient release, has attracted much recent attention. It has been established that released nutrient is patchy on the microscale (Lehman and Scavia, 1982). It is proposed that microscale patchiness of nutrient release allows resource-saturated growth because cells are occasionally exposed to high resource concentration in between time periods spent at extremely low resource concentration (Goldman et al., 1979; McCarthy and Goldman, 1979). In the ocean, microaggregates of organic material are suggested to play a similar role in creating microscale heterogeneity (Goldman, 1984). This idea has led at least some phytoplanktologists to rule out resource limitation out of hand (Harris, 1986). Closer examination of these arguments, however, indicates that this is premature. It is still very unclear what effect this scale of heterogeneity has on algal populations (see Lehman, 1984, and Lehman and Scavia, 1984). Because it is difficult to quantify algal-zooplankton encounter rates (Section 4.2; Scavia et al., 1984), the frequency of algal encounters with microscale patches is completely unknown. Exploitation of patches experienced by cells for time intervals much shorter than the life span of the cell may be made possible by uptake rates that are much greater than rates of cell growth, especially for cells that are somewhat depleted in nutrients, i.e., are resource-limited (Goldman and Glibert, 1982) but also for cells of some species growing near resource-saturation (Goldman and Glibert, 1982; Goldman, 1984). However, actual incorporation of nutrient transported into the cell is much slower, occurring at rates closer to the cell's growth rate (Zehr et al., 1988). Thus, cells tend to average over small-scale patchiness more than is indicated in studies of short-term nutrient transport. It has already been noted that small-scale (i.e., short term) heterogeneity has less effect on the diversity of algal communities than does heterogeneity occurring on a larger (longer) scale (see Figure 3.12). It should also be noted that heterogeneity is not itself reason to abandon models based on equilibria (Levins, 1979; Tilman, 1982). The ecological importance of microscale patchiness has drawn much debate (Jackson, 1980; Currie, 1984; McCarthy and Altabet, 1984; Reynolds, 1984; Lehman and Sandgren, 1985; Levitan, 1987; see also Chapter 3), and pending some resolution of the arguments pro and con, wholehearted adoption of the view that microscale patches preclude resource limitation of phytoplankton is unwise.

Grazer gradients—experimental results   A number of experiments have recorded the effects of experimentally established gradients of grazer density on algal population dynamics. These gradients have probed the short-term dynamics of zooplankton/phytoplankton interactions by determining the sensitivity of algal populations to changes in the density of grazers. In addition, by establishing the functional form of this dependence, these experiments have helped to unravel the mechanisms at work. As a set, these gradient experiments have been conducted at various times throughout the growing season in a sample of lakes that differ widely in their algal communities as well as in other characteristics, though nearly all of them have been performed during dominance (or at least prevalence) by *Daphnia*. In the following discussion, each algal taxon within each gradient that was judged to have sufficient sample size for statistical analysis will be called a "case."

   In each of these experiments, a natural phytoplankton community was incubated with various densities of the natural zooplankton community; the response variables were the rates of change of the phytoplankton population. Enclosure volumes ranged from 0.85 to 36 liters (l). Incubation times (ranging from 3 to 6 days) reflect methodological constraints—allowing sufficient time for effects to develop while minimizing enclosure effects—rather than any primacy of importance of one particular time scale. The time scales examined are thus in some respects arbitrary, and may themselves affect the magnitudes and functional forms of the responses (Carpenter and Kitchell, 1987b).

   These gradient experiments have demonstrated that a large fraction of phytoplankton populations are affected by grazers. Statistically significant effects of grazer density on algal net growth were found in 41 out of 98 cases (42 percent) (Lehman and Sandgren, 1985), 9 out of 16 cases (56 percent) (Bergquist and Carpenter, 1986), and 67 out of 137 cases (49 percent) (Elser et al., 1987). These percentages are probably underestimates of the degree to which grazers influence algal population since there must have been some species that responded too weakly for statistical analysis to achieve significance. Also, a net response to grazers is the sum of two responses, one positive and the other negative, so that lack of a net response does not mean no response at all. Sterner (1986a) showed a net response composed of two opposing individual responses so that the net effect on rates of change was weak. When this occurs, grazers increase the turnover rate of the algal populations but not the net rates of change. One cannot yet judge to what extent this particular effect pervades the algal community, but one must conclude from these studies that the dynamics of many algal populations are influenced partly by the density of the herbivores.

   The nature of the dynamics seen in these experiments suggests several possible mechanisms. The most common response to increasing grazer density is for the net growth rate to decrease, i.e., the net effect of the herbivores is negative. This type of response accounts for 88 percent and 52 percent of significant responses in the two studies with large numbers of cases (Lehman and Sandgren, 1985, and Elser et al., 1987, respectively), though it constitutes only 22 percent of the study with a smaller number of cases (Bergquist and Carpenter, 1986). A simple, linear decline in net rates of change with increasing grazer density is expected theoretically from the direct effect of grazing alone (see below). However, in many empirical cases the decline in net rates of change is not linear. Curvature was searched for by quadratic regression and found in 25 percent (Bergquist and Carpenter, 1986) and 37 percent of the significant grazer responses (Elser et al., 1987). All curvilinear responses so far reported have been convex, i.e., their second derivative with grazer density was negative. In some cases net growth increased with increasing grazer density at low grazer density, but at higher grazer density it decreased with increasing grazer density, i.e., the response was unimodal, or hump-shaped.

Algal net growth rates may also increase with increasing grazer density, though this seems to be less common, being found in 12 percent (Lehman and Sandgren, 1985), 56 percent where the number of cases is not as large (Bergquist and Carpenter, 1986), and 6 percent (Elser et al., 1987) of significant responses. Algal size seems to be strongly associated with this type of response—nearly all of the algae that have showed positive net effects from increasing grazer density have been greater than 100 μm in their longest linear dimension. Figure 4.5 illustrates taxa that show this response. (This figure also illustrates the difficulty of summarizing the size of large, colonial algae by any single morphometric parameter.) Taken together, these studies demonstrate that *Daphnia*-dominated zooplankton assemblages tend to have negative net effects on algal species that are efficiently grazed, but can often have positive net effects on species with a refuge from *Daphnia* grazing. This probably occurs by a transfer of limiting nutrient from the edible to the inedible species (Bergquist et al., 1985).

Grazer gradients—theoretical results   Grazer gradient experiments have exposed a rich set of herbivore/producer dynamics: algal growth rates may increase or decrease with increasing grazer density, and the relationship may be linear or convex. These different types of dynamics are somehow indicative of grazing rates, selectivity, nutrient limitation, and nutrient regeneration, but lack of theoretical study has so far hindered interpretation.

The direct effect of zooplankton on phytoplankton levels is easily analyzed. It was shown earlier that the clearance rate is directly proportional to algal mortality rate. Thus, when clearance rate per animal does not vary with animal density, algal mortality will be a simple linear function of grazer density with a slope equal to the specific clearance rate, as shown by the dotted lines in Figure 4.6A. Mortality from grazing will be zero when grazer density is zero, and will increase as grazer density increases within the entire range of grazer density. Mortality from sources other than grazing would simply add a constant (they do not vary with grazer density), or they could

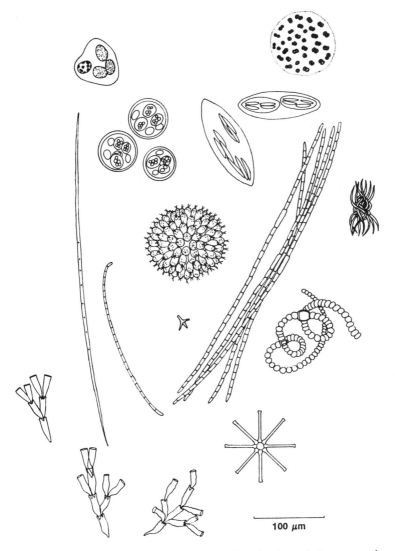

Figure 4.5   Phytoplankton taxa which have been found to have their net growth rates positively affected by increasing grazer density. With only one exception, these taxa are all large enough to have some degree of refuge from grazing. Drawings adapted from those in Prescott (1962).

be subsumed into the reproduction function. Highly edible species will have mortality rates that are sensitive to grazer gradients, whereas mortality of less selected species will be relatively insensitive. If the only effect of grazer density on algal per capita rates of change occurred via the direct effect of grazing,

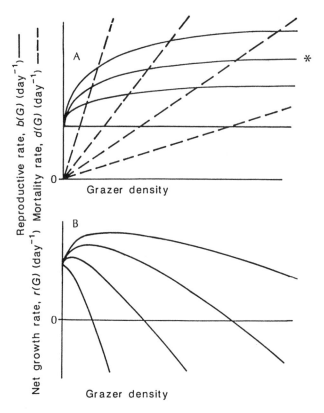

Figure 4.6   Algal population dynamics in gradients of grazer density. A. Per capita mortality (dotted) vs. grazer density [$d(G)$] for different clearance rates. Per capita reproduction (solid) vs. grazer density [$b(G)$] for varying degrees of enhancement of productivity. B. Net per capita rates of change [$r(G)$] when the $b(G)$ function labeled with a star (Figure 4.6A) and the four different clearance rates are used. Net response goes from decreasing to unimodal as susceptibility to grazing decreases.

grazer gradients would result in the net rates of change (reproduction minus mortality) of all algal species having a negative, linear dependence on grazer density, with the slope determined by selectivity. As already discussed, other dynamics do occur.

When algal reproduction is limited by a resource made more available by the grazer, reproduction will increase with increasing grazer density until resource saturation occurs (see Figure 4.6A, solid lines). Resource availability may increase either by nutrient regeneration or by cropping of the algal community, precluding monopolization of any resource, including light. Grazer effects on algal per capita productivity are stimulatory, and when the grazer gradient is broad enough they must be curvilinear: no organism can reproduce at an infinite rate. The net effect of a grazer gradient is the difference between reproduction and mortality. Reproduction functions may be either flat (no effect) or be hyperbolically saturating (positive reproduction stimulated by grazing). Mortality functions are linear. There is no concave component (positive second derivative with grazer density). The total stimulation in per capita reproduction (achieved at high grazer density) relates to the degree of nutrient limitation (actual growth rate divided by resource-saturated growth rate). A severely nutrient-limited taxon may be strongly stimulated when grazer density is very high.

The above relationships help explain the various dynamics observed in grazer gradient experiments. Phytoplankton responses to grazer gradients fall into one of four categories (Table 4.3) according to each species' sensitivity to grazing and its degree of resource limitation. Another way to summarize this is by the following equation:

$$r(G) = \mu' + \mu\,(G/K + G) - C{\cdot}G \qquad (4.5)$$

where: $r(G)$ is defined as the net per capita rate of change of algae as a function of grazer density; μ' is the per capita reproductive rate at grazer density of zero, $\mu + \mu'$ is the resource-saturated growth rate, K is a half-saturation constant for growth (with dimension equal to grazer density), C is the specific clearance rate, and G is the grazer density or biomass. The direct effect of grazing depends upon C and thus only on relative edibility. The indirect effect depends upon the parameters μ' and μ. Where $\mu/(\mu' + \mu)$ (which is closely related to the intensity of nutrient limitation discussed in Chapter 3) is large, the sensitivity to the indirect effect may also be large. The actual magnitudes of these different effects then depends upon the actual grazer density, G, and the release rate of nutrient from the grazer pool compared to the population size of the phytoplankton. Sterner (1986a) shows an application of this approach.

Unless nutrient recycling by grazers can decrease cell numbers, conservation of mass stipulates that more than one unit of algal biomass cannot be made from recycling the nutrients contained in one unit. However, when nutrients come from sources other than recycling from algal material, nutrient regeneration may cause the net effects of grazers to produce positive functions of grazer density for the phytoplankton as a whole. Nutrients may be transferred to algae from nonalgal sources like bacteria, detritus, and protozoa.

Table 4.3 Algal net per capita rates of change should exhibit any of these four types of responses to increasing grazer density

Algal type	Resource limited	Not resource limited
Edible	Curvilinear, unimodal, or negative	Linear, negative
Inedible	Curvilinear, positive	No response

Release of phosphatase enzymes by zooplankton (Wynne and Gophen, 1981; Boavida and Heath, 1984) may make an otherwise refractory resource available. Finally, short term, nonequilibrium experiments might find stimulatory effects because of nutrient assimilated by the animals before the experiment began. In this last case, total nutrient within the grazer biomass would decrease during the experiment, and thus this should be restricted either to starving, unhealthy or senescent animals or to animals incurring significant mortality by carnivores whose nutrient regeneration may also be significant aspect of the trophic web (Tátrai and Istvánovics, 1986; Threlkeld, 1987). Of course, for individual phytoplankton species, net effects can be positive because of nutrient release by other species that are being eaten and digested by grazers.

The particular parameters of productivity that are enhanced by grazing herbivores are easily confused. The choices are numerous, and include per capita biomass incorporation (e.g., carbon fixation), per capita reproductive rate, areal or volumetric biomass incorporation, and evolutionary fitness. Though for phytoplankton the first two choices are at least not too distantly related (Turpin et al., 1985; Williams and Turpin, 1985), distinctions among these dimensions can be critical. Volumetric or areal productivity may be unimodal with grazer density (increase with grazer density at low grazer density, but decrease with grazer density at higher grazer density) (e.g., Cooper, 1973; Carpenter and Kitchell, 1984) simply because of density dependence, in the same way that dN/dt is maximal at intermediate N in the logistic equation (Carpenter and Kitchell, 1985). This is unrelated to the unimodal response of per capita productivity (dN/Ndt) to grazer density that can occur because of a dynamic interaction between grazing losses and nutrient transfer from competing species.

The grazing/excretion model   It may be useful at this point to construct a numerical model for the effects of grazing and nutrient regeneration on the dynamics of an algal population. The model simulates a grazer gradient experiment. For the first equation the following assumptions were made:

- Nutrients were added to the dissolved pool via regeneration of some fraction of the cleared particulate nutrient.
- Clearance rate was a rectilinear function of algal concentration.
- Grazer density was treated as a constant.
- Nutrients were removed from the dissolved pool by algal uptake, which was Michaelis-Menten in form (S is the concentration of the dissolved nutrient).

$$dS/dt = a \cdot C(N) \cdot G \cdot N \cdot Q - \rho[S/(k_\rho + S)] \cdot N \qquad (4.6)$$

Table 4.4 contains descriptions and definitions of the symbols used.

Table 4.4   Descriptions and definitions of the reference parameter set used in the grazing/excretion model

Parameter	Description	Value in reference run	Dimension
a	Fraction of cleared nutrient that is rereleased	0.50	Dimensionless
$C(N)$	Clearance rate[1] (varies with N)	$N \leq 10^4$: 1.77 $N > 10^4$: $(1.77 \cdot 10^4)/N$	ml·animal^{-1}·hr^{-1}
G	Grazer density	0–100	Animals/l
k_q	Subsistence quotient	$0.25 \cdot 10^{-9}$	μmoles/cell
k_p	Half-saturation constant for uptake	0.05	μmoles·cell^{-1} ·hr^{-1}
N	Algal density	Variable[2]	Cells/ml
Q	Cell quotient	Variable[2]	μmoles/cell
r	Per capita rate of change	Variable[2]	Day^{-1}
ρ	Maximum uptake rate	$0.25 \cdot 10^{-9}$	μmoles·cell^{-1} ·hr^{-1}
S	Dissolved nutrient concentration	Variable[2]	μmoles/l
μ	Maximum growth rate	1.20	Day^{-1}

[1] Clearance rate parameters taken from Porter et al., 1982.
[2] Value varied during simulation. Initial, dissolved nutrient concentration ($S_{t=0}$) was 0.005 μmoles/l. Initial cell quotient ($Q_{t=0}$) was $0.25 \cdot 10^{-9}$ μmoles/cell.

The particulate nutrient pool increased via algal uptake and decreased via grazing:

$$dNQ/dt = \rho[S/(k_p + S)] \cdot N - C(N) \cdot G \cdot N \cdot Q \qquad (4.7)$$

Algal population density began at a prespecified initial level, and the per capita reproductive rate was determined by internal stores, following Droop's (1973) cell quota formulation. Algal mortality occurred solely from grazing:

$$r = dN/Ndt = \mu[(Q - k_q)/Q] - C(N) \cdot G \qquad (4.8)$$

The equation for the rate of change of cell quota was found by using the chain rule on eq. 4.7, giving:

$$dQ/dt = \rho[S/(k_p + S)] - \mu(Q - k_q) \qquad (4.9)$$

In this model, grazers filter a portion of the algal standing crop and regenerate a fraction of that nutrient into the dissolved pool. Nutrient may flow into or out of the grazer compartment, depending upon the value of the assimilation parameter. To see this, note that the rate of change of nutrient in the dissolved pool and algal pool is:

$$dS/dt + d(N \cdot Q)/dt = (a - 1) \cdot C(N) \cdot G \cdot N \cdot Q \qquad (4.10)$$

This rate of change will be negative when grazers assimilate nutrient (nutrients can be incorporated into grazer biomass).

Simultaneous analytical solution of eqs. 4.6, 4.8, and 4.9 was intractable analytically because of the nonlinear terms. A modified Euler method of solution was utilized for numerical solution. The functions $b(G)$, $d(G)$, and $r(G)$ (as defined in Figure 4.6), were calculated over simulated periods of six days (time step $= 0.3$ hr, lowered automatically when dynamics were rapid. Further reduction of the time step did not significantly alter the results).

The reference set of parameters (see Table 4.4) was chosen to be realistic for small, edible algae being grazed by *Daphnia magna*, with the sole limiting nutrient being phosphorus. Initial algal density was varied in logarithmic steps from 100 to 1,000,000 cells per ml. Grazer density varied in arithmetic steps from zero to 100 animals per liter. Rates of change were plotted over the initial algal density/grazer density plane. Vertical slices of these surfaces parallel to the "grazer density" axes are $b(G)$, $d(G)$, and $r(G)$ functions (see Figure 4.6). One can examine these slices to determine the algal densities where algal mortality, reproduction, and net rates of change are sensitive to grazer density. In doing so, one can see where the direct effect of grazing is important and where the indirect effect mediated via the dissolved nutrient pool is important.

In the reference run, the sensitivity of reproductive rates, $b(G)$, to grazer density was highly dependent on the initial algal density, as shown in Figure 4.7A. At very low algal density, algae grew at rates determined by the initial nutrient concentration, that were nearly independent of grazer density. Near the incipient limiting concentration, ILC, (10^4 cells per ml) $b(G)$ was very sensitive to grazer density. Then, at very high algal density, algae quickly depleted the dissolved nutrient so that the reproduction rate was nearly zero, grazers stimulated reproduction only negligibly, and growth over the six-day period was very low regardless of grazer density.

Likewise, Figure 4.7B shows that the sensitivity of mortality rates, $d(G)$, to grazer density depended on initial algal density. Mortality rates were linear functions of grazer density (indicative of constant clearance rates) at low and intermediate algal density, but at high algal density, because clearance rates decline, mortality rates became insensitive to grazer density. The sharp bends in the $d(G)$ surface around the ILC delineate the combinations of initial algal density and grazer density where algal density was reduced below the ILC during the course of the simulations. Changes in mortality far exceeded changes in reproduction in the range of parameters studied in these simulations, so the surface of the net growth rates curves (Figure 4.7C) closely mirrored the mortality rate surface.

The results of these simulations of the grazing/excretion model suggest two important conclusions. First, some (perhaps most) of the sen-

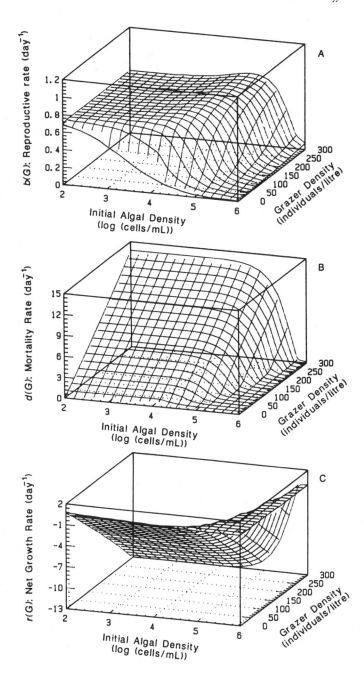

Figure 4.7   Solutions of the grazing/excretion model for the reference parameter set.
A. Per capita reproduction, $b(G)$, versus initial algal density and grazer density, showing
greatest dependence on grazers at intermediate algal density. B. Per capita mortality,
$d(G)$. C. Net rates of change, $r(G)$.

sitivity of algal reproductive rates to grazer density occurs not because of nutrient release by grazers, but because grazers—by thinning the algal population—prevent algae from monopolizing the resources that are in the dissolved pool at the start of the simulations. This explains why algal reproductive rates at zero grazer density decline with increasing initial algal abundance and why the presence of grazers prevents this density dependence. Figure 4.8A shows that this "thinning" effect can easily be seen by setting the assimilation efficiency ($a - 1$) equal to 100 percent, instead of the 50 percent used in the reference run and rerunning the simulations. Here, nutrient release is zero, but a grazer gradient experiment might find positive or curvilinear responses anyway. Comparison with identical parameter settings, except with assimilation efficiency equal to 0 percent (all nutrient ingested is excreted) illustrates the effect of nutrient release (Figure 4.8B). Nutrient release makes stimulation of algal reproduction by grazers greater than when no nutrients were released, especially near the ILC. The results shown in Figure 4.8 indicate that it is not true that all increases in algal productivity by grazing animals seen in experiments which manipulate grazer density are a result of nutrient regeneration.

Secondly, these simulations suggest that grazer stimulation of per capita productivity by grazers is highest near the ILC. At very low algal density, feeding rates and therefore excretion rates were very low (Figure 4.9B and C). Thinning was relatively ineffective because algae at low density cannot monopolize much resource. Well below the ILC, algae grew at rates determined mainly by the initial concentration of the dissolved nutrient. At very high algal density, feeding rates were the same as at intermediate algal density because the functional response has satiated (Figure 4.9B). Since nutrients were regenerated in direct proportion to the nutrient ingested, at high algal density there were more phytopankton among whom the same amount of nutrient must be distributed than at the ILC. Also, because clearance rates are very low at high algal density, there is little algal mortality (Figure 4.7B) and thus little thinning. For these reasons grazers were ineffective at enhancing productivity at algal density well above the ILC. Changing the model so that animals filtered at constant rate regardless of the algal density showed the importance of satiation (Figure 4.10). Mortality effects no longer dampened at high algal density.

To summarize, the context in which interactions occur influences how strong they are. In this case, algal density strongly influenced algal responses to grazing and nutrient regeneration. Maximal effects of grazers were seen around the ILC for several reasons:

1. Grazers prevent resource monopolization, which would otherwise occur.

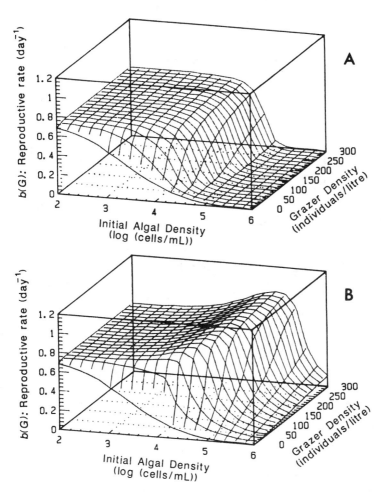

Figure 4.8 Solutions of the model when only the assimilation efficiency is changed compared to the reference run. A. Complete assimilation, i.e., no nutrient regeneration. B. No assimilation, i.e., complete regeneration.

2. Grazers regenerate the greatest amount of nutrient per animal per algal biomass.
3. Feeding rates per algal biomass are maximal.

Such effects, admittedly in a highly simplified model of natural populations, may explain some of the differences seen in grazer experiments.

Nitrogen to phosphorus regeneration ratios   Studies of algal competition have shown that the ratio of supply of potentially limiting nutrients strongly

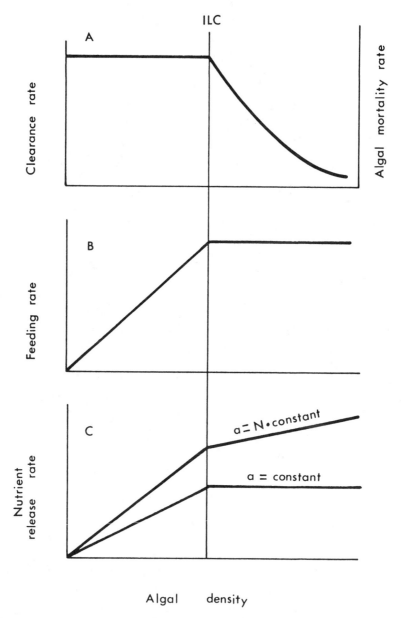

Figure 4.9   A, B. The rectilinear model of zooplankton feeding. A. The clearance rate is high and independent of algal density below the ILC; above that concentration, clearance rate decreases with increasing algal density. B. A Holling Type 1 functional response. C. Presumed relationship between nutrient release rate and algal density. Bottom curve would be seen if the grazers, assimilation efficiency (*a*) is constant with algal density. Top curve is produced if the assimilation efficiency steadily decreases with increasing algal density.

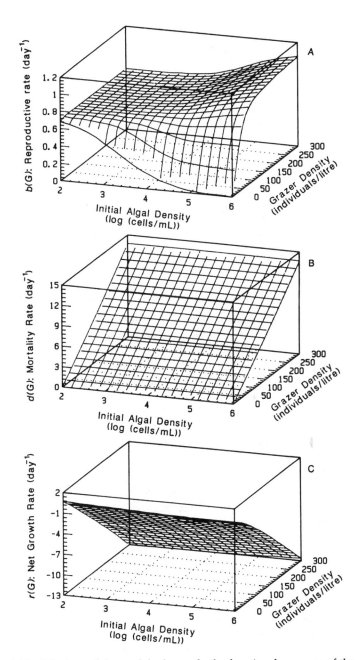

Figure 4.10 Solutions of the model when only the functional response of the grazers is eliminated. Clearance rate was held at the maximal rate, even above the ILC. This modification dramatically alters the predictions of the model and suggests that the functional response of the animals has a major effect on the results seen in grazer gradient experiments.

influences which algal species will be competitively dominant (Kilham and Kilham, 1984; Chapter 3). Thus, where zooplankton herbivores are quantitatively important as agents of resource supply, grazers should affect the algal competitive arena by the ratio of resources released into the water. The ratio of nutrients recycled by the animals does not have to equal the ratio in the food but may be influenced by the stoichiometries of digestion, metabolism, growth, and production of refractory materials (Lehman, 1984).

The ratio of silicon to phosphorus supply by grazers has already been discussed (Chapter 3): grazers provide low Si:P supply ratios because Si is not released in an available form, but P is. Herbivores should therefore disfavor diatom competitive dominance. The influence of zooplankton on another important nutrient supply ratio, that of nitrogen to phosphorus, is much less obvious because both nitrogen and phosphorus can be released at high rates. Therefore we will review the stoichiometries of nitrogen and phosphorus release in some detail.

Early methodology for measuring nutrient release involved incubating pre-starved animals (to purge gut contents) in filtered water to which antibiotics were added to inhibit nutrient uptake by bacteria (e.g., Corner et al., 1972; Corner et al., 1976). Measurements of this sort may determine the basal metabolic release of starving animals, but they seriously underestimate the actual supply rate from actively feeding animals where nutrients may be released from cell breakage, egestion, and metabolic processes associated with feeding itself in addition to basal metabolism. When algal food is present, however, uptake of released nutrient is rapid. Attempts to measure nutrient release by zooplankton in the presence of food have taken many forms: flux of ^{32}P from heavily labeled animals (Peters and Rigler, 1973), estimating algal uptake with Michaelis-Menten kinetics (Takahashi and Ikeda, 1975), or with nutrient-saturated kinetics (Lehman, 1980b); feeding animals heat-killed algae (Vanderploeg et al., 1986); extrapolating high resolution time courses obtained on animals without food back to the time of removal from food (Gardner and Scavia, 1981); assaying glutamate dehydrogenase (thought to be a regulatory enzyme of ammonium production (Bidigare and King, 1981); and calculating a mass balance among grazers, algae, and the dissolved pool (Olsen and Østgaard, 1985). Of course, none of these methods is free of limitations. Highly specialized flow-through devices may be needed (time courses), or simultaneous collection of data from algae, grazers, and the dissolved pool may be necessary for just one measurement (kinetic saturation and mass balance). Some methods alter the food base to some degree (heat-killed algae and kinetic saturation), and one method can be affected by particulate nutrient release (mass balance). Care must be taken to choose the technique to best match the experimental design.

Ammonium is the major form of released nitrogen (Lehman, 1984, and Bidigare, 1983) although release of urea (Mitamura and Saijo, 1986) and of amino acids (Johannes and Webb, 1965; Gardner and Miller, 1981; Gardner and Paffenhöfer, 1982; Riemann et al., 1986) can be measured. Organic as well as inorganic nitrogen can be used by many algae (Wheeler, 1983), but some algal species specialize on certain forms of nitrogen (e.g., Wehr et al., 1987). The chemical speciation of released nitrogen may have some subtle effects on algal populations.

Ammonium is produced primarily from the deamination of protein (Bidigare, 1983). The resultant keto acids enter the TCA cycle and may supply much of the energy budget of the animal (Harris, 1959; Blažka, 1966, 1967; Ganf and Blažka, 1974; Blažka et al., 1982). Ammonium would be toxic were it to build up in the tissues. Bidigare (1983) provides a review of the physiology of nitrogen release. Time courses have shown a rapid decline of ammonium release after a well-fed animal is removed from food (Gardner and Scavia, 1981; Scavia and Gardner, 1982; Vanderploeg et al., 1986), which suggests that egestion of catabolized but unassimilated organic nitrogen is a significant source of release as well. These same studies show ammonium release to be steady rather than pulsed. On the contrary, amino acid release was extremely episodic (Gardner and Paffenhöfer, 1982).

Most of the dissolved phosphorus released by grazing animals seems to be inorganic orthophosphate (Rigler, 1961; Peters and Lean, 1973; Taylor and Lean, 1981). The physiological pathways culminating in phosphorus excretion are less characterized than for nitrogen, perhaps because they are less specific. As with ammonium, time courses show a rapid decline in phosphorus release when well-fed animals are taken away from food (Scavia and Gardner, 1982), but, in contrast to ammonium release, occasional pulses superimposed upon a background level have been observed (Taylor and Lean, 1981; Scavia and McFarland, 1982). These pulses have been attributed to release of fecal material, possibly contributing to small-scale patchiness of phosphorus.

Though the major fractions of released nitrogen and phosphorus are available for algal uptake, not all released nutrient is in a dissolved form. Particulate nutrient is more refractory to algal uptake. Copepods distinguish themselves by enclosing their fecal materials in membrane-covered pellets which quickly sink out of the trophogenic upper waters. Cladocerans, on contrast, live in a perennial state of "diarrhea," and although most carbon is released by *Daphnia* in soluble form (Lampert, 1978), some too is particulate. For example, Olsen et al. (1986a) observed that 80 percent does not pass a Whatman GF/F filter. Some nitrogen and phosphorus is probably associated with this particulate carbon, and the magnitude of this form of release is probably sensitive to the food source. For example, the cell membranes of *Scenedesmus acutus* are not digested by *Daphnia* (Sterner, pers. obs.) probably because assimilation of cellulose by cladocerans is very low (Schoenberg et al., 1984). Particulate matter enters into slow recycling cycles (days to years) rather than the fast cycles (seconds to hours) that characterize dissolved nutrient.

Some direct measurements of the stoichiometry of nitrogen and phosphorus release have been made. A series of marine studies was carried out by Harris and Riley (1956), who examined the ratio of nitrogen and phosphorus within zooplankton, algae, and dissolved pools in Long Island Sound. They concluded that N:P ratios within the zooplankton were less variable during the season than were ratios in the phytoplankton, which were, in turn, less variable than those in the dissolved pools. In his lucid review, Ketchum (1962) pointed out that Harris and Riley's data showed a higher N:P ratio in the tissues of the zooplankton than in the phytoplankton they ate so the N:P ratio in the excreta must have been relatively low. Later, studies on the nitrogen and phosphorus balances of *Calanus* (Butler et al., 1969; Corner et al., 1972; Corner et al., 1976) measured higher N:P ratios in the animals (typical value of 20) and lower N:P ratios in excretion (typical value of 11) than Redfield proportions value of 16 (reviewed further with more examples in Corner and Davies, 1971). Similar conclusions were reached by Blažka et al. (1982) and Ejsmont-Karabin (1983, 1984) working on freshwater zooplankton species. The latter reported extremely low N:P release ratios (range about 4 to 13). These marine and freshwater studies suggest that recycling lowers N:P supply ratios. However, the metabolic state of the animal and the quantity and quality of the food have been recognized to affect rates of N and P release: excreted N:P varies with the quantity and quality of the food. This effect was seen by Scavia and Gardner (1982) who reported N:P release ratios between 25 to 51 for *Daphnia magna* consuming a soy flour/bacteria mixture, but ratios of 9 to 19 for animals consuming chlorococcalean algae.

Recent measurements have underscored the influence of the phosphorus content of the food of the rate of P release. Measurements conducted on *Daphnia* exposed to food with a range of carbon:phosphorus between 23 and 170 (by atoms) (Olsen et al., 1986b) found phosphorus release decreased as C:P became higher (i.e., when the percent phosphorus in the food became lower). Phosphorus release was almost zero at the highest values of C:P, which nevertheless are considered diagnostic of only modest P limitation in the algae (see Chapter 3). Similarly, other measurements on *Daphnia* have shown that net phosphorus assimilation is higher (Lehman and Naumoski, 1985) and excretion is lower (Lehman, 1980b) for animals fed algae with low cell quotae than for animals fed P-sufficient food. Animals feeding on P-limited algae husband P efficiently whereas ingestion of P-rich algae results in higher P release. The ratio of assimilated C:P is apparently more constant than the ratio that is found in the food. When phosphorus appears in the food in excess of this assimilation ratio, it is excreted rapidly. Though zooplankton do not hold their body burdens of phosphorus rigidly constant, they appear to maintain them relatively constant compared to phytoplankton. For example, Lehman and Naumoski (1985) found that when algae whose phosphorus contents varied by a factor of 2.3 to 2.7 were the nutrient resource for *Daphnia*, the P content of the *Daphnia* varied by 1.6-fold.

Stoichiometric considerations for nitrogen release are slightly different than these of phosphorus. As discussed above, nitrogen is a necessary metabolic byproduct of catabolizing proteins for energy; a zooplankter that must burn protein releases nitrogen. This can be seen in the change in the ratio of nitrogen excretion to nitrogen ingestion which *decreases* as food becomes more abundant (Corner et al., 1976; Gardner and Paffenhöfer, 1982). Starving animals must catabolize a greater share of their dietary protein, making it difficult to know how nitrogen excretion varies with food level. There is a conflict between the need for structural material during growth and reproduction and the need for energy. Indeed, some workers have concluded that animals facing low food abundance release more nitrogen than well-fed animals (Martin, 1968, but note that the methods in this study made no allowance for nutrient uptake by algae during the measurement). Several studies reached the opposite conclusion: nitrogen release increases as food density increases (Corner et al., 1976; Gardner and Paffenhöfer, 1982). The laboratory studies reviewed by Corner and Davies (1971) revealed no consistent relationship between food level and nitrogen-assimilation efficiency, which says that below the ILC, increasing food density should increase N release rates. Blažka et al. (1982) took the point of view that N excretion (normalized for oxygen consumption) increases with increasing food levels for carnivorous zooplankton, but decreases for herbivores.

The diversity of opinions and the scarcity of freshwater studies make further interpretations speculative, but regardless of these physiological complexities, Roman (1983), Corner and Davies (1971), as well as the previously mentioned study by Harris and Riley (1976) all concluded that nitrogen content is more constant for marine zooplankton than for phytoplankton. This suggests that the animals maintain their stoichiometry by regulating the net incorporation of N into their tissues—when N is present in excess of the amount needed in the diet, its concentration in the released materials must be greater than when the food is N deficient. This is similar to the pattern documented for phosphorus. The relationship between the C:N:P needed for metabolism and structural material and the C:N:P ingested determine the C:N:P released. In this view, grazers ingest phytoplankton with C:N:P ratios that vary within wide limits; they assimilate a C:N:P ratio that is relatively constant, and they release the difference between the two. Zooplankton structural material may either build up in grazer biomass, flow further up the food chain, or if physiological death is important, be removed to the hypolimnetic environment. Each of these fates is a drain of a relatively constant C:N:P proportion from the algal pool.

The implications of this nutrient stoichiometry on the seasonal dynamics of phytoplankton are intriguing. Consider a population of grazers feeding on a diet of mildly P-limited phytoplankton. If phosphorus appears in lower proportion in the algae than is optimal for the grazers, P release, as compared to release of nitrogen and presumably other elements as well, will be low. The ratio of N:P released is more P deficient, increasing the degree of P limitation in the phytoplankton. With the ingestion of these more P-limited algae, the animals further lower P release compared to release of other nutrients even more. With every iteration of this process, the phytoplankton become more P limited. The same argument could be repeated for nitrogen limitation: mild N limitation would become severe N limitation. Grazers might cause alterations in the nutrient limitation patterns of the phytoplankton. This effect would produce the counter-intuitive result that, over the course of seasonal succession, grazers intensify competition for a single nutrient by causing nutrient supply ratios to diverge. This contrasts somewhat with their more obvious role in amelioration of nutrient limitation that occurs on shorter times scales. A finding that N and P availability were inversely related and that transitions between N and P limitations were abrupt (Elser et al., 1988) provides strong support for this hypothetical process.

4.6   Results of Field Manipulations

Field experiments (including so-called "natural experiments") that test some aspect of the influence of grazers on algal succession are common. The following discussion will focus on five such studies chosen to illustrate the evidence available from field studies in general. These also serve as examples of several themes of this chapter. Each case allows a test of the premise that inedible phytoplankton succeed edible ones because of grazing pressure. In these experiments, grazers were manipulated either directly or indirectly by altering the fish communities.

Round Lake, Minnesota case study   In 1980 Shapiro and Wright (1984) applied rotenone to this small dimictic lake which had a planktivorous fish fauna (bluegill sunfish and black crappie) and an herbivore community dominated by *Bosmina*. The lake was then restocked with piscivorous largemouth bass and walleye about half as abundant as bluegills. During the following two years *Daphnia* became the dominant zooplankter, and grazing pressure (estimated by allometric equations) approximately doubled compared to pretreatment levels. Posttreatment chlorophyll was lower (though not for every sample) and secchi depths were deeper than in the pretreatment year. The pretreatment algal succession was *Cryptomonas* to Chlorophyceae, including *Chlamydomonas, Selenastrum, Gloeocystis, Schroederia,* and *Staurastrum,* as well as the cyanobacterium *Aphanizomenon.* In both posttreatment years the succession was *Cryptomonas* to mixed dominance by *Cryptomonas, Schroederia,* and filamentous cyanobacteria (*Aphanizomenon* in colonies small enough to be grazed and *Anabaena*). Though the alteration in fish communities resulted in several hundred *Daphnia* per liter with estimated grazing rates as high as 1.4 day^{-1}, the algal community did *not* become heavily dominated by inedible species. An indirect effect of the fish manipulation on nutrient concentration was suggested, as both total epilimnetic phosphorus and nitrogen were lower in posttreatment years than in pretreatment years. The possibility that nutrients were removed from the epilimnion by a net downward flux owing to vertical migration was suggested. Thus, though it appeared that the trophic manipulation influenced this succession in important ways, the mechanisms involved were unclear, and a tight link between grazing pressure and dominance by inedible species was not seen.

Lake St. George, Ontario case study   Three limnocorrals were placed in this lake in 1982 (Post and McQueen, 1987) and different densities of fish were added to two of them, the third serving as a fishless control. The experiment included weekly supplements of nitrogen (2 μmoles per l) and phosphorus (0.07 μmoles per l) to all enclosures. Daphnids and *Diaptomus* were the dominant grazers in the fishless enclosure whereas small cladocerans were dominant in the enclosures containing fish. Chlorophytes and Cryptophytes dominated the algal communities at all times in all enclosures. In June, at the beginning of the experiment, edible species, e.g., *Chlorella*, *Chlamydomonas*, and *Cryptomonas*, dominated all three enclosures. Succession in the presence of the *Daphnia* differed from the other treatments, with the September phytoplankton in the *Daphnia* enclosure being heavily dominated by *Oocystis*, but with *Gloeococcus*, *Tetraedron*, and *Scenedesmus* sharing dominance in both of the other two. This is a case of high *Daphnia* grazing pressure apparently directing a succession toward relatively inedible (digestion-resistant) gelatinous greens. Perhaps this trend was aided by a weakening of nutrient limitation because of the additions of N and P.

Dynamite Lake, Illinois case study   In 1980 and 1981 Vanni (1987) performed factorial enclosure experiments in this meso-oligotrophic lake. The treatments—presence or absence of fish, and weekly nutrient addition or no addition—were triplicated. The exclusion of fish affected the abundance and size structure of the cladocerans that dominated (primarily *Bosmina*, *Ceriodaphnia*, and *Diaphanasoma*) with increased total biomass and larger body size in fishless enclosures. The "normal" succession was from dominance by *Mallomonas*, *Chrysochromulina*, and *Cryptomonas* to dominance by greens and diatoms, with sheathed or coated species, especially *Oocystis* and *Coenococcus*, being common. The absence of fish (where grazing rates were presumably higher) resulted in increased proportional abundance of inedible, digestion-resistant algae, consistent with the premise that resistance to grazing contributes to their late season success. This effect was noticeable even though the phytoplankton were strongly nutrient limited: nutrient additions led to great increases in densities of many algal species.

Lake George, New York case study   A historical shift in algal dominance from diatoms, *Cryptomonas*, and *Dinobryon*, to colonial coccoid cyanobacteria in the south basin of this large lake was analyzed by Siegfried (1987). For some years prior to 1975, seasonal succession did not lead toward cyanobacterial dominance, but during the period 1976 to 1981, late-season dominance by cyanobacteria (*Anacystis* and *Aphanothece*) developed. Siegfried argued that nutrient loading to the lake was not responsible for this shift, but that changes in the fish stock, notably the establishment of rainbow smelt, were responsible. Owing to shoreline topography, smelt did not establish themselves in the north basin, and large zooplankton (*Daphnia pulicaria* and *Epischura lacustris*) were more common there than in the south basin, though copepods (especially *Diaptomus sicilis*) dominated both basins. Allometric calculations of grazing rate based on zooplankton census data showed that grazing rates in the north exceeded those in the south in April and May, but that rates in the south exceeded those in the north in June. Grazing rates were always low, the maximum being 0.08 day^{-1} (80 l·m^{-3}·d^{-1}), in keeping with observations earlier in this chapter that high grazing rates are associated only with dominance by large cladocerans.

Allometric calculations of P release by zooplankton—which must be viewed with extreme caution because the support for this approach (Peters, 1975) is not as extensive as for corresponding grazing-rate calculations—suggested that P release in June in the south basin exceeded that in the north. Siegfried went on to argue that this enhanced P release resulted in Si depletion and the resulting cyanobacteria dominance, which is similar to the autogenic changes hypothesized for Lake Michigan (Schelske and Stoermer, 1971; but see Shapiro and Swain, 1983). More studies of this mechanism of changing patterns of seasonal succession would be valuable, but this example illustrates how shifts in zooplankton communities from large cladocerans to smaller ones may be associated with dominance by cyanobacteria.

Heney Lake, Quebec case study   In 1976 and 1977 enclosures were established in this deep, hardwater lake (Briand and McCauley, 1978; McCauley and Briand, 1979), and the amount of herbivorous zooplankton was reduced by vertical tows of a plankton net. Zooplankton compositional data are not given, but the species list includes primarily small-bodied species. The major result reported was that lowered herbivore levels may have caused intensified exploitative competition which was said to favor edible species and a few inedible species at the expense of many inedible species that were eliminated. The manipulation of herbivore levels produced only a weak shift in the balance between species classified as edible and species classified as inedible after the 4-week experiment; Table 2 of McCauley and Briand, 1979, shows that by numbers and by biomass, "inedible" species accounted for >92 percent in all samples. However, the percent contribution to the inedible group shifted to *Synedra* in the enclosures with lowered zooplankton abundance, this genus accounting for less than 10 percent initially and less than 30 percent at the end of the control enclosures, but accounting for more than 50 percent at the end in the treatment enclosures. *Asterionella* dominated the "edible" fraction but was not always overwhelming. This study supported the premise that reduced grazing pressure enhances succession toward edible algal species (i.e., grazing promotes inedible dominance) though the method of classification as edible or inedible was not well supported.

Lake Michigan case study   In the past decades, the offshore epilimnetic phytoplankton community has undergone dramatic shifts in its successional trajectories that have been interpreted recently by Fahnenstiel and Scavia (1986) and Scavia et al. (1986). Superimposed upon long-term chemical changes, namely decreased epilimnetic phosphorus and silicon, have been more abrupt changes in the trophic web. The characteristic succession of the late 1960's and 1970's was for diatoms (including *Melosira*, *Asterionella*, and *Fragilaria*) to succeed to greens and cyanobacteria that made up more than 50 percent of algal carbon. This successional pattern apparently persisted through 1982 when greens and cyanobacteria (especially *Anabaena flos-aquae*) made up about 30 percent of the mid-summer phytoplankton. From 1983 to 1985 (the last year reported), though total algal carbon was not greatly changed, greens and cyanobacteria were very rare, and chrysophytes and cryptophytes dominated during summer. The timing of this switch in successional trends is significant, because between 1982 and 1983 the dominant crustacean zooplankton switched from *Diaptomus* to *Daphnia*—probably as a result of decreasing alewife predation. The abruptness of the change in algal succession suggests that this sudden trophic web alteration is the cause, rather than the more gradually occurring chemical changes, though plausible nutrient-based arguments could be constructed too. It thus appeared that the dominance of *Daphnia* rather than *Diaptomus* facilitated succession toward flagellates rather than *Anabaena*. This might then be a case of increasing grazing pressure facilitating dominance by edible algae. Alternatively, *Anabaena* may be cleared with greater efficiency by *Daphnia* than by *Diaptomus*, so that changes in differential mortality on the different algal groups were responsible.

One can look at these case studies to see how the relative proportion of two general classes of relatively inedible algal species, digestion-resistant greens and colonial cyanobacteria, respond to grazing pressure. Grazing facilitation of late season dominance by digestion-resistant greens appeared to be fairly clear (Lake St. George and Dynamite Lake), whereas a connection between grazing pressure and colonial cyanobacterial dominance was not (Round Lake, Lake George, and Lake Michigan). Changes in grazing rates should determine succession more strongly when the inferior competitors (assuming mortality rates are equal) are relatively resistant to grazing. Digestion-resistant greens have never been observed to dominate whole-community chemostats even though they can strongly dominate the plankton of lakes (e.g., Kerfoot, 1987). Thus, though independent measurements of competitive ability are needed for final proof, gelatinous greens seem to be inferior nutrient competitors.

Differential mortality (grazing on their competitors) may be necessary to explain their dominance.

In contrast, when grazers focus upon inferior competitors, such as on flagellates under strong nitrogen limitation, they may accelerate or enhance a succession already occurring (Sterner, 1989). It seems significant that the three case studies of *Daphnia*/filamentous cyanobacteria interactions found that the presence of the *Daphnia* was neutrally or negatively associated with the percent dominance of filamentous cyanobacteria (Round Lake, Lake George, and Lake Michigan), contrary to a simple premise of grazing enhancing dominance by filamentous cyanobacteria. Spencer and King (1984, 1987) made similar observations and concluded that an absence of *Daphnia* fostered spring algal blooms which resulted in a light climate that favored the development of buoyant cyanobacteria. Grazers feeding on algae that are otherwise unfavored are not confined to cyanobacteria; for example, the dinoflagellate *Peridinium inconspicuum* is relatively inedible compared to its competitors, and it commonly dominates acid lakes. Grazing seems only to intensify its ability to dominate (Havens and DeCosta, 1985). A clearer understanding of the competitive abilities of the algal species under the nutrient regines in force during studies of grazing would greatly facilitate the interpretation of studies of grazing.

4.7   Conclusions

Seasonal trajectories of grazing and nutrient release   No species of zooplankton herbivore is so completely unselective that all algal species are cleared at precisely the same rates. When grazing rates are high, i.e., when zooplankton density becomes high, even slight selectivity will have strong effects on algal dominance patterns. This is especially important during dominance by large *Daphnia* species, where strong peaks of grazing rates may occur. During these peaks, edible algal species may experience loss rates that exceed maximal growth rates. At these times, the algal community moves toward dominance by inedible species. In addition, nutrient release by herbivores can stimulate the growth of inedible species. At other times, even during the growing season, grazing loss rates may be very low so that grazing does not play a strong role in determining algal succession. Thus, for seasonal environments, discussions about the importance of grazing in lake communities is integrally and inextricably embedded in both the size structure and the seasonality of zooplankton populations.

It would be desirable to make parallel observations on seasonal variations of nutrient release. However, I am aware of no published study using modern methodology to determine nutrient release by grazing herbivores through a seasonal cycle. Although allometric relationships of excretion versus body size have been used for this purpose (Bartell and Kitchell, 1978; Ejsmont-

Karabin et al., 1983), the support for such relationships is not strong. Excretion per unit body length is highly variable as it is influenced greatly by both the quantity and quality of the ingested seston. This is a major gap in basic information about planktonic interactions.

Grazing–competition interactions   The "competition vs. predation" paradigm of community structure, which surfaces in the limnological literature as the debate over top-down versus bottom-up influence has been fastidiously avoided in this chapter. Although grazers do exert significant influence on successional trajectories, this does not mean that algal succession is necessarily and consistently controlled from the "top down." The influence of zooplankton on algal communities often comes indirectly through bottom-up factors, i.e., through changed algal resource relations.

Furthermore, the top-down effect of grazing depends upon the nature of algal competition, because grazing on competitive dominants produces a different succession series than does grazing on competitive subordinates. The idea that the implications of predation depend very much on whether the competitive dominant or the competitive subordinate is preferred by the predator is an old idea in general community ecology, but it is still not well integrated into discussions of plankton/grazing interactions. In a network of interacting components, "effects" may propagate in one direction or another but not all control comes from either the top or the bottom (McQueen et al., 1986). Community ecologists are becoming more and more aware that interactions themselves interact (Roughgarden and Diamond, 1986). Competition and predation are not mutually exclusive factors, and a search for dominance of either top-down or bottom-up effects in determining planktonic community structure may distract us from the truly important interactions.

Acknowledgments
This chapter was written while I was in residence at the Max Planck Institute for Limnology, Plön, FRG. I thank W. Lampert and U. Sommer for stimulating discussion. W.R. DeMott, Z.M. Gliwicz, S.S. Kilham, C.S. Reynolds, K.-O. Rotthaupt, and U. Sommer suggested improvements of the manuscript. R. Kiesling, J. Lehman, and M. Leibold read an earlier description of the numerical simulations of grazer gradients. J. Sterner contributed the artwork in Figure 4.5. Financial assistance was from a Max Planck Stipendium.

References

Abrams, P. 1987. Indirect interactions between species that share a predator: varieties of indirect effects, pp. 38–54, in Kerfoot, W.C. and Sih, A. (editors), *Predation: Direct and Indirect Impacts on Aquatic Communities.* University Press of New England, Hanover, New Hampshire.

Axler, R.P., Redfield, G.W., and Goldman, C.R. 1981. The importance of regenerated nitrogen to phytoplankton productivity in a subalpine lake. *Ecology* 62: 345–354.

Barlow, J.P. and Bishop, J.W. 1965. Phosphate regeneration by zooplankton in Cayuga Lake. *Limnology and Oceanography* 10 (supplement): R15–R25.

Bartell, S.M. and Kitchell, J.F. 1978. Seasonal impact of planktivory on phosphorus release by Lake Wingra zooplankton. *Internationale Vereinigung für Theoritische und Angewandte Limnologie, Verhandlungen* 20: 466–474.

Bartram, W.L. 1980. Experimental development of a model for the feeding of neritic copepods on phytoplankton. *Journal of Plankton Research* 3: 1525–1551.

Bergquist, A.M. and Carpenter, S.R. 1986. Limnetic herbivory: effects on phytoplankton populations and primary production. *Ecology* 67: 1351–1360.

Bergquist, A.M., Carpenter, S.R., and Latino, J.C. 1985. Shifts in phytoplankton size structure and community composition during grazing by contrasting zooplankton assemblages. *Limnology and Oceanography* 30: 1037–1045.

Berman, M.S. and Richman, S. 1974. The feeding behavior of *Daphnia pulex* from Lake Winnebago, Wisconsin. *Limnology and Oceanography* 19: 105–109.

Bidigare, R.R. 1983. Nitrogen excretion by marine zooplankton, pp. 385–409, in Carpenter, E.J. and Capone, D.G. (editors), *Nitrogen in the Marine Environment*. Academic Press, New York.

Bidigare, R.R. and King, F.D. 1981. The measurement of glutamate dehydrogenase activity in *Praunus flexuosus* and its role in the regulation of ammonium excretion. *Comparative Biochemistry and Physiology* 70B: 409–413.

Blažka, P. 1966. The ratio of crude protein, glycogen and fat in the individual steps of the production chain. *Hydrobiological Studies* 1: 395–408.

Blažka, P. 1967. Physiological basis of secondary production, pp. 222–228, in Edmondson, W.T. and Winberg, G.C. (editors), *Secondary Productivity in Fresh Waters*. IBP Handbook 17. Blackwell, Oxford.

Blažka, P., Brandl, Z., and Procházková, L. 1982. Oxygen consumption and ammonia and phosphate excretion in pond zooplankton. *Limnology and Oceanography* 27: 294–303.

Bleiwas, A.H. and Stokes, P.M. 1985. Collection of large and small particles by *Bosmina*. *Limnology and Oceanography* 30: 1090–1092.

Boavida, M.J. and Heath, R.T. 1984. Are the phosphatases released by *Daphnia magna* components of its food? *Limnology and Oceanography* 29: 641–644.

Bogdan, K.G. and Gilbert, J.J. 1982. Seasonal patterns of feeding by natural populations of *Keratella, Polyarthra*, and *Bosmina*: Clearance rates, selectivities, and contributions to community grazing. *Limnology and Oceanography* 27: 918–934.

Bogdan, K.G. and Gilbert, J.J. 1984. Body size and food size in freshwater zooplankton. *Proceedings of the National Academy of Sciences, U.S.A.* 81: 6427–6431.

Bogdan, K.G. and Gilbert, J.J. 1987. Quantitative comparison of food niches in some freshwater zooplankton. A multi-tracer cell approach. *Oecologia* (Berlin) 72: 331–340.

Bogdan, K.G. and McNaught, D.C. 1975. Selective feeding by *Daphnia* and *Diaptomus*. *Internationale Vereinigung für Theroetische und Angewandte Limnologie. Verhandlungen*. 19: 2935–2942.

Børsheim, K.Y. and Anderson, S. 1987. Grazing and food size selection by crustacean zooplankton compared to production of bacteria and phytoplankton in a shallow Norwegian mountain lake. *Journal of Plankton Research* 9: 367–379.

Boyd, C.N. 1976. Selection of particle sizes by filter-feeding copepods: a plea for reason. *Limnology and Oceanography* 21: 175–180.

Brendelberger, H. 1985. Filter mesh-size and retention efficiency for small particles: comparative studies with Cladocera. *Archiv für Hydrobiologie Beiheft. Ergebnisse der Limnologie* 21: 135–146.

Brendelberger, H. and Geller, W. 1985. Variability of filter structures in eight *Daphnia* species: mesh sizes and filtering areas. *Journal of Plankton Research* 7: 473–486.

Brendelberger, H., Herbeck, M., Lang, H., and Lampert, W. 1987. *Daphnia*'s filters are not solid walls. *Archiv für Hydrobiologie* 107: 197–202.

Briand, F. and McCauley, E. 1978. Cybernetic mechanisms in lake plankton systems: how to control undesirable algae. *Nature* 273: 228–230.

Burns, C.W. 1968. The relationship between body size of filter-feeding cladocera and the maximum size of particle ingested. *Limnology and Oceanography* 13: 675–678.

Burns, C.W. and Rigler, F.H. 1967. Comparison of filtering rates of *Daphnia rosea* in lake water and suspensions of yeast. *Limnology and Oceanography* 12: 492–502.

Butler, E.I., Corner, E.D.S., and Marshall, S.M. 1969. On the nutrition and metabolism of zooplankton. VI. Feeding efficiency of *Calanus* in terms of nitrogen and phosphorus. *Journal of the Marine Biological Association of the U.K.* 49: 977–1001.

Canfield, D.E. Jr. and Watkins, C.E. III. 1984. Relationships between zooplankton abundance and chlorophyll *a* concentrations in Florida lakes. *Journal of Freshwater Ecology* 2: 335–344.

Carpenter, S.R. and Kitchell, J.F. 1984. Plankton community structure and limnetic primary production. *The American Naturalist* 124: 159–172.

Carpenter, S.R. and Kitchell, J.F. 1985. Cascading trophic interactions and lake productivity. *Bioscience* 35: 634–639.

Carpenter, S.R. and Kitchell, J.F. 1987a. The temporal scale of variance in limnetic primary productivity. *The American Naturalist* 129: 417–433

Carpenter, S.R. and Kitchell, J.F. 1987b. Analysis of temporally variable processes in lake ecosystems, pp. 141–153, in *Basic Issues in Great Lakes Research*. Special Report No. 123, Great Lakes Research Division, Ann Arbor, MI.

Chow-Frazer, P. 1986. An empirical model to predict in situ grazing rates of *Diaptomus minutus* Lilljeborg on small algal particles. *Canadian Journal of Fisheries and Aquatic Sciences* 43: 1065–1070.

Chow-Fraser, P. and Knoechel, R. 1985. Factors regulating in situ filtering rates of Cladocera. *Canadian Journal of Fisheries and Aquatic Sciences* 42: 567–576.

Christoffersen, K. and Jespersen, A.-M. 1986. Gut evacuation rates and ingestion rates of *Eudiaptomus graciloides* measured by means of the gut fluorescence method. *Journal of Plankton Research* 8: 973–983.

Conover, R.J., Durvasula, R., Roy, S., and Wang, R. 1986. Probable loss of chlorophyll-derived pigments during passage through the gut of zooplankton and some of the consequences. *Limnology and Oceanography* 31: 878–887.

Cooper, D.C. 1973. Enhancement of net primary productivity by herbivore grazing in aquatic laboratory microcosms. *Limnology and Oceanography* 18: 31–37.

Corner, E.D.S. and Davies, A.G. 1971. Plankton as a factor in the nitrogen and phosphorus cycles in the sea. *Advances in Marine Biology* 9: 101–204.

Corner, E.D.S., Head, R.N., and Kilvington, C.C. 1972. On the nutrition and metabolism of zooplankton. VIII. The grazing of *Biddulphia* cells by *Calanus helgolandicus*. *Journal of the Marine Biological Association of the U.K.* 52: 847–861.

Corner, E.D.S., Head, R.N., Kilvington, C.C., and Pennycuick, L. 1976. On the nutrition and metabolism of zooplankton. X. Quantitative aspects of *Calanus heloglandiscus* feeding as a carnivore. *Journal of the Marine Biological Association of the U.K.* 56: 345–358.

Coughlan, J. 1969. The estimation of filtering rate from the clearance of suspension. *Marine Biology* 2: 356–358.

Crumpton, W. and Wetzel, R.G. 1982. Effects of differential growth and mortality in the seasonal succession of phytoplankton populations in Lawrence Lake, Michigan. *Ecology* 63: 1729–1739.

Currie, D.J. 1984. Microscale nutrient patches: do they matter to the phytoplankton? *Limnology and Oceanography* 29: 211–213.

Cushing, D.H. 1976. Grazing in Lake Erken. *Limnology and Oceanography* 21: 349–356.

Dawidowicz, P. and Gliwicz, Z.M. 1987. Biomanipulation. III. The role of direct and indirect relationship between phytoplankton and zooplankton. *Wiadomosci Ekologiczne* 33: 259–277.

DeMott, W.R. 1982. Feeding selectivities and relative ingestion rate of *Daphnia* and *Bosmina*. *Limnology and Oceanography* 27: 518–527.

DeMott, W.R. 1985. Relations between filter mesh-size, feeding mode, and capture efficiency for cladocerans feeding on ultrafine particles. *Archiv für Hydrobiologie Beiheft. Ergebnisse der Limnologie* 21: 125–134.

DeMott, W.R. 1986. The role of taste in food selection by freshwater zooplankton. *Oecologia* (Berlin) 69: 334–340.

DeMott, W.R. and Kerfott, W.C. 1982. Competition among cladocerans: nature of the interaction between *Bosmina* and *Daphnia*. *Ecology* 63: 1949–1966.

Downing, J.A. 1981. In situ foraging responses of three species of littoral cladocerans. *Ecological Monographs* 51: 85–103.

Droop, M.R. 1973. Some thoughts on nutrient limitation in algae. *Journal of Phycology* 9: 264–272.

Duarte, C.M., Agusti, S., and Peters, H. 1987. An upper limit to the abundance of aquatic organisms. *Oecologia* (Berlin) 74: 272–276.

Dumont, H.J. 1977. Biotic factors in the population dynamics of rotifers. *Archiv für Hydrobiologie Beiheft. Ergebnisse der Limnologie* 8: 98–122.

Ejsmont-Karabin, J. 1983. Ammonia nitrogen and inorganic phosphorus excretion by the planktonic rotifers. *Hydrobiologia* 104: 231–236.

Ejsmont-Karabin, J. 1984. Phosphorus and nitrogen excretion by lake zooplankton (rotifers and crustaceans) in relationship to individual body weights of the animals, ambient temperature and presence or absence of food. *Ekologia Polska* 32: 3–42.

Ejsmont-Karabin, J., Bownik-Dylinska, L., and Godlewska-Lipowa, W.A. 1983. Biotic structure and processes in the lake system of R. Jorka watershed (Masurian Lakeland, Poland) VII. Phosphorus and nitrogen regeneration by zooplankton as the mechanism of the nutrient supplying for bacterio- and phytoplankton. *Ekologia Polska* 31: 719–746.

Elder, R.G. and Parker, M. 1984. Growth response of a nitrogen fixer (*Anabaena flosaquae*, Cyanophyceae) to low nitrate. *Journal of Phycology* 20: 296–301.

Elser, J.J., Elser, M.M., MacKay, N., and Carpenter, S.R. 1988. Zooplankton-mediated transitions between N- and P-limited algal growth. *Limnology and Oceanography* 33: 1–14.

Elser, J.J., Goff, N.C., MacKay, N.A., St. Amand, A.L., Elser, M.M., and Carpenter, S.R. 1987. Species-specific algal responses to zooplankton: experimental and field observations in three nutrient-limited lakes. *Journal of Plankton Research* 9: 699–717.

Fahnenstiel, G.L. and Scavia, D. 1986. Dynamics of Lake Michigan phytoplankton: recent changes in surface and deep communities. *Canadian Journal of Fisheries and Aquatic Sciences* 44: 499–508.

Friedman, M.M. 1980. Comparative morphology and functional significance of copepod receptors and oral structures, pp. 185–197, in Kerfoot, W.C. (editor), *Evolution and Ecology of Zooplankton Communities*. University Press of New England, Hanover.

Friedman, M.M. and Strickler, J.R. 1975. Chemoreception and feeding in calanoid copepods. *Proceedings of the National Academy of Sciences USA* 72: 4185–4188.

Frost, B.W. 1972. Effects of size and concentration of food particles on the feeding behavior of the marine planktonic copepod *Calanus pacificus*. *Limnology and Oceanography* 17: 805–815.

Frost, B.W. 1975. A threshold feeding behavior in *Calanus pacificus*. *Limnology and Oceanography* 20: 263–266.

Fryer, G. 1987. The feeding mechanisms of the Daphniidae (Crustacea: Cladocera):

recent suggestions and neglected considerations. *Journal of Plankton Research* 9: 419–432.

Fulton, R.S. III and Paerl, H.W. 1987. Effects of colonial morphology on zooplankton utilization of algal resources during blue-green algal (*Microcystis aeruginosa*) blooms. *Limnology and Oceanography* 32: 634–644.

Ganf, G.G. and Blažka, P. 1974. Oxygen uptake, ammonia and phosphate excretion by zooplankton of a shallow equatorial lake (Lake George, Uganda). *Limnology and Oceanography* 19: 313–325.

Ganf, G.G. and Shiel, R.J. 1985a. Particle capture by *Daphnia carinata*. *Australian Journal of Marine and Freshwater Research* 36: 371–381.

Ganf, G.G. and Shiel, R.J. 1985b. Feeding behavior and limb morphology of two cladocerans with small intersetular distances. *Australian Journal of Marine and Freshwater Research* 36: 69–86.

Gardner, W.S. and Miller, W.H. III. 1981. Intracellular composition and net release of free amino acids in *Daphnia magna*. *Canadian Journal of Fisheries and Aquatic Sciences* 38: 157–162.

Gardner, W.S. and Paffenhöfer, G.-A. 1982. Nitrogen regeneration by the subtropical marine copepod *Eucalanus pileatus*. *Journal of Plankton Research* 4: 725–734.

Gardner, W.S. and Scavia, D. 1981. Kinetic examination of N release by zooplankters. *Limnology and Oceanography* 26: 801–810.

Geller, W. and Müller, H. 1981. The filtration apparatus of cladocera: filter mesh-sizes and their implications of food selectivity. *Oecologia* (Berlin) 49: 316–321.

Gerritsen, J. and Porter, K.G. 1982. The role of surface chemistry in filter feeding by zooplankton. *Science* 216: 1225–1227.

Gilbert, J.J. and Bogdan, K.G. 1984. Rotifer grazing: in situ studies on selectivity and rates, pp. 97–133, in Meyers, D.G. and Strickler, J.R. (editors), *Trophic Interactions Within Aquatic Ecosystems*. AAAS Selected Symposium 85, Westview Press, Boulder, Colorado.

Glasser, J.W. 1984. Analysis of zooplankton feeding experiments: some methodological considerations. *Journal of Plankton Research* 6: 553–569.

Gliwicz, Z.M. 1969. Studies on the feeding of pelagic zooplankton in lakes of varying trophy. *Ekologia Polska A* 17: 663–707.

Gliwicz, Z.M. 1975. Effect of zooplankton grazing on photosynthetic activity and composition of phytoplankton. *Internationale Vereinigung für Theroetische und Angewandte Limnologie. Verhandlungen* 19: 1490–1497.

Gliwicz, Z.M. 1977. Food size selection and seasonal succession of filter feeding zooplankton in an eutrophic lake. *Ekologia Polska* 25: 179–225.

Gliwicz, Z.M. 1980. Filtering rates, food size selection, and filtering rates in cladocerans—another aspect of interspecific competition in filter-feeding zooplankton, pp. 282–291, in Kerfoot, W.C. (editor), *Evolution and Ecology of Zooplankton Communities*. University Press of New England, Hanover.

Gliwicz, Z.M. and Siedlar, E. 1980. Food size limitation and algae interfering with food collection in *Daphnia*. *Archiv für Hydrobiologie* 88: 155–177.

Goldman, J.C. 1984. Oceanic nutrient cycles, pp. 137–170, in Fasham, M.J. (editor), *Flows of Energy and Materials in Marine Ecosystems: Theory and Practice*. Plenum Press, New York.

Goldman, J.C. and Glibert, P.M. 1982. Comparative rapid ammonium uptake by four species of marine phbytoplankton. *Limnology and Oceanography* 27: 814–827.

Goldman, J.C., McCarthy, J.J., and Peavey, D.G. 1979. Growth rate influence on the chemical composition of phytoplankton in oceanic waters. *Nature* 279: 210–215.

Gophen, M. and Geller, W. 1984. Filter mesh size and food particle uptake by *Daphnia*. *Oecologia* (Berlin) 64: 408–412.

Grant, P.R. 1986. Interspecific competition in fluctuating environments, pp. 173–191 in Diamond, J. and Case, T.J. (editors), *Community Ecology*. Harper and Row, New York.

Gulati, R.D. 1975. A study on the role of herbivorous zooplankton community as primary consumers of phytoplankton in Dutch lakes. *Internationale Vereinigung für Theoritische und Angewandte Limnologie. Verhandlungen* 19: 1202–1210.

Gulati, R.D. 1984. The zooplankton and its grazing rate as measures of trophy in the Loosdrecht Lakes. *Internationale Vereinigung für Theoritische und Angewandte Limnologie. Verhandlungen* 22: 863–867.

Gulati, R.D. 1985. Zooplankton grazing methods using radioactive tracers: technical problems. *Hydrobiological Bulletin* 19: 61–69.

Gulati, R.D., Siewertsen, K., and Postema, G. 1982. The zooplankton: its community structure, food and feeding, and role in the ecosystem of Lake Vechten. *Hydrobiologia* 95: 127–163.

Haney, J.F. 1971. An in situ method for the measurement of zooplankton grazing rates. *Limnology and Oceanography* 16: 970–977.

Haney, J.F. 1973. An in situ examination of the grazing activities of natural zooplankton communities. *Archiv für Hydrobiologie* 72: 87–132.

Haney, J.F. 1985. Regulation of cladoceran filtering rates in nature by body size, food concentration, and diel feeding patterns. *Limnology and Oceanography* 30: 397–411.

Hanson, J.M. and Peters, R.H. 1984. Empirical prediction of zooplankton and profundal macrobenthos biomass in lakes. *Canadian Journal of Fisheries and Aquatic Sciences* 41: 439–445.

Hargrave, B.T. and Geen, G.H. 1968. Phosphorus excretion by zooplankton. *Limnology and Oceanography* 13: 332–342.

Hargrave, B.T. and Geen, G.H. 1970. Effects of copepod grazing on two natural phytoplankton communities. *Journal of the Fisheries Research Board of Canada* 27: 1395–1403.

Harris, E. 1959. The nitrogen cycle in Long Island Sound. *Bulletin of the Bingham Oceanographic Collection* 17: 31–65.

Harris, E. and Riley, G.A. 1956. Oceanography of Long Island Sound, 1952–1954. VIII. Chemical composition of the plankton. *Bulletin of the Bingham Oceanographic Collection* 15: 315–323.

Harris, G.P. 1986. *Phytoplankton Ecology. Structure, Function, and Fluctuation*. Chapman and Hall, New York.

Hart, R.C. 1986. Aspects of the feeding ecology of turbid water zooplankton. In situ studies of community filtration rates in silt-laden Lake le Roux, Orange River, South Africa. *Journal of Plankton Research* 8: 401–426.

Havens, K. and DeCosta, J. 1985. An analysis of selective herbivory in an acid lake and its importance in controlling phytoplankton community structure. *Journal of Plankton Research* 7: 207–222.

Hessen, D.O. 1985. Filtering structures and particle size selection in coexisting cladocera. *Oecologia* (Berlin) 66: 368–372.

Holm, N.P., Ganf, G.G., and Shapiro, J. 1983. Feeding and assimilation rates for *Daphnia pulex* fed *Aphanizomenon flos-aquae*. *Limnology and Oceanography* 28: 677–687.

Horn, W. 1981. Phytoplankton losses due to zooplankton grazing in a drinking water reservoir. *Internationale Revue der Gesamten Hydrobiologie* 66: 787–810.

Horn, W. 1985a. Investigations in the food selectivity of the planktic crustaceans *Daphnia hyalina, Eudiaptomus gracilis,* and *Cyclops vicinus. Internatinale Revue der Gesamten Hydrobiologie* 70: 603–612.

Horn, W. 1985b. Results regarding the food of the planktic crustaceans *Daphnia hyalina* and *Eudiaptomus gracilis. Internationale Revue der Gesamten Hydrobiologie* 70: 703–709.

Hrbáček, J., Dvořakova, M., Kořinek, M., and Prochźkóva, L. 1961. Demonstration of the effect of the fish stock on the species composition of zooplankton and the intensity of metabolism of the whole plankton association. *Internationale Vereinigung für Theoretische und Angewandte Limnologie. Verhandlungen* 14: 192–195.

Infante, A. and Edmondson, W.T. 1985. Edible phytoplankton and herbivorous zooplankton in Lake Washington. *Archiv für Hydrobiologie Beiheft. Ergebnisse der Limnologie* 21: 161–171.

Infante, A. and Riehl, W. 1984. The effect of Cyanophyta upon zooplankton in a eutrophic tropical lake (Lake Valencia, Venezuela). *Hydrobiologia* 113: 293–198.

Jackson, G.A. 1980. Phytoplankton growth and zooplankton grazing in oligotrophic oceans. *Nature* 284: 439–441.

Jacobs, J. 1974. Quantitative measurement of food selection: A modification of the forage ratio and Ivlev's selectivity index. *Oecologia* (Berlin) 14: 413–417.

Jarvis, A.C. 1986. Zooplankton community grazing in a hypertrophic lake (Hartbeespoort Dam, South Africa). *Journal of Plankton Research* 8: 1065–1078.

Johannes, R.W. and Webb, K.L. 1965. Release of dissolved amino acids by marine zooplankton. *Science* 150: 76–77.

Kalff, J. and Knoechel, R. 1978. Phytoplankton and their dynamics in oligotrophic and eutrophic lakes. *Annual Review of Ecology and Systematics* 9: 475–495.

Kerfoot, W.C. 1987. Cascading effects and indirect pathways, pp. 57–70, in Kerfoot, W.C. and Sih, A. (editors), *Predation. Direct and Indirect Impacts on Aquatic Communities*. University Press of New England, Hanover, New Hampshire.

Kerfoot, W.C. and DeMott, W.R. 1984. Food web dynamics: dependent chains and vaulting, pp. 347–382, in Meyers, D.G. and Strickler, J.R. (editors), *Trophic Interactions Within Aquatic Ecosystems*. AAAS Selected Symposium 85, Westview Press, Boulder, Colorado.

Kerfoot, W.C., DeMott, W.R., and DeAngelis, D.L. 1985. Interactions among cladocerans: food limitation and exploitative competition. *Archiv für Hydrobiologie Beiheft. Ergebnisse der Limnologie* 21: 161–171.

Kerfoot, W.C. and Sih, A. 1987. *Predation. Direct and Indirect Impacts on Aquatic Communities*. University Press of New England, Hanover, New Hampshire.

Ketchum, B.H. 1962. Regeneration of nutrients by zooplankton. *Rapport et Procés-Verbaux des Réunions. Consiel Permanent International pour l'exploration de la Mer* 152: 142–146.

Kibby, H.V. 1971. Energetics and population dynamics of *Diaptomus gracilis*. *Ecological Monographs* 41: 311–327.

Kilham, S.S. 1988. Phytoplankton responses to changes in mortality rates. *Internationale Vereinigung für Theoretische und Angewandte Limnologie. Verhandlungen* 23: 677–682.

Kilham, S.S. and Kilham, P. 1984. The importance of resource supply rates in determining phytoplankton community structure, pp. 7–27, in Meyers, D.G. and Strickler, J.R. (editors), *Trophic Interactions Within Aquatic Ecosystems*. AAAS Selected Symposium 85, Westview Press, Boulder, Colorado.

Kiørboe, T. and Tiselius, P.T. 1987. Gut clearance and pigment destruction in a herbivorous copepod, *Arcartia tonsa*, and the determination of in situ grazing rates. *Journal of Plankton Research* 9: 525–534.

Knisely, K. and Geller, W. 1986. Selective feeding of four zooplankton species on natural lake phytoplankton. *Oecologia* (Berlin) 69: 86–94.

Knoechel, R. 1977. Analyzing the significance of grazing in Lake Erken. *Limnology and Oceanography* 22: 967–969.

Knoechel, R. and Holtby, L.B. 1986. Construction and validation of a body-length-based model for the prediction of cladoceran community filtering rates. *Limnology and Oceanography* 31: 1–16.

Koehl, M.A.R. 1984. Mechanisms of particle capture by copepods at low Reynolds numbers: possible modes of selective feeding, pp. 135–166, in Meyers, D.G. and

Strickler, J.R. (editors), *Trophic Interactions Within Aquatic Ecosystems*. AAAS Selected Symposium 85, Westview Press, Boulder, Colorado.

Koehl, M.A.R. and Strickler, J.R. 1981. Copepod feeding currents: Food capture at low Reynolds number. *Limnology and Oceanography* 26: 1062–1073.

Lack, T.J. and Lund, J.W.G. 1974. Observations and experiments on the phytoplankton of Blelham Tarn, English Lake District. I. The experimental tubes. *Freshwater Biology* 4: 399–415.

Lampert, W. 1978. Release of dissolved organic carbon by grazing zooplankton. *Limnology and Oceanography* 23: 831–834.

Lampert, W. 1986. Wer bestimmt die Struktur von pelagischen Biocoenosen? DieRolle von Phyto- und Zooplankton-Interactionen, pp. 66–73, in Siebeck, O. (editor), *Elemente der Steuerung und Regulation in der Pelagialbiozönose*, Akademie für Naturschutz und Landschaftspflege, Laufen/Salzac, FRG.

Lampert, W. 1987. Vertical migration of freshwater zooplankton: indirect effects of vertebrate predators on algal communities, pp. 291–299, in Kerfoot, W.C. and Sih, A. (editors), *Predation. Direct and Indirect Impacts on Aquatic Communities*. University Press of New England, Hanover, New Hampshire.

Lampert, W., Fleckner, W., Rai, H., and Taylor, B.E. 1986. Phytoplankton control by grazing zooplankton: A study on the spring clear water phase. *Limnology and Oceanography* 31: 478–490.

Lampert, W. and Taylor, B.E. 1984. In situ grazing rates and particle selection by zooplankton: effects of vertical migration. *Internationale Vereinigung für Theoretische und Angewandte Limnologie. Verhandlungen* 22: 943–946.

Lampert, W. and Taylor, B.E. 1985. Zooplankton grazing in a eutrophic lake: implications of diel vertical migration. *Ecology* 66: 68–82.

Lehman, J.T. 1976. The filter-feeder as an optimal forager, and the predicted shapes of feeding curves. *Limnology and Oceanography* 21: 501–516.

Lehman, J.T. 1980a. Nutrient cycling as an interface between algae and grazers in freshwater communities, pp. 251–263, in Kerfoot, W.C. (editor), *Evolution and Ecology of Zooplankton Communities*. University Press of New England, Hanover, New Hampshire.

Lehman, J.T. 1980b. Release and cycling of nutrients between planktonic algae and herbivores. *Limnology and Oceanography* 25: 620–632.

Lehman, J.T. 1984. Grazing, nutrient release, and their impacts on the structure of phytoplankton communities, pp. 49–72, in Meyers, D.G. and Strickler, J.R. (editors), *Trophic Interactions Within Aquatic Ecosystems*. AAAS Selected Symposium 85, Westview Press, Boulder, Colorado.

Lehman, J.T. and Naumoski, T. 1985. Content and turnover rates of phosphorus in *Daphnia pulex*: effect of food quality. *Hydrobiologia* 128: 119–125.

Lehman, J.T. and Sandgren, C. 1985. Species-specific rates of growth and grazing loss among freshwater algae. *Limnology and Oceanography* 30: 34–46.

Lehman, J.T. and Scavia, D. 1982. Microscale patchiness of nutrients in plankton communities. *Science* 216: 729–730.

Lehman, J.T. and Scavia, D. 1984. Measuring the ecological significance of microscale nutrient patches. *Limnology and Oceanography* 29: 214–216.

Levins, R. 1979. Coexistence in a variable environment. *The American Naturalist* 114: 765–783.

Levitan, C. 1987. Formal stability analysis of a planktonic freshwater community, pp. 71–100, in Kerfoot, W.C. and Sih, A. (editors), *Predation. Direct and Indirect Impacts on Aquatic Communities*. University Press of New England, Hanover, New Hampshire.

Lewis, W.M. Jr. 1977. Comments on the analysis of grazing in Lake Erken. *Limnology and Oceanography* 22: 966–967.

Lynch, M. 1980. *Aphanizomenon* blooms: Alternate control and cultivation by *Daphnia pulex*, pp. 299–304, in Kerfoot, W.C. (editor), *Evolution and Ecology of Zooplankton Communities*. University Press of New England, Hanover, New Hampshire.

Lynch, M. and Shapiro, J. 1981. Predation, enrichment, and phytoplankton community structure. *Limnology and Oceanography* 26: 86–102.

Mackas, D. and Bohrer, R. 1976. Fluorescence analysis of zooplankton gut contents and an investigation of diel feeding patterns. *Journal of Experimental Marine Biology and Ecology* 25: 77–85.

Martin, J.H. 1968. Phytoplankton-zooplankton relationships in Narragansett Bay. 3. Seasonal changes in zooplankton excretion rates in relation to phytoplankton abundance. *Limnology and Oceanography* 13: 63–71.

McCarthy, J.J. and Altabet, M.A. 1984. Patchiness in nutrient supply: implications for phytoplankton ecology, pp. 29–47, in Meyers, D.G. and Strickler, J.R. (editors), *Trophic Interactions Within Aquatic Ecosystems*. AAAS Selected Symposium 85, Westview Press, Boulder, Colorado.

McCarthy, J.J. and Goldman, J.C. 1979. Nitrogenous nutrition of marine phytoplankton in nutrient depleted waters. *Science* 203: 670–672.

McCauley, E. and Briand, F. 1979. Zooplankton grazing and phytoplankton species richness: Field tests of the predation hypothesis. *Limnology and Oceanography* 24: 243–252.

McCauley, E. and Downing, J.A. 1985. The prediction of cladoceran grazing rate spectra. *Limnology and Oceanography* 30: 202–212.

McCauley, E. and Kalff, J. 1981. Empirical relationships between phytoplankton and zooplankton biomass in lakes. *Canadian Journal of Fisheries and Aquatic Sciences* 38: 458–463.

McCauley, E. and Murdoch, W.W. 1987. Cyclic and stable populations: plankton as paradigm. *The American Naturalist* 129: 97–121.

McMahon, J.W. and Rigler, F.H. 1965. Feeding rate of *Daphnia magna* Straus in different foods labelled with radioactive phosphorus. *Limnology and Oceanography* 10: 105–114.

McQueen, D.J. 1970. Grazing rates and food selection in *Diaptomus oregonesis* (Copepoda) from Marion Lake, British Columbia. *Journal of the Fisheries Research Board of Canada* 27: 13–20.

McQueen, D.G., Post, J.R., and Mills, E.L. 1986. Trophic relationships in freshwater pelagic ecosystems. *Canadian Journal of Fisheries and Aquatic Sciences* 43: 1571–1581.

Meise, C.J., Munns, W.R. Jr., and Hairston, N.G. Jr. 1985. An analysis of the feeding behavior of *Daphnia pulex*. *Limnology and Oceanography* 30: 862–870.

Miller, T.E. and Kerfoot, W.C. 1987. Redefining indirect effects, pp. 33–37, in Kerfoot, W.C. and Sih, A. (editors), *Predation. Direct and Indirect Impacts on Aquatic Communities*. University Press of New England, Hanover, New Hampshire.

Mitamura, O. and Saijo, Y. 1986. Urea metabolism and its significance in the nitrogen cycle in the euphotic layer of Lake Biwa. IV. Regeneration of urea and ammonia. *Archiv für Hydrobiologie* 107: 425–440.

Muck, P. and Lampert, W. 1980. Feeding of freshwater filter-feeders at very low food concentrations: Poor evidence for "threshold feeding" and "optimal foraging" in *Daphnia longispina* and *Eudiaptomus gracilis*. *Journal of Plankton Research* 2: 367–379.

Muck, P. and Lampert, W. 1984. An experimental study on the importance of food conditions for the relative abundance of calanoid copepods and cladocerans. 1. Comparative feeding studies with *Eudiaptomus gracilis* and *Daphnia longispina*. *Archiv für Hydrobiologie. Supplement* 66: 157–179.

Murdoch, W.W. and McCauley, E. 1985. Three distinct types of dynamic behavior shown by a simple planktonic system. *Nature* 316: 628–630.

Murtaugh, P.A. 1985. The influence of food concentration and feeding rate on the gut residence time of Daphnia. Journal of Plankton Research 7: 415–420.

Okamoto, K. 1984. Size-selective feeding of Daphnia longispina hyalina and Eudiaptomus japonicus on a natural phytoplankton assemblage with the fractionizing method. Memoirs of the Faculty of Science, Kyoto University, Series of Biology 9: 23–40.

Olsen, Y. and Østgaard, K. 1985. Estimating release rates of phosphorus from zooplankton: Model and experimental verification. Limnology and Oceanography 30: 844–852.

Olsen, Y., Vřum, K.M., and Jensen, A. 1986a. Some characteristics of the carbon compounds released by Daphnia. Journal of Plankton Research 8: 505–517.

Olsen, Y., Jensen, A., Reinertsen, H., Børsheim, K.Y., Heldal, M., and Langeland, A. 1986b. Dependence of the rate of release of phosphorus by zooplankton on the P:C ratio in the food supply, as calculated by a recycling model. Limnology and Oceanography 31: 34–44.

Pace, M.L. 1986. Zooplankton community structure, but not biomass influences the phosphorus-chlorophyll a relationship. Canadian Journal of Fisheries and Aquatic Sciences 41: 1089–1096.

Paffenhöfer, G.-A. 1984. Calanoid copepod feeding: grazing on small and large particles, pp. 75–95, in Meyers, D.G. and Strickler, J.R. (editors), Trophic Interactions Within Aquatic Ecosystems. AAAS Selected Symposium 85, Westview Press, Boulder, Colorado.

Paffenhöfer, G.-A., Strickler, J.R., and Alcaraz, M. 1982. Suspension-feeding byherbivorous calanoid copepods: a cinematographic study. Marine Biology 67: 193–199.

Paloheimo, J.E. 1979. Indices of food preference by a predator. Journal of the Fisheries Research Board of Canada 36: 470–473.

Persson, G. 1985. Community grazing and the regulation of in situ clearance and feeding of planktonic crustaceans in lakes in the Kuskkel area, northern Sweden. Archiv für Hydrobiologie Supplement 70: 197–238.

Peters, R.H. 1975. Phosphorus regeneration by natural populations of limnetic zooplankton. Internationale Vereingung für Theoretische und Angewandte Limnologie. Verhandlungen 19: 273–279.

Peters, R.H. 1983. The Ecological Implications of Body Size. Cambridge University Press, Cambridge. 329 pp.

Peters, R.H. 1984. Methods for the study of feeding, filtering and assimilation by zooplankton, pp. 336–412, in Downing, J.A. and Rigler, F.H. (editors), A Manual for the Assessment of Secondary Productivity in Fresh Waters. IBP Handbook 17, Blackwell, Oxford.

Peters, R.H. and Downing, J.A. 1984. Empirical analysis of zooplankton filtering and feeding rates. Limnology and Oceanography 29: 763–784.

Peters, R.H. and Lean, D. 1973. The characterization of soluble phosphorus released by limnetic zooplankton. Limnology and Oceanography 18: 270–279.

Peters, R.H. and Rigler, F.H. 1973. Phosphorus release by Daphnia. Limnology and Oceanography 18: 821–839.

Porter, K.G. 1973. Selective grazing and differential digestion of algae by zooplankton. Nature 244: 179–180.

Porter, K.G. 1976. Enhancement of algal growth and productivity by grazing zooplankton. Science 192: 1332–1334.

Porter, K.G. 1977. The plant-animal interface in freshwater ecosystems. American Scientist 65: 159–170.

Porter, K.G., Gerritsen, J., and Orcutt, J.D. Jr. 1982. The effect of food concentration on swimming patterns, feeding behavior, ingestion, assimilation, and respiration by Daphnia. Limnology and Oceanography 27: 935–949.

Porter, K.G. and McDonough, R. 1984. The energetic cost of response to blue-green algal filaments by cladocerans. *Limnology and Oceanography* 29: 365–369.
Porter, K.G. and Orcutt, J.D. Jr. 1980. Nutritional adequacy, manageability, and toxicity as factors that determine the food quality of green and blue-green algae for *Daphnia*, pp. 268–281, in Kerfoot, W.C. (editor), *Evolution and Ecology of Zooplankton Communities*. University Press of New England, Hanover, New Hampshire.
Post, J.R. and McQueen, D.J. 1987. The impact of planktivorous fish on the structure of a plankton community. *Freshwater Biology* 17: 79–89.
Poulet, S.A. 1973. Grazing of *Pseudocalanus minutus* on naturally occurring particulate matter. *Limnology and Oceanography* 18: 564–573.
Poulet, S.A. and Marsot, P. 1980. Chemosensory feeding and food-gathering by omnivorous marine copepods, pp. 198–218, in Kerfoot, W.C. (editor), *Evolution and Ecology of Zooplankton Communities. University Press of New England, Hanover, New Hampshire.*
Pourriot, R. 1977. Food and feeding habits of Rotifera. *Archiv für Hydrobiologie Beiheft. Ergebnisse der Limnologie* 8: 243–260.
Prescott, G.W. 1962. *Algae of the Western Great Lakes Area*. Willam C. Brown Company, Dubuque, Iowa. 977 pp.
Price, H.J. and Paffenhöfer, G.-A. 1985. Perception of food availability by calanoid copepods. *Archiv für Limnologie Beiheft. Ergebnisse der Limnologie* 21: 115–124.
Redfield, G.W. 1980. The effect of zooplankton on phytoplankton productivity in the epilimnion of a subalpine lake. *Hydrobiologia* 70: 217–224.
Reynolds, C.S. 1984. *The Ecology of Freshwater Phytoplankton*. Cambridge University Press. 384 pp.
Reynolds, C.S., Thompson, J.M., Ferguson, A.J.D., and Wiseman, S.W. 1982. Loss processes in the population dynamics of phytoplankton maintained in closed systems. *Journal of Plankton Research* 4: 561–600.
Richman, S., Bohon, S.A., and Robbins, S.E. 1980. Grazing interactions among freshwater calanoid copepods, pp. 219–233, in Kerfoot, W.C. (editor), *Evolution and Ecology of Zooplankton Communities*. University Press of New England, Hanover, New Hampshire.
Richman, S. and Dodson, S.I. 1983. The effect of food quality on feeding and respiration by *Daphnia* and *Diaptomus*. *Limnology and Oceanography* 28: 948–956.
Riemann, B., Jorgensen, N.O.G., Lampert, W., and Fuhrman, J.A. 1986. Zooplankton induced changes in dissolved free amino acids and in production rates of freshwater bacteria. *Microbial Ecology* 12: 247–258.
Rigler, F.H. 1961. The uptake and release of inorganic phosphorus by *Daphnia magna* Straus. *Limnology and Oceanography* 6: 165–174.
Robarts, R.D. and Zohary, T. 1984. *Microcystis aeruginosa* and underwater light attenuation in a hypertrophic lake (Hartbeespoort Dam, South Africa). *Journal of Ecology* 72: 1001–1017.
Roman, M.R. 1983. Nitrogenous nutrition of marine invertebrates, pp. 345–383 in Carpenter, E.J. and Capone, D.G. (editors), *Nitrogen in the Marine Environment*. Academic Press, New York.
Roughgarden, J. and Diamond, J. 1986. Overview: the role of species interactions in community ecology, pp. 333–343, in Diamond, J. and Case, T.J. (editors), *Community Ecology*. Harper and Row, New York.
Scavia, D., Fahnenstiel, G.L., Davis, J.A., and Kreiss, R.G. Jr. 1984. Small-scale nutrient patchiness: Some consequences and a new encounter mechanism. *Limnology and Oceanography* 29: 785–793.
Scavia, D., Fahnenstiel, G.L., Evans, M.S., Jude, J.T., and Lehman, J.T. 1986. Influence of salmonine predation and weather on long-term water quality trends in Lake Michigan. *Canadian Journal of Fisheries and Aquatic Sciences* 43: 435–443.

Scavia, D. and Gardner, W.S. 1982. Kinetics of nitrogen and phosphorus release in varying food supplies by *Daphnia magna*. *Hydrobiologia* 96: 105–111.

Scavia, D. and McFarland, M.J. 1982. Phosphorus release patterns and the effects of reproductive stage and ecdysis in *Daphnia magna*. *Canadian Journal of Fisheries and Aquatic Sciences* 39: 1310–1314.

Schelske, C.L. and Stoermer, E.F. 1971. Eutrophication, silica depletion, and predicted changes in algal quality in Lake Michigan. *Science* 173: 423–424.

Schoenberg, S.A. and Carlson, R.E. 1984. Direct and indirect effects of zooplankton grazing on phytoplankton in a hypertrophic lake. *Oikos* 42: 291–302.

Schoenberg, S.A., Maccubbin, A.E., and Hodson, R.E. 1984. Cellulose digestion by freshwater microcrustacea. *Limnology and Oceanography* 29: 1132–1136.

Shapiro, J. 1980. The importance of trophic-level interactions to the abundance and species composition of algae in lakes, pp. 105–116, in Barica, J. and Mur, L.R. (editors), *Hypertrophic Ecosystems*. Junk, The Hague.

Shapiro, J. and Swain, E.B. 1983. Lessons from the silica "decline" in Lake Michigan. *Science* 221: 457–459.

Shapiro, J. and Wright, D.I. 1984. Lake restoration by biomanipulation: Round Lake, Minnesota—the first two years. *Freshwater Biology* 14: 371–383.

Siegfried, C.A. 1987. Large-bodied crustacea and rainbow smelt in Lake George, New York: trophic interactions and phytoplankton community composition. *Journal of Plankton Research* 9: 27–39.

Sommer, U. 1985. Seasonal succession of phytoplankton in Lake Constance. *Bioscience* 35: 351–357.

Sommer, U. 1986. Phytoplankton competition along a gradient of dilution rates. *Oecologia* (Berlin) 68: 503–506.

Sommer, U. 1988. Phytoplankton succession in microcosm experiments under simultaneous grazing pressure and resource limitation. *Limnology and Oceanography* 33: 1037–1054.

Sommer, U., Gliwicz, Z.M., Lampert, W., and Duncan, A. 1986. The PEG-Model of seasonal successional events in fresh waters. *Archiv für Hydrobiologie* 106: 433–471.

Spencer, C.N. and King, D.L. 1984. Role of fish in regulation of plant and animal communities in eutrophic ponds. *Canadian Journal of Fisheries and Aquatic Sciences* 41: 1851–1855.

Spencer, C.N. and King, D.L. 1987. Regulation of blue-green algal buoyancy and bloom formation by light, inorganic nitrogen, CO_2, and trophic level interactions. *Hydrobiologia* 144: 183–192.

Starkweather, P.L. 1980. Aspects of the feeding behavior and trophic ecology of suspension feeding rotifers. *Hydrobiologia* 73: 63–72.

Sterner, R.W. 1986a. Herbivores' direct and indirect effects on algal populations. *Science* 231: 605–607.

Sterner, R.W. 1986b. *Nutrients, Algae and Zooplankton: A Mechanistic Consideration of Direct and Indirect Effects*. Ph.D. thesis, University of Minnesota, Minneapolis, Minnesota.

Sterner, R.W. 1989. Resource competition during seasonal succession toward cyanobacteria. *Ecology* 70: 229–245.

Stockner, J.G. and Antia, N.J. 1986. Algal picoplankton from marine and freshwater ecosystems: a multidisciplinary perspective. *Canadian Journal of Fisheries and Aquatic Sciences* 43: 2472–2503.

Strickler, J.R. 1984. Sticky water: a selective force in copepod evolution, pp. 187–239, in Meyers, D.G. and Strickler, J.R. (editors), *Trophic Interactions Within Aquatic Ecosystems*. AAAS Selected Symposium 85, Westview Press, Boulder, Colorado.

Takahashi, M. and Ikeda, T. 1975. Excretion of ammonia and inorganic phosphorus by *Euphausia pacifica* and *Metridia pacifica* at different concentrations of phytoplankton. *Journal of the Fisheries Research Board of Canada* 32: 2189–2195.

Tátrai, I. and Istvánovics, V. 1986. The role of fish in the regulation of nutrient cycling in Lake Balaton, Hungary. *Freshwater Biology* 16: 417–424.

Taylor, W.D. 1984. Phosphorus flux through epilimnetic zooplankton from Lake Ontario: relationship with body size and significance to phytoplankton. *Canadian Journal of Fisheries and Aquatic Sciences* 41: 1702–1712.

Taylor, W.D. and D.R. Lean. 1981. Radiotracer experiments on phosphorus uptake and release by limnetic microzooplankton. *Canadian Journal of Fisheries and Aquatic Sciences* 38: 1316–1321.

Threlkeld, S.T. 1987. Experimental evaluation of trophic-cascade and nutrient-mediated effects of planktivorous fish on plankton community structure, pp. 161–173, in Kerfoot, W.C. and Sih, A. (editors), *Predation, Direct and Indirect Impacts on Aquatic Communities*. University Press of New England, Hanover, New Hampshire.

Thompson, J.M., Ferguson, A.J.D., and Reynolds, C.S. 1982. Natural filtration rates of zooplankton in a closed system: the derivation of a community grazing index. *Journal of Plankton Research* 4: 545–560.

Tilman, D. 1982. *Resource Competition and Community Structure*. Princeton University Press, Princeton, New Jersey.

Tilman, D., Kiesling, R., Sterner, R., Kilham, S.S., and Johnson, F.A. 1986. Green, bluegreen and diatom algae: Taxonomic differences in competitive ability for phosphorus, silicon and nitrogen. *Archiv für Hydrobiologie* 106: 473–485.

Turpin, D.H. Miller, A.G., Parslow, J.S., Elrifi, I.R., and Canvin, D.T. 1985. Predicting the kinetics of dissolved inorganic carbon limited growth from the short-term kinetics of photosynthesis in *Synechococcus leopoliensis* (Cyanophyta). *Journal of Phycology* 21: 409–418.

Vanderploeg, H.A. 1981. Seasonal particle-size selection by *Diaptomus sicilis* in offshore Lake Michigan. *Canadian Journal of Fisheries and Aquatic Sciences* 38: 504–517.

Vanderploeg, H., Laird, G.A., Leibig, J.R., and Gardner, W.S. 1986. Ammonium release rates by zooplankton in suspensions of heat-killed algae and an evaluation of the flow cell method. *Journal of Plankton Research* 8: 341–352.

Vanderploeg, H.A. and Ondricek-Fallischeer, R.L. 1982. Intersetule distances are a poor predictor of particle retention efficiency in *Diaptomus sicilis*. *Journal of Plankton Research.* 4: 237–244.

Vanderploeg, H.A. and Paffenhöffer, G.-A. 1985. Modes of algal capture by the freshwater copepod *Diaptomus sicilis* and their relation to food-size selection. *Limnology and Oceanography* 30: 871–885.

Vanderploeg, H. and Scavia, D. 1979a. Two electivity indices for feeding with special reference to zooplankton grazing. *Journal of the Fisheries Research Board of Canada* 36: 362–365.

Vanderploeg, H.A. and Scavia, D. 1979b. Calculation and use of selectivity coefficients of feeding: zooplankton grazing. *Ecological Modelling* 7: 135–149.

Vanderploeg, H.A. and Scavia, D. 1983. Misconceptions about estimating prey preference. *Canadian Journal of Fisheries and Aquatic Sciences* 40: 248–250.

Vanderploeg, H.A., Scavia, D., and Liebig, J.R. 1984. Feeding rate of *Diaptomus sicilis* and its relation to selectivity and effective food concentration in algal mixtures in Lake Michigan. *Journal of Plankton Research* 6: 919–941.

Vanni, M.J. 1987. Effects of nutrients and zooplankton size on the structure of a phytoplankton community. *Ecology* 68: 624–635.

Vogel, S. 1981. *Life in Moving Fluids: The Physical Biology of Flow*. Willard Grant Press, Boston.

Watts, E.C. and Young, S. 1980. Components of *Daphnia* feeding behavior. *Journal of Plankton Research* 2: 203–212.

Wehr, J.D., Brown, L.M., and O'Grady, K. 1987. Highly specialized nitrogen metabolism in a freshwater phytoplankter, *Chrysochromulina breviturrita*. *Canadian Journal of Fisheries and Aquatic Sciences* 44: 736–742.

Werner, E.E. 1986. Species interactions in freshwater fish communities, pp. 344–358, in Diamond, J. and Case, T.J., *Community Ecology*. Harper and Row, New York.

Wetzel, R.G. 1983. *Limnology*. Second Edition. Saunders College Publishing, New York.

Wheeler, P.A. 1983. Phytoplankton nitrogen metabolism, pp. 307–346, in Carpenter, E.J. and Capone, D.G. (editors), *Nitrogen in the Marine Environment*. Academic Press, New York.

Wiens, J.A. 1977. On competition and variable environments. *American Scientist* 65: 590–597.

Williams, T.G. and Turpin, D.H. 1987. Photosynthetic kinetics determine the outcome of competition for dissolved inorganic carbon by freshwater microalgae: implications for acidified lakes. *Oecologia* (Berlin) 73: 307–311.

Wilson, D.S. 1973. Food size selection among copepods. *Ecology* 54: 909–914.

Wynne, D. and Gophen, M. 1981. Phosphatase activity in freshwater zooplankton. *Oikos* 37: 369–376.

Zánkai, P.N. and Ponyi, J.E. 1986. Composition, density and feeding of crustacean zooplankton community in a shallow, temperate lake (Lake Balaton, Hungary). *Hydrobiologia* 135: 131–147.

Zaret, R.E. 1980. The animal and its viscous environment, pp. 3–9 in Kerfoot, W.C. (editor), *Evolution and Ecology of Zooplankton Communities*. University Press of New England, Hanover, New Hampshire.

Zehr, J.P., Falkowski, P.G., Fowler, J., and Capone, D.G. 1988. Coupling between ammonium uptake and incorporation in a marine diatom: experiments with the short-lived radioisotope [13]N. *Limnology and Oceanography* 33: 518–527.

5

The Role of Fungal Parasites in Phytoplankton Succession

Ellen Van Donk

Provincial Waterboard of Utrecht
Utrecht, The Netherlands

5.1 Introduction

Parasitism is a one-sided organism relationship in which one of the organisms benefits at the expense of the other. The parasite uses the host as a source or supply of food. Parasites that cause disease and possibly death of the host are called pathogens (Ahmadjian and Paracer, 1986). It is typical of parasitism that the host populations are greatly reduced in numbers. Selective parasitism on one phytoplankton species will favor the development of other species, and in this way it can be one of the factors influencing seasonal succession (Canter and Lund, 1953; Reynolds, 1973). Although parasites are known for both marine and freshwater phytoplankton species, and also for nonplanktonic algae, the emphasis in this chapter is placed upon the parasites that infect freshwater phytoplankton.

Knowledge of the large number of pathogenic organisms to which phytoplankton species are susceptible has been accumulated over many years. The effect of pathogens on the ecology of phytoplankton, however, is still poorly understood. In their study of lakes, limnologists have concentrated nearly exclusively on physicochemical factors, and on grazing by the herbivores responsible for alterations in the number of phytoplankton. The possible importance of a biological factor such as parasitism has remained virtually unexplored. This is primarily due to the fact that parasites usually escape the notice of most limnologists. Species dyes, for example, are often necessary to identify parasitic structures on phytoplankton (Müller and v. Sengbusch, 1983). Further, epidemics of parasitism may have been apparent enough, but their hitherto unpredictible development, combined with their evident host specificity, has made them difficult to study in relation to other factors affecting host populations (Reynolds, 1984).

171

Viral pathogens, algal-lysing bacteria, protozoa, and fungal parasites are the main pathogens of phytoplankton species. Phycoviruses can actively lyse laboratory strains of cyanobacteria (Safferman and Morris, 1963; Luftig and Haselkorn, 1967; Shilo, 1971). All these organisms exhibit some degree of host specificity (Fay, 1983). Goryushin and Chaplinskaya (1968) attributed the formation of clearwater spots in natural cyanobacteria blooms to viral activity. A similar effect of viral activity was also found in several Scottish lochs (Daft et al., 1970). Some experimental work on the use of viruses to control cyanobacterial blooms has been undertaken (Shilo, 1971). However, the appearance of cyanophage-resistant mutants of cyanobacteria may be an important limitation to any practical application of the cyanophage in large-scale control (Shilo, 1971).

Bacterial pathogens of various phytoplankton species all seem to belong to the Myxobacteriales (Shilo, 1970). They can cause rapid lysis of a wide range of unicellular and filamentous phytoplankton species, especially cyanobacteria. However, the available evidence does not suggest that lysis of blooms is a frequent occurrence.

Protozoa demonstrate all the transition phases between parasitism and grazing in their interaction with phytoplankton. Some, such as amoeboid forms, ingest algae by flowing around them, others enter the cells and engulf the contents. Still other species of protozoa attach themselves to algae and either extract the contents through filiform appendages or digest the contents while remaining outside the cells (Lund, 1965; Canter, 1973, 1979). Size-selective feeding is particularly conspicuous among the protozoans. Certain protozoans feeding on small colonial algae have been shown to eliminate over 99 percent of the population of certain chlorophycean planktonic algae in time periods as short as 7 to 14 days (Canter and Lund, 1968).

Many phytoplankton species are parasitized by aquatic fungi. The groups mainly involved are the uniflagellate Chytridiales (Chytridiomycetes) and simple biflagellate forms (Oomycetes). Most of the Chytridiales—or Chytrids as they are commonly called—are partly or wholly external parasites, while the biflagellate forms are internal parasites. Because most of the knowledge of the effect of pathogens on phytoplankton growth is with aquatic fungi, the remaining part of this chapter will be restricted to this group of organisms composed of uni- and biflagellate fungi.

Many fungal parasites of phytoplankton have been identified to species level, and their life cycles have been described (e.g., Canter and Lund, 1953; Patterson, 1958; Sparrow, 1960; Barr, 1978; Canter, 1972, 1979). In this group there is a stage in the life cycle of the fungus when numerous free-swimming flagellate bodies called zoospores are formed (Figure 5.1). These zoospores are the principle dispersive agents. The zoospores of the uniflagellates settle on phytoplankton cells, penetrate the cell wall with their flagellum and then form an internal rhizoidal system. Via these rhizoids, nourishment is conveyed back to the zoospore-body outside the phytoplankton cell, which en-

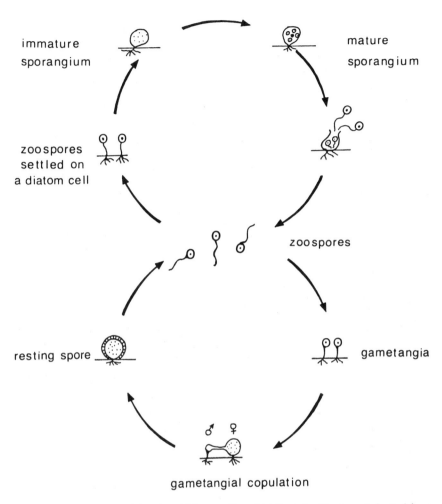

immature
sporangium

mature
sporangium

zoospores
settled on
a diatom cell

zoospores

resting spore

gametangia

gametangial copulation

Figure 5.1   Life cycle of the chytrid fungus *Zygorhizidium planktonicum* (adapted from Van Donk and Ringelberg, 1983).

larges and becomes a sporangium. The sporangium matures and new zoospores are produced. The mechanism of dehiscense varies diagnostically among species. The sporangia of the uniflagellate fungi form a well-defined operculum, a number of small holes, or a portion of the sporangium may dissolve away (see Figures 5.1 and 5.2).

The sporangia of biflagellate fungi do not possess rhizoids and are entirely located within the algal host cell. They often form narrow tubular structures or are sack-like in shape. In many species, when the sporangium is fully grown, its contents—surrounded by a vesicle—pass out of the alga via a narrow exit tube. Within this vesicle the zoospores undergo their final maturation

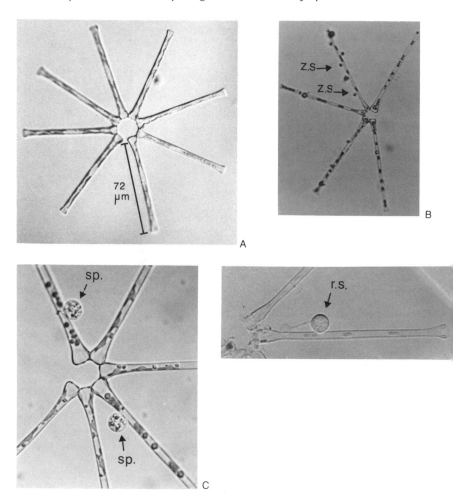

Figure 5.2   A. Uninfected cells of *A. formosa* forming a colony of 8 cells. B. Cells of *A. formosa* infected by zoospores (z.s) of *Z. planktonicum*. C. Cells of *A. formosa* infected by sporangia (sp.) of *Z. planktonicum*. D. A cell of *A. formosa* infected by *Z. planktonicum*, showing gametangial copulation, resulting in the formation of a resting spore (r.s) (adapted from Van Donk, 1983).

and are liberated when it bursts. The production of thick-walled asexually or sexually formed resting spores occurs in both these groups of fungi (Figure 5.2D) (Sussman, 1966). The prerequisite for resting spore formation is an ability to survive environmental conditions that would be lethal to the organism at other stages of its development. However, for some frequently encountered species, such resting-spore stages have not yet been discovered. When well-developed fungal sporangia or resting spores are present, the chromatophores of the host algae are either considerably disorganized or completely destroyed (Figure 5.2C and 5.2D).

5.2   The Effect of Parasites on Phytoplankton Populations

Most knowledge of the effect of pathogens in relation to fluctuations in the numbers of algae and the ecological significance of pathogens is derived from the aquatic fungi. The first investigators (e.g., Wesenberg-Lund, 1908; de Wildeman, 1900, 1931; Huber-Pestalozzi, 1946) described the occurrence of chytridiaceous fungi parasitizing planktonic algae, but they gave no indication about whether the fungi exerted any marked effect on the algal population. Reynolds (1940) showed that a chytrid fungus reduced the numbers of a natural population of *Staurastrum paradoxum* Meyer. Weston (1941) reviewed the role of aquatic fungi in limnology and pointed to the ability of fungal parasites to control the number of plankton algae. However, no attempts were made to quantitatively assess the importance of the phenomenon until the studies of Canter and Lund (1948, 1951, and 1953). Their earliest work was carried out in the English lake district and centered around the diatoms *Asterionella formosa, Fragilaria crotonensis, Tabellaria fenestrata,* and *Melosira italica,* which were all parasitized by chytrids.

Canter and Lund (1951) stated that the chytrid parasites may delay the time of appearance of the maximum algal number, or may decrease the size of this maximum. Many of the chytrids appeared to be highly specific for a single algal species or a small range of related species. Later investigations, over a period of 20 years, showed that the desmid population in these lakes may also be controlled to a large extent by fungal epidemics (Canter and Lund, 1969). They further noted that the parasites that attacked desmids were more frequently biflagellate than uniflagellate. Another long-term study of the effects of fungal parasites on algae was that of Pongratz (1966). This paper reported the results of ten years of study on Lake Geneva. However, only dates of occurrences of fungus blooms were given, without host concentrations, and often without an estimate of the percentage of host cells infected. Subsequently, epidemics involving both chytrids and biflagellate fungi have been widely reported for planktonic diatoms, desmids, green algae, dinoflagellates, chrysophycean algae, and cyanobacteria (e.g., Koob, 1966; Soeder and Maiweg, 1969; Canter, 1963, 1972; Blinn and Button, 1973; Reynolds, 1973; Masters, 1971, 1976; Youngman et al., 1976; Müller and Sengbusch, 1983; Van Donk and Ringelberg, 1983; Canter and Heaney, 1984; Sommer et al., 1984; Sommer, 1984).

The course of epidemics caused by fungi usually takes the following pathway. At the outset, there is an increase in the number of young, recently encysted zoospores on the living algal cells. Subsequently, the proportion of sporangia in the fungal population rises, and the algal cells bearing them die. In the last stage of an epidemic, the number of encysted zoospores is relatively small, and many dead cells bear the empty cases of dehisced sporangia. In addition, sexually formed resting spores may now be produced (see Figures 5.1 and 5.2). Epidemics are usually of short duration—on the order of two to three weeks (Lund, 1957). Lund further stated that the minimum algal-cell density for epidemics of the diatom *Asterionella formosa* is approximately 10 cells/ml. Low host concentrations require that infective zoospores travel relatively further to find suitable cells.

Fungal epidemics of one phytoplankton species may favor the development of other algal species, and in this way parasitism can be one of the factors controlling seasonal succession. One of the first examples of parasitism as a factor influencing seasonal succession was the report of the replacement of a highly infected *Asterionella formosa* population by *Fragilaria crotonensis* and *Tabellaria fenestrata* in Esthwaite Water in 1949 (Canter and Lund, 1951). Canter and Lund (1969) found that fungal parasitism of desmid populations did not alter the overall seasonal pattern of periodicity of these algae, but that it did have a marked effect on interspecific competition. The proportional abundance of species changed as a result of parasitism.

Reynolds (1973) stated that one important effect of the epidemic of the chytrid fungus *Zygorhizidium affluens* on *Asterionella* in Crose Mere in 1968 was to permit the dominance of *Stephanodiscus astraea*. This species was typically subdominant in other nonepidemic years. Youngman et al. (1976) found that the growth of *A. formosa* in Farmoor Reservoir was interrupted when 45.5 percent of the cells were parasitized by *Zygorhizidium affluens*. This infection favored the development of *S. astraea* and *S. hantzschii*. Van Donk and Ringelberg (1983) also showed that severe parasitism on *A. formosa* favored the development of other diatoms—primarily *F. crotonensis, S. hantzschii*, and *S. astraea*.

However, all these observations on natural populations have not clarified the conditions under which the fungi multiply quickly enough to start an epidemic. Contemporary understanding of the factors that enable parasites to grow at the expense of their algal host is still very sketchy. The host-parasite relationship for phytoplankton is further complicated by: 1) hypersensitivity of hosts to infective spores, such that algal cells die so soon after infection that sporangia fail to develop (Canter and Jaworski, 1979), and 2) hyperparasitism—the chytrid *Zygorhizidium affluens*, for example, is itself frequently parasitized by another chytrid, *Rozella* (Canter, 1969; Held, 1974). Experimental studies now are essential for the elucidation of the questions raised from field studies.

5.3   Factors Influencing Fungal Parasitism

Largely because of the difficulty of isolating parasitic fungi (Canter and Jaworski, 1978; Van Donk and Ringelberg, 1983; Bruning and Ringelberg, 1987), little experimental work has been done on the pathogenicity of these organisms. Furthermore, information on the factors that influence chytrid parasitism is scarce. The change in the fraction of infected cells in an infected algal population is the resultant of the reproductive rates of both host and parasite (Sparrow, 1968; Bruning and Ringelberg, 1987). The fraction of infected cells I will change over time t according to:

$$I(t) = I_0 \cdot \exp[(\mu_p - \mu_h)t] \qquad (5.1)$$

μ_p and μ_h being the reproductive rates of parasite and infected host populations respectively. The expression $(\mu_p - \mu_h)$ is a measure of the rate at which an epidemic will develop. An increase of fungal reproductive rate and a decrease of algal reproductive rate will both facilitate the development of a parasitic epidemic. Therefore, environmental factors that retard the algal reproductive rates—e.g., nutrient limitation or low light intensities—may be expected to be favorable for the development of an epidemic, insofar as the parasitic reproductive rate is not negatively influenced by the same factor (Bruning and Ringelberg, 1987).

Nutrient deficiency   Reynolds (1984) stated that a nutrient-deficient algal population may be more susceptible to bacterial or fungal attack than a healthy one. It might appear reasonable to suppose that the level of parasitism would be higher if a phytoplankton population is not in "good condition." Under such conditions the reproductive rate of the algae would be relatively slow, and a given rate of infection and reproduction of a parasite will be relatively fast. However, many investigators have reported that severe parasitism usually arises when the algae are growing relatively fast—often at their maximum annual rate and while nutrient concentrations are still high (Canter and Lund, 1969; Youngman et al., 1976; Van Donk and Ringelberg, 1983).

Van Donk (1983) observed great differences in maximum spring abundance of *Asterionella formosa* in Lake Maarsseveen (the Netherlands) over a five-year period, in spite of the fact that the concentrations of nutrients (P, Si, and N) at the start of each of the spring blooms were almost identical. Compared with 1978, 1979, and 1982, the maximum abundance of *A. formosa* in 1980 and 1981 was low, and P, Si, and N concentrations were relatively high at the time of the decline of *A. formosa*. In 1980 and 1981, high rates of infection of *A. formosa* by the fungus *Zygorhizidium planktonicum* were observed (Figures 5.3 to 5.7).

In the laboratory, Bruning and Ringelberg (1987) investigated the influence of phosphorus limitation of *A. formosa* on the rate of reproduction of one of its parasites, *Rhizophydium planktonicum*. The alga and the parasite were both isolated from Lake Maarsseveen. At saturated phosphorus concentrations (at 16°C), the host cells reached a reproductive rate of 0.95 d⁻¹. Growing on these phosphorus-saturated host cells, the mean rate of parasite zoospore production was 26 spores per sporangium, and the mean development time of a sporangium was 45 hours. Growing on phosphorus-limited hosts, the zoospore production decreased to less than 9 spores per sporan-

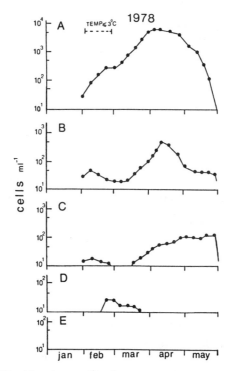

Figure 5.3

Figures 5.3 to 5.7   Abundance of *A. formosa* in various stages of infection were observed in the upper 10 meters of Lake Maarsseveen during the spring of 1978, 1979, 1980, 1981, and 1982, respectively. For each year, the number of uninfected cells (A) and of cells in various stages of infection by *Z. planktonicum* are shown. For infected cells, B indicates cells with zoospores; C, cells with sporangia; D, cells with thick-walled spores; and E, cells with resting spores (adapted from Van Donk, 1983).

gium, and the development time decreased slightly to 40 hours. In Figure 5.8 the mean number of spores per sporangium (N) is plotted as a function of the phosphorus limited reproductive rate of the host. In Figure 5.9 the development time of the sporangia (D) is plotted as a function of the reproductive rate of the host. The shorter development time of the sporangia on phosphorus-limited hosts is reasonable, since fewer spores are produced. Also, spore production proceeds at a lower rate on phosphorus-limited hosts (Figure 5.10). Here the doubling time (Td) of the spores within the sporangia ($D \times \ln2/\ln N$) is plotted against the phosphorus-limited reproductive rate of the host. On a host growing at high phosphorus concentrations, the number of spores doubles approximately every 9.5 hours, but on phosphorus-limited hosts the doubling time increases to 13 hours or more. In summary, phosphorus limitation of the host causes the parasite to produce fewer spores and at a lower rate.

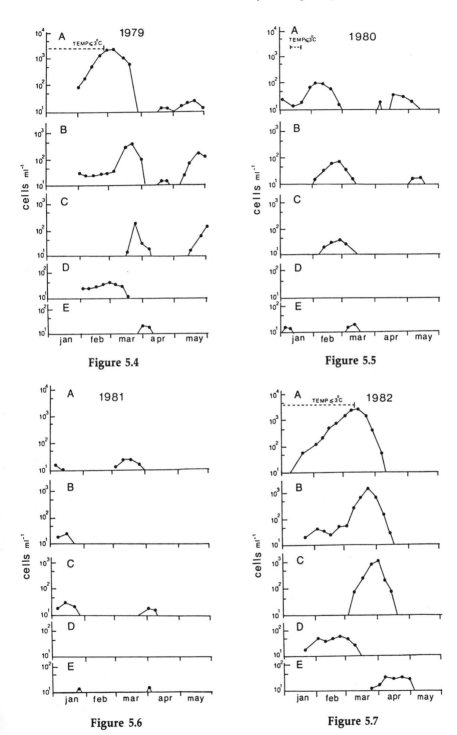

Figure 5.4

Figure 5.5

Figure 5.6

Figure 5.7

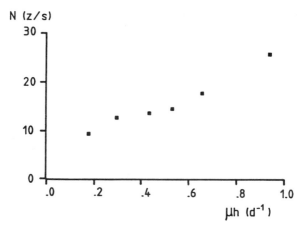

Figures 5.8   The mean number of zoospores per sporangium (*N*) produced by the parasite as a function of the phosphorus-limited reproductive rate of its host (μ_h) (adapted from Bruning and Ringelberg, 1987).

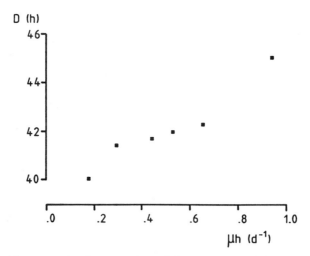

Figure 5.9   The mean development time of the sporangia of the parasite (*D*) as a function of the phosphorus-limited reproductive rate of its host (μ_h) (adapted from Bruning and Ringelberg, 1987).

Having established the influence of phosphorus limitation on the algal reproductive rate and on the rate of fungal spore production, the question of the influence of phosphorus limitation on the development of a parasitic epidemic arises. At high host densities—a situation in which nearly every zoospore finds a host after a relatively short searching time—the reproductive

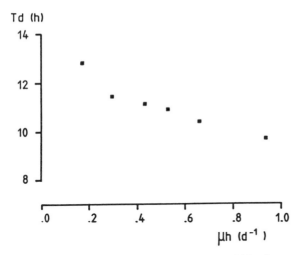

Figure 5.10   The mean doubling time of the zoospores within the sporangia of the parasite (Td) as a function of the phosphorus-limited reproductive rate of its host (μ_h) (adapted from Bruning and Ringelberg, 1987).

rate of the parasite is a function only of the development time (D) and number of spores per sporangium (N):

$$\mu_p = (\ln N)/D \tag{5.2}$$

At low host densities, the searching time will become significant, and zoospores will have no effect if they can not find a host during the infective period. Consequently, if host densities are low the reproductive rate of a parasite as calculated using eq. 5.2 will be an overestimate. Since the experiments related to the searching time and the infective period performed by Bruning have not been finished yet, the influence of phosphorus limitation on epidemic development can only be evaluated at high host densities. In Figure 5.11, calculated values of $(\mu_p - \mu_h)$ at high host densities are plotted against the phosphorus-limited reproductive rate of the host (Bruning and Ringeberg, 1987). The figure shows that $(\mu_p - \mu_h)$ is always positive. This suggests that, at high host densities, an epidemic can occur regardless of a possible phosphorus limitation of the host. However, since $(\mu_p - \mu_h)$ increases with decreasing reproductive rates of the host, an epidemic will develop at a higher rate when the host cells become phosphorus limited.

Thus, it can be concluded that phosphorus limitation of *Asterionella* will facilitate the development of an epidemic of its parasite *Rhizophydium*, at least at a high host density. It must be noted that this conclusion should not be interpreted to mean that nutrient-limited algae are more susceptible to fungal attack than healthy cells (Reynolds, 1984). On the contrary, the parasite seems

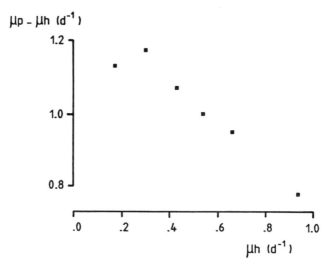

Figure 5.11 The difference in reproductive rate of parasite and host at high host density ($\mu_p - \mu_h$) as a function of the phosphorus-limited reproductive rate of its host (μ_h) (adapted from Bruning and Ringelberg, 1987).

to grow better on "healthy," nonlimited algae, producing more zoospores at a higher rate. Phosphorus limitation has a negative effect on both host and parasite, but the impact on the reproductive rate of the host is greater than that of the parasite (Bruning and Ringelberg, 1987).

Light   The infection process of chytrids has been found to require light (Canter and Jaworski, 1981). While zoospores can remain alive in darkness, rates of adhesion to host cells are low in the dark and increase dramatically under illumination. For *Rhizophydium planktonicum*, Canter and Jaworski (1981) showed that absolute darkness was not necessary to reduce the numbers of zoospores that adhered to *Asterionella* cells. It appeared that even in dim light—up to approximately 3.5 μEm^{-2}sec^{-1}—a similar reduction in attachment rate takes place. There was evidence that zoospores remained alive in the presence of *Asterionella*, because they had the ability to adhere to host cells once they had been transferred from the darkened condition into an illumination of 35 μEm^{-2}sec^{-1}. Canter and Jaworski (1979) also reported that a zoospore, once encysted on a host cell, could continue its development in the darkness after a brief inoculation period in the light (10–30 min). After 2.5 days in total darkness at 15°C, well-developed sporangia with an occasional branched endobiotic rhizoidal thread were found.

In laboratory studies, Blinn and Button (1973) found that increasing the length of the photoperiods may increase infection rates of the chytrid *Dangeardia mammillata* on *Pandorina* sp. as much as 4- to 5-fold. Multiple cell infections were also induced under continuous light regimes within the optimum range of temperature. Other experimental work concerned with the effects of light and darkness on parasitism of algae has been conducted by Barr and Hickman (1967), Abeliovich and Dikbuck (1977), and Kumar (1978). In all cases where host and parasite were subjected to experimentation in darkness, reduced or negligible percentages of infection were recorded.

Barr and Hickman (1967) noted that although many zoospores of *Rhizophydium sphaerocarpum* had encysted on *Spirogyra* cells in the dark, few had germinated. At the present there is, however, no evidence that the responses here described also exist in the natural environment. The primary role of light might be to influence the rate of zoospore swimming, since some zoospores among the various types of chytrids are known to exhibit phototaxis (e.g., Strasburger, 1878; Kazama, 1972). Alternatively, the influence of light might be indirect, perhaps caused by the attraction of zoospores to organic substances produced by algae growing in the light. Both possibilities require further study.

Extracellular algal products   Lund (1957) has already suggested the significance of extracellular algal products in the stage of chytrid zoospores finding a new host. The possible importance of extracellular algal products in the process of infection by chytrid parasites has been further emphasized by several other authors (Masters, 1971; Blinn and Button, 1973; Reynolds, 1973; Stewart and Daft, 1976). Differences in the composition or amount of exudate might help to account for differences in the susceptibility of an algal population.

Role of temperature   For a number of host-parasite interactions there is evidence that temperature is the most important factor controlling chytrid parasitism (Barr and Hickman, 1967; Masters, 1971; Blinn and Button, 1973; Van Donk and Ringelberg, 1983). Barr and Hickman (1967) reported that the optimum temperature for the chytrid *Rhizophydium sphaerocarpum* to infect *Spirogyra* in the laboratory was 30°C, with no infection occurring below 10°C. These results are supported by the observation that natural epidemics of *Rhizophydium sphaerocarpum* on *Spirogyra* occur in summer only. Masters (1971) found that *Chytridium deltanum* did not appear on species of *Oocystis* in Lake Manitoba until the water had reached a temperature of 25°C. Furthermore, the ability of the chytrid *Dangenardia mammillata* to infect a *Pandorina*-strain at the pathogenic level under controlled laboratory conditions appears to be closely correlated with temperature (Blinn and Btton, 1973). Here, the percentage of chytrid infection increased directly with water temperature from 3.5°C to 14°C, but the infection rate then declined steadily with increasing water temperature between 14°C and 17°C. With an increase in temperature, levels of infection after an incubation time of 14 days decreased from 78 percent at 4–8°C to less than 1 percent at temperatures above 20°C. These laboratory results correlate with the high level of infection during the winter in Upper Lake Mary.

In Lake Maarsseveen, Van Donk and Ringelberg (1983) found that the fungus *Zygorhizidium planktonicum* was capable of affecting the spring bloom of *A. formosa* and bringing about a premature end to the bloom (see also Section 5.4). Only when the fungus was temporarily inhibited in its parasitic activity, during times coinciding with periods of ice cover, was *A. formosa* able to reach a high abundance. In the five years of the study (Van Donk, 1983) the formation of thick-walled fungal spores and a decrease in number of cells with zoospores and sporangia was observed during cold periods at temperature ≤3°C (see Figures 5.3 to 5.7). In contrast to 1978 an enormous increase in the infection rate was observed in 1979 and 1982 as soon as the temperature rose above 3°C. The cause of this difference remains to be determined. Perhaps the time required for ripening of the thick-walled spores played an important role. In 1978, the period of ice cover was short. After the frost period the temperature rose to 4°C within 10 days, but another 10 days were required for a perceptible reinfection to begin. Thick-walled spores persisted for over a month. If this period of one month actually represents the time required for maturation, the fungus might not have been able to overtake the fast-growing diatom population. At the end of the cold periods in 1979 and in 1982, the thick-walled spores had already been present for at least a month, and within one week after the ice melted (as soon as the temperature had risen above 3°C), *Z. planktonicum* showed enormous activity. In January 1980, resting spores of an epidemic that must have occurred in the previous month were still present. At that time the number of *A. formosa* cells was low. A short period of ice cover in January had no appreciable effect on the temperature of the water column. In February, the parasite increased rapidly and soon overtook *A. formosa*.

Host/parasite studies performed in the laboratory showed that temperature can be an important environmental factor associated with epidemics of *Z. planktonicum* on *A. formosa* (Van Donk and Ringelberg, 1983). At very low temperatures (1.5 ± 1°C) fungal activity was inhibited although *A. formosa* still grew well. However, at 5, 10, and 18°C the fungus manifested a high infection rate and was able to overtake *A. formosa* (Figure 5.12). At these three temperatures the fungus was capable of destroying an *A. formosa* population within one week, even in the case when the population was not limited in its growth by any other factor. Bruning (personal communication) recently performed laboratory experiments on *Rhizophydium planktonicum* parasitizing *A. formosa*. He found an increase in development time of the sporangia from 2 days at 16°C to 19 days at 2°C. It is unknown whether the observed inhibition of fungal parasitism at very low temperatures is not only a direct but also an indirect effect of temperature. A possible indirect effect may be the result of a change in the structure of the membrane and/or the wall of the diatoms, which often become thicker (and perhaps less permeable for

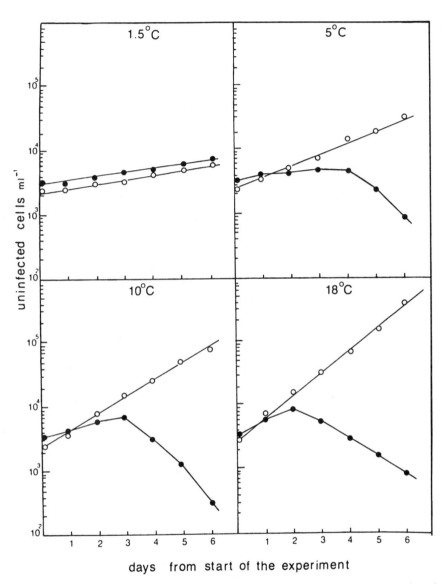

Figure 5.12   The number of uninfected cells during six days of incubation of uninfected (○) and infected (•) cultures of A. *formosa* at four different temperatures. At day zero the (•) cultures were inoculated with infected A. *formosa* cells (adapted from Van Donk and Ringelberg, 1983).

fungal parasites) at very low temperatures. In this context, perhaps species-specific infections of the fungus can be attributed to differences in the membrane/wall structures of the various algal species.

5.4   Interaction of Parasitism and Abiotic Factors in Controlling Phytoplankton Succession in Lake Maarsseveen

In order to analyze the factors controlling the spring succession of phytoplankton in Lake Maarsseveen both abiotic and biotic factors have been studied, and a linkage has been made between field and laboratory studies (Van Donk, 1983).

As a first step, the seasonal variations of the dominant species of algae were described over a five-year period. Simultaneously, the concentrations of some important nutrients were measured. During this time (1978–1982), diatoms were predominant in the spring. Blooms of five different diatom species—*Asterionella formosa* Hass., *Stephanodiscus astraea* (Ehr.) Grun., *Stephanodiscus hantzschii* Grun., *Fragilaria crotonensis* Kitt., and *Cyclotella comta* Ehr. Kütz were observed. Over these five years the similarities in succession patterns were striking. Each year, in winter and early spring, *A. formosa*, *S. hantzschii*, and *S. astraea* were the first to develop (Figure 5.13). However, great differences were observed in the level of the maximal spring abundance of *A. formosa*. In 1980, when *A. formosa* failed to reach a high abundance it was followed by a bloom of *F. crotonensis* (see Figure 5.13C).

The silicate, nitrate, and orthophosphate concentrations were nearly identical each year at the start of the spring bloom. The orthophosphate concentration was always very low (Figure 5.14). The highest concentration was measured in December (6 μg/l), declining during the spring to undetectable values. From August onward, the orthophosphate concentration again rose above the detection level. The nitrate concentrations remained high, and varied only slightly throughout the year between 0.3 and 0.5 mg/l (Van Donk, 1983). The dissolved silicate concentrations were also relatively high (Figure 5.14); the highest value was again measured in December (2 mg/l). In the epilimnion the silicate concentration declined during the spring, reaching its lowest values in June and July (0.6 mg/l).

These closely timed observations, however, were by themselves insufficient to decipher the mechanisms responsible for the apparent succession because the observed net change in population abundance could have been produced by a widely divergent combination of gains and losses. Dissection of the interplay of abiotic (e.g., nutrients, temperature, light) and biotic factors (e.g., grazing, parasitism) in shaping the behavior of the populations in situ demanded not only an intensive sampling scheme but also a meaningful experimental scheme.

To determine the role of abiotic factors, natural-community bioassays were carried out and physiological parameters, such as cell nutrient contents and nutrient uptake kinetics, were used as indicators. The bioassays were performed under natural light and temperature conditions in a newly devel-

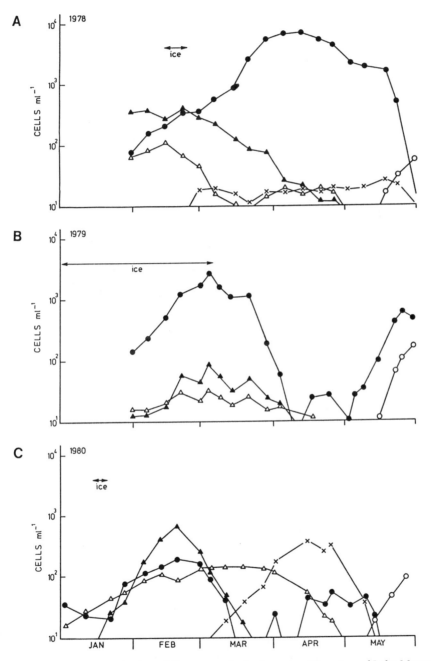

Figure 5.13 The succession of diatom species in the upper 10 meters of Lake Maarsseveen during the spring of A, 1978; B, 1979; and C, 1980. •, *Asterionella formosa*; ▲, *Stephanodiscus hantzschii*; △, *Stephanodiscus astraea*; X, *Fragilaria crotonensis*; ○, *Cyclotella comta*. (Adapted from Van Donk and Ringelberg, 1983).

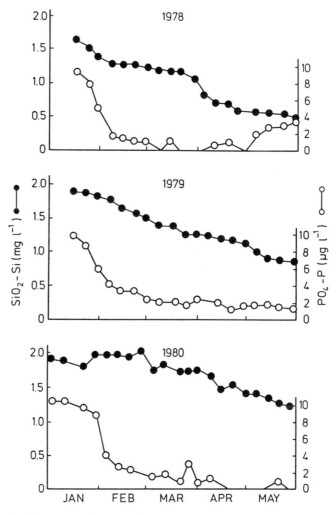

Figure 5.14   The mean values of silicate (●) and phosphate (○) concentrations measured in the upper 10 meters of Lake Maarsseveen during the spring of 1978, 1979, and 1980. (Adapted from Van Donk and Ringelberg, 1983).

oped bioassay apparatus outside the laboratory (Van Donk et al., 1988). The experiments were conducted over a two-year period.

Phosphate was the major growth-rate-limiting nutrient. From January through June, during the decline of the phosphorus concentration, the diatoms became successively phosphate limited in the following order: *S. hantzschii*, *S. astraea*, *F. crotonensis*, *A. formosa*, and *C. comta*. Light limitation was probably the major cause of the relatively late start in early spring of the growth of *F. crotonensis* (Van Donk, et al., 1988). Grazing by zooplankton was not an important loss factor for the diatom populations since zooplankton were scarce during both the winter and the spring (Hulsmann and Van Rijswijk, 1980).

The role of nutrient competition was also studied in the laboratory. The growth kinetics for silicon and phosphorus of *A. formosa, F. crotonensis,* and *S. hantzschii* were determined in batch cultures using uninfected clones isolated from the lake (Van Donk, 1983). The experiments were performed at four different temperatures (5°, 10°, 15°, and 20°C). *A. formosa* had the highest affinity for phosphorus at all four temperatures. *S. hantzschii* was shown to have the highest affinity for silicon at all four temperatures, but its growth rate was inferior to that of both *A. formosa* and *F. crotonensis* under phosphorus limitation. *F. crotonensis* and *A. formosa* were similar in their growth responses to silicon limitation. The predicted outcome of competition when all three species were limited by phosphorus or silicon was tested in continuous-culture, competition experiments. The results were consistent with the predictions: *A. formosa* dominated under phosphorus limitation and *S. hantzschii* under silicon limitation (Van Donk, 1983; Van Donk and Kilham, in prep.).

In performing the bioassays and the nutrient-uptake experiments, it was noted that occasionally *A. formosa* was heavily infected by the parasitic chytrid fungus *Zygorhizidium planktonicum* Canter. The infected *A. formosa* population did not respond with growth stimulation to the addition of any nutrient. In order to elucidate the significance of this biotic factor in the seasonal succession, a detailed quantitative study of the course of the infection was made (see Figures 5.3 to 5.7). *A. formosa* was the only species with a high rate of fungal infection (>70 percent). The fungus seemed capable of affecting the abundance of *A. formosa,* and bringing about a premature end to the bloom. Only when the fungus was temporarily inhibited in its parasitic activity (this phenomenon always coincided with periods of frost), was *A. formosa* able to reach a high abundance (see also Section 5.3). Severe parasitism on *A. formosa* further favored the development of *F. crotonensis* and both *Stephanodiscus* species. As was previously noted, temperature had a marked effect upon the rate of parasitism of algal cells (see Section 5.3); fungal infection is inhibited only at the very low water temperatures found under ice cover.

Coupling the results of the laboratory experiments to the seasonal succession and chytrid infection events observed in Lake Maarsseveen, it may be concluded that in growing seasons when parasitism of *A. formosa* was of minor importance due to inhibition of fungal activity (1978, 1979, and 1982), *A. formosa* was able to outcompete *F. crotonensis* and both *Stephanodiscus* species (*S. hantzschii* and *S. astraea*), because of its higher affinity for phosphorus. However, in years when *A. formosa* was heavily infected by *Z. planktonicum* (1980 and 1981), *F. crotonensis* and both *Stephanodiscus* species were able to bloom because of the absence of competitive pressure from *A. formosa* (Figure

5.15). Although *F. crotonensis* was initially limited by light, it ultimately dominated both *Stephanodiscus* species under P-limited conditions.

In summary it may be concluded that the growth and succession of the various diatom species in Lake Maarsseveen during spring is regulated by several interacting biotic and abiotic factors. Phosphorus limitation, parasitism, light, and temperature were the important factors for the diatoms. From these data it is obvious that the commonly used concept of a *single* limiting factor is a great oversimplification when applied to the natural phytoplankton population. Biotic and abiotic factors continuously interact and the limiting factors change seasonally from one year to the next.

5.5   Conclusions

Fungal parasitism can play a significant role in the control of phytoplankton succession. Fungal parasites are able to prevent phytoplankton species from blooming or may delay bloom development and reduce the level of the max-

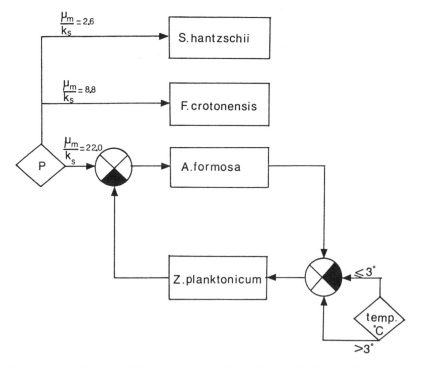

Figure 5.15   A diagram of the relationship of three diatoms (*A. formosa, F. crotonensis,* and *S. hantzschii*) and the fungus *Z. planktonicum*. μ/K_s is a measure of the competitive ability of each diatom for phosphorus (P) at 5°C—the higher the value, the greater the competitive ability. A white quadrant stands for stimulation of the growth rate and a black quadrant for inhibition. *Z. planktonicum* inhibits the growth of *A. formosa* at temperatures higher than 3°C (adapted from Van Donk, 1983).

imum abundance. Because of the host specificity of most fungi, infection of one algal species often favors successional development of other, competing phytoplankton species. On the other hand, fungal parasitism is influenced by a number of environmental factors, including light, temperature, nutrient limitation, and probably the production of extracellular substances by the host. The losses and loss rates of phytoplankton populations caused by parasites can not yet be quantified because of the complexity of the infection process. In most field studies, there is no information available about the time elapsed from infection to host-cell death, so it is not possible to convert the percent values of parasitized cells into exponential death rates. Although parasitism is likely to be just as important as other loss processes, many of the dynamic models of phytoplankton-biomass changes have incorporated factors for the rate of removal by herbivores and for sedimentation, but losses due to parasitic attacks have not been considered (Sommer et al., 1986). Whether phytoplankton succession is predominantly controlled by the algal growth rate or by the algal loss rate is still uncertain and the relative importance of these processes must vary both temporally and among lakes. The rate with the higher temporal variability has the greater influence on the temporal pattern of population development (Kalff and Knoechel, 1978). Only after the factors influencing both growth and mortality have been ascertained will a full understanding of phytoplankton dynamics be achieved (e.g., Kalff and Knoechel, 1978). Finally, a significant amount of additional research (both field and laboratory studies) is necessary to quantify the role of parasitism in the control of phytoplankton succession.

Acknowledgments
The author thanks Dr. K. Bruning, Prof. Dr. J. Ringelberg, Prof. Dr. W.R. Swain, and the referees for reading the manuscript critically and thanks Monica Hoppenbrouwers for typing the manuscript. The investigations on fungal parasitism carried out by E. Van Donk and K. Bruning were supported by the Foundation for Fundamental Biological Research (BION), which is subsidized by the Netherlands Organization for Advancement of Pure Research (ZWO).

References

Abeliovich, A. and Dikbuck, S. 1977. Factors affecting infection of *Scenedesmus obliquus* by a *Chytridium* sp. in sewage oxidation ponds. *Applied and Environmental Microbiology* 6: 832–836.
Ahmadjian, V. and Paracer, S. 1986. *Symbiosis. An Introduction to Biological Associations*. University Press of New England, Hanover and London. 212 pp.
Barr, D.J.S. 1978. Taxonomy and phylogeny of chytrids. *Biosystems* 10: 153–165.
Barr, D.J.S. and Hickman, C.J. 1967. Chytrids and algae. II. Factors influencing parasitism of *Rhizophydium sphaerocarpum* on *Spirogyra*. *Canadian Journal of Botany* 45: 431–440.
Blinn, D.W. and Button, K.S. 1973. The effect of temperature on parasitism of *Pandorina* sp. by *Dangeardia mammillata* B. Schröder in an Arizona mountain lake. *Journal of Phycology* 9: 323–326.
Bruning, K. and Ringelberg, J. 1987. The influence of phosphorus limitation of the diatom *Asterionella formosa* on the zoospore production of its fungal parasite *Rhizophydium planktonicum*. *Hydrobiological Bulletin* 21: 49–54.

Canter, H.M. 1963. Studies on British chytrids. 23 new species of chrysophysean algae. *Transactions British Mycological Society* 46: 305–320.

Canter, H.M. 1969. Studies on British chytrids. XXIX. A taxonomic revision of certain fungi found on the diatom *Asterionella*. *Botanical Journal of the Linnean Society* 62: 267–278.

Canter, H.M. 1972. A guide to the fungi occurring on planktonic blue-green algae, pp. 148–158, in Desikachary, T.V. (editor), *Proceedings of the Symposium on Taxonomy and Biology of Blue-green Algae*. Madras, University of Madras.

Canter, H.M. 1973. A new primitive protozoan devouring centric diatoms in the plankton. *Zoological Journal of the Linnean Society* 52: 63–83.

Canter, H.M. 1979. Fungal and protozoan parasites and their importance in the ecology of phytoplankton. *Freshwater Biological Association* Annual Report 47: 43–50.

Canter, H.M. and Heaney, S.I. 1984. Observations on zoosporic fungi of *Ceratium* spp. in lakes of the English Lake District: Importance for phytoplankton population dynamics. *New Phytologist* 97: 601–612.

Canter, H.M. and Jaworski, G.H.M. 1978. The isolation, maintenance and host range studies of a chytrid, *Rhizophydium planktonicum* Canter emend., parasitic on *Asterionella formosa* Hass. *Annals of Botany* 42: 967–979.

Canter, H.M. and Jaworski, G.H.M. 1979. The occurrence of a hypersensitive reaction in the planktonic diatom *Asterionella formosa* Hass. parasitized by the chytrid *Rhizophydium planktonicum* Canter emend., in culture. *New Phytologist* 82: 187–206.

Canter, H.M. and Jaworski, G.H.M. 1981. The effect of light and darkness upon infection of *Asterionella formosa* Hass. by the chytrid *Rhizophydium planktonicum* Canter emend. *Annals of Botany* 47: 13–30.

Canter, H.M. and Lund, J.W.G. 1948. Studies on plankton parasites. I. Fluctuations in the numbers of *Asterionella formosa* Hass. in relation to fungal epidemics. *New Phytology* 47: 238–261.

Canter, H.M. and Lund, J.W.G. 1951. Studies on plankton parasites. III. Examples of interaction between parasitism and other factors determining the growth of diatoms. *Annals of Botany* 15: 359–371.

Canter, H.M. and Lund, J.W.G. 1953. Studies on plankton parasites. II. The parasitism of diatoms with special reference to lakes in the English Lake District. *Transactions British Mycological Society* 36: 13–37.

Canter, H.M. and Lund, J.W.G. 1968. The importance of protozoa in controlling the abundance of planktonic algae in lakes. *Proceedings of the Linnean Society*, London 179: 203–219.

Canter, H.M. and Lund, J.W.G. 1969. The parasitism on planktonic desmids by fungi. *Österreichische Botanische Zeitung* 116: 351–377.

Daft, N.J., Begg, J., and Stewart, W.D.P. 1970. A virus of blue-green algae from freshwater habitats in Scotland. *New Phytologist* 69: 1029–1038.

Fay, P. 1983. *The Blue-Greens*. The Institute of Biology's Studies in Biology no. 160, Edward Arnold, London. 88 pp.

Goryushin, V.A. and Chaplinskaya, S.M. 1968. Finding the viruses lysing blue-green algae, pp. 171–174, in Topacheveskii et al. (editors), *Tsveteniye Vody* (in Russian), Kiev Naukova Dumkoi.

Held, A.A. 1974. Attraction and attachment of zoospores of the parasitic chytrid *Rozella allomycis* in response to host-dependent factors. *Archiv für Mikrobiologie* 95: 97–114.

Huber-Pestalozzi, G. 1946. Der Walensee und sein Plankton. *Zeitung für Hydrologie* 10: 1–200.

Hulsmann, A.D. and Van Rijswijk, P. 1980. Grazing in Lake Maarsseveen, pp. 127–155, in Ringelberg, J. (editor), *Limnological Research in the Maarsseveen Lakes*. Intern Report, University of Amsterdam.

Kalff, J. and Knoechel, R. 1978. Phytoplankton and their dynamics in oligotrophic and trophic lakes. *Annual Review of Ecology and Systematics* 9: 475–495.

Kazama, F.Y. 1972. Ultrastructure of phototaxis of the zoospores of *Phlyctochytrium* sp., an estuarine Chytrid. *Journal of General Microbiology* 71: 555–566.

Koob, D.D. 1966. Parasitism of *Asterionella formosa* Hass. by a chytrid in two lakes of the Rawah Wild Area of Colorado. *Journal of Phycology* 2: 41–45.

Kumar, C.R. 1978. Physiology of infection of the marine diatom *Licmophora* by the fungus *Ectrogella perforans*. *Veröffentlichunger des Instituts für Meeresforschung, Bremerhoven* 17: 1–14.

Luftig, R. and Haselkorn, R. 1967. Morphology of a virus of blue-green algae and properties of its deoxyribonucleic acid. *Journal of Virology* 1: 344–361.

Lund, J.W.G. 1957. Fungal diseases of plankton algae, pp. 19–24, in Horton-Smith, C. (editor), *Biological Aspects of the Transmission of Disease*. Oliver and Boyd, Edinburgh.

Lund, J.W.G. 1965. The ecology of the freshwater phytoplankton. *Biological Reviews of the Cambridge Philosophical Society* 40: 231–293.

Masters, M.J. 1971. The ecology of *Chytridium deltanum* and other fungus parasites on *Oocystis* spp. *Canadian Journal of Botany* 49: 75–87.

Masters, M.J. 1976. Freshwater *Phycomycetes* on Algae, pp. 489–512, in Jones, E.B.G. (editor), *Recent Advances in Aquatic Mycology*. Elek Science, London.

Müller, U. and Sengbusch, P. 1983. Visualization of aquatic fungi (Chytridiales) parasitizing on algae by means of induced fluorescence. *Archiv für Hydrobiologie* 97: 471–485.

Paterson, R.A. 1958. Parasitic and saprophytic phycomycetes which invade planktonic organisms. I. New taxa and records of chytridiaceous fungi. *Mycologia* 50: 85–96.

Pongratz, E. 1966. De quelques champignons parasites d'organismes planctoniques du Léman. *Schweizer Zeitung Hydrologie Rev. Suisse d'Hydrologie* 28: 104–132.

Reynolds, C.S. 1973. The seasonal periodicity of planktonic diatoms in a shallow eutrophic lake. *Freshwater Biology* 3: 89–110.

Reynolds, C.S. 1984. *The Ecology of Freshwater Phytoplankton*. Cambridge Studies in Ecology. Cambridge University Press. 384 pp.

Reynolds, N. 1940. Seasonal variations in *Staurastrum paradoxum* Meyen. *New Phytologist* 39: 86–89.

Safferman, R.S. and Morris, M.E. 1963. Algal virus: isolation. *Science* 140: 679–686.

Shilo, M. 1970. Lysis of blue-green algae by myxobacter. *Journal of Bacteriology* 104: 453–461.

Shilo, M. 1971. Biological agents which cause lysis of blue-green algae. *Mitteilungen der internationale Vereinigung für theoretische und angewandte Limnologie* 19: 206–213.

Soeder, C.J. and Maiweg. D. 1969. Einfluss pilzlicher Parasiten auf unsterile Massenkulturen von *Scenedesmus*. *Archiv für Hydrobiologie* 66: 48–55.

Sommer, U. 1984. Population dynamics of three planktonic diatoms in Lake Constance. *Holarctic Ecology* 7: 257–261.

Sommer, U., Wedemeyer, C., and Lowsky, B. 1984. Comparison of potential growth rates of *Ceratium hirundinella* with observed population density changes. *Hydrobiologia* 109: 159–164.

Sommer, U., Gliwicz, Z.M., Lampert, W., and Duncan, A. 1986. The PEG-model of seasonal succession of planktonic events in freshwaters. *Archiv für Hydrobiologie* 106: 433–471.

Sparrow, F.K. 1960. *Aquatic Phycomycetes*, 2nd edition. University of Michigan, Ann Arbor.

Sparrow, F.K. 1968. Ecology of freshwater fungi, pp. 41–93, in Ainsworth, G.C. and Sussman, A.S. (editors), *The Fungi*, Vol. 3. Academic Press, New York.

Strasburger, E. 1878. Wirkung des Lichtes und der Wärme auf *Schwamsporen. Jena, Zeitung für Naturwissenschaft* 12: 551–625.

Stewart, W.D.P. and Daft, M.J. 1976. Microbial pathogens of cyanophycean blooms, pp. 177–218, in Droop, M.R. and Jannasch, H.W. (editors), *Advances in Aquatic Microbiology.* Academic Press, London.

Sussman, A.S. 1966. Dormancy of spore germination, pp. 41–93, in Ainsworth, G.C. and Sussman, A.S. (editors), *The Fungi,* Vol. 2. Academic Press, New York.

Van Donk, E. 1983. Factors influencing phytoplankton growth and succession in Lake Maarsseveen. *Thesis, University of Amsterdam,* 148 pp.

Van Donk, E. and Kilham, S.S. In press. Temperature effects on silicon-and phosphorus-limited growth and competitive interactions among three diatoms. *Journal of Phycology.*

Van Donk, E. and Ringelberg, J. 1983. The effect of fungal parasitism on the succession of diatoms in Lake Maarsseveen (The Netherlands). *Freshwater Biology* 13: 241–251.

Van Donk, E., Veen, A., and Ringelberg, J. 1988. Natural community bioassays to determine the abiotic factors that control phytoplankton growth and succession. *Freshwater Biology* 20: 199–210.

Wesenberg-Lund, C. 1908. *Plankton Investigations of the Danish Lakes,* Copenhagen.

Weston, W.H. 1941. The role of aquatic fungi in hydrobiology. *In a Symposium of Hydrobiology.* University of Wisconsin Press, Madison.

Wildeman, E. de. 1900. Observations sur quelques Chytridinées nouvelles ou peu connues. *Mém. Herb. Boissier* pp. 1–2.

Wildeman, E. de. 1931. Sur quelques Chytridinées parasites d'algues. *Bull. Acad. Belg. Cl. Sci.* 5: 281–295.

Youngman, R.E., Johnson, D., and Farley, M.R. 1976. Factors influencing phytoplankton growth and succession in Farmoor Reservoir. *Freshwater Biology* 6: 253–263.

6

The Role of Competition in Zooplankton Succession

William R. DeMott

*Department of Biological Sciences
and Crooked Lake Biological Station
Indiana-Purdue University at Fort Wayne
Fort Wayne, Indiana, USA*

6.1 Introduction

Debates over the role of competition in natural communities often consider two broad alternatives. The first view, based on the Lotka Volterra model, is that populations are consistently food-limited, and species that coexist in nature do so by virtue of niche partitioning (Schoener, 1982). In this and related equilibrium models, temporal changes in environmental conditions and poulation sizes are assumed to be unimportant. In contrast, nonequilibrium models emphasize the role of changing conditions in stabilizing species' coexistence (for reviews of nonequilibrium concepts see Chesson and Case, 1986; DeAngelis and Waterhouse, 1987). Most commonly, environmental fluctuations or predation are considered to keep populations at low densities, where exploitative competition is unimportant (Wiens, 1977; Strong, 1986). As pointed out by Hutchinson (1961) in his discussion of the "paradox of the plankton," however, changing environmental conditions could cause shifts in competitive ability, thus promoting species' persistence despite continuous resource limitation and consistently strong competition.

Early studies on zooplankton communities described patterns that are consistent with equilibrium competition theory. For example, Hutchinson (1951) noted that coexisting copepods often differ considerably in body size, presumably an indication of food-size partitioning. Pennak (1957) remarked on the low diversity of zooplankton in Colorado lakes in comparison to the local species pool, inferring that competitive exclusion was a strong force in zooplankton communities. He noted, however, that when congeners coexisted, they usually exhibited population peaks during different seasons of the year.

Circumstantial evidence for strong food limitation among herbivorous zooplankton is both widespread and easy to obtain. Because many freshwater zooplankters carry their eggs until hatching, evidence that food limits reproductive rates can be obtained directly from field samples. For example, clutch sizes in *Daphnia* are typically high during the spring peak in phytoplankton biomass but are much reduced during the remainder of the year (Hebert, 1978). Reductions in fecundity can usually be attributed to food limitation, although selective predation on larger, more fecund females can also be important (Threlkeld, 1979; Gliwicz et al., 1981). Recently, low reserves lipids (Tessier and Goulden, 1982) and decreased slopes in length/body mass relationships (Geller and Müller, 1985; Duncan, 1985) have also been used as indicators of food limitation in field zooplankton populations.

Although zooplankton populations often show symptoms of strong food limitation, they frequently undergo distinct seasonal successions, which confound equilibrium models. There are at least four alternative mechanisms whereby environmental change, predation, competition, or interactions between these factors may alternately favor one species over another, causing species successions. Coexistence may depend upon a balance between competition and predation. For example, in their "size-efficiency hypothesis," Brooks and Dodson (1965) argued that larger species were superior competitors but also more vulnerable to predation by visually-feeding fishes. According to this mechanism, changes in predator abundance or activity would explain the seasonal succession and long-term persistence of a better "escaper" and a superior competitor (Jacobs, 1977a, b). Intense predation could, of course, keep populations below levels where exploitative competition is important (Zaret, 1980; also see Chapter 7).

Even in the absence of predation, changing environmental conditions might prevent competitive exclusion and drive seasonal succession. Two distinct mechanisms have been postulated. As pointed out by Hutchinson (1961), changing environmental conditions could cause reversals in competitive advantage before exclusion occurs. Competitive ability is assumed to vary with environmental factors such as resource productivity, food quality, and temperature. Interspecific competition is continuous and intense, but competitive rankings of species vary through time. Succession could also result from an alternation between density-independent and density-dependent population growth (Koch, 1974; Huston, 1979). In this case, seasonal succession would reflect the replacement of a fast-growing exploiter (*r*-selected species) by a slow-growing but efficient competitor (*K*-selected species).

The fourth and final alternative to resource partitioning focuses on the dynamics of phytoplankton, the major food resource of herbivorous zooplankton, and provides a mechanism for food limitation without competition. With the exception of interference between large *Daphnia* and certain rotifers (Gilbert and Stemberger, 1985), competition among zooplankton is regarded as an indirect interaction mediated through resource exploitation (Levitan,

1987). Thus, competition can occur only when reductions in food quantity or quality are brought about by zooplankton feeding. Studies of phytoplankton, however, have long emphasized the importance of nutrient limitation and physical factors in causing seasonal changes in productivity and composition (see Chapters 2 and 3). In a recent attempt to classify broad categories of resources, Price (1984) described phytoplankton as a "pulsing ephemeral resource" for zooplankton. According to this point of view, food limitation would be a consequence of seasonal changes in resources, which are largely independent of exploitation by grazers. Zooplankton seasonal succession would thus reflect a specialized tracking of phytoplankton succession, with little competition.

Many of the common patterns of zooplankton seasonal succession and proposed mechanisms are summarized in the PEG-model (Sommer et al., 1986). The effects of predation are analyzed in Chapter 7 of this book. The emphasis of this chapter is on mechanisms of interspecific competition. Both niche partitioning and nonequilibrium alternatives are considered. The emphasis on interspecific interactions leads me to ignore many studies that show seasonal patterns of resource limitation in single species populations. On the other hand, studies of food limitation and competition that emphasize differences between lakes are considered here because of their implications for seasonal succession. The evidence for grazer impacts on phytoplankton dynamics, obviously a critical part of an understanding of zooplankton competition, is the focus of Chapter 4. Predatory zooplankton are excluded from consideration. Although some taxa, particularly omnivorous cyclopoids, undergo distinctive seasonal successions (see Hutchinson, 1967), evidence of competition is too limited to warrant discussion.

This review can be divided into two parts. The first part emphasizes the responses of individuals to food concentration, food quality, temperature, and other physicochemical factors under laboratory conditions. The goals of this part are to consider how seasonal events could cause reversals in competitive ability and to evaluate the potential for food resource partitioning. The second part evaluates the nature of competition among zooplankton, using the results of laboratory competition experiments and studies of natural communities.

6.2   Responses to Food Concentration

Threshold food concentrations   The cornerstone of the size-efficiency hypothesis proposed by Brooks and Dodson (1965) is the notion that the rate of food collection and assimilation should increase more rapidly with increased body size than the respiration rate. Thus, larger species were thought to be able to grow and reproduce at lower food concentrations than could smaller species. The evidence originally presented for this hypothesis was scanty and based primarily on size-specific rates of feeding and respiration within species. In their detailed review of the size-efficiency hypothesis, Hall et al. (1976) analyzed the size-dependency of 11 components of population growth. Although their analysis generally supported the notion of a positive relationship between body size and growth efficiency, it also pointed to serious problems in trying to predict population responses from many physiological components. Until recently, the technical difficulties of maintaining a low and constant food level discouraged direct tests of growth under low food conditions.

Lampert (1977a, b, c) used improved radioisotope techniques and a continuous flow culture system to make the first estimates of threshold food levels for freshwater zooplankton. He defined two critical food concentrations:

1. The individual threshold, where metabolic losses are just equalled by assimilation, so that the body mass of the animal remains constant.

2. The population threshold, which allows enough production to compensate for losses due to senescence and predation.

Using this own data on assimilation rates and literature values for respiration, Lampert (1977c) showed that individual threshold values for *Daphnia pulicaria* were sensitive to body size, food species, and temperature. For readily ingested and assimilated algae (e.g., *Scenedesmus*), the threshold for individual growth was about 50 µg C (Carbon)/liter (l) over a temperature range from 10–20°C for animals from 1.0–3.0 mm. Temperatures above 20°C caused sharp increases in threshold concentrations, especially for larger, older individuals. Using the same basic approach with simultaneous estimates of ingestion, assimilation, and respiration, Kersting (1983) directly determined a threshold food concentration for individual *Daphnia magna* of 60 µg C/l of *Chlorella*.

Because growth and reproduction must offset mortality due to senescence, the minimal population threshold concentration is necessarily higher than the individual threshold. The minimal requirement to sustain a population is that the animals grow to maturity and reproduce. More formally, a food-limited population threshold can be defined as the food concentration that allows a zero net population growth rate in the absence of predators. Extending earlier work, Lampert and Schober (1980) estimated the threshold concentration for egg production in continuous flow experiments with *Daphnia pulicaria* and compared their results to an unpublished study by Peters that employed a plankton wheel and *D. rosea*. As expected, thresholds for egg production were slightly higher than individual thresholds. In addition, the threshold for *D. rosea* appeared to be lower than the threshold for the larger species, *D. pulicaria*. The threshold for reproduction for *Daphnia* spp. in Lake Constance during spring

(Lampert, 1978) was about twice the concentration determined in laboratory experiments with *Scenedesmus*, presumably due to the lower food value of ingestible seston (particulate organic carbon, POC <50 μm).

In the absence of direct, between-species comparisons, two theoretical approaches predicted, contrary to the size-efficiency hypothesis, that small size at maturity should be advantageous when food is limiting (Lynch, 1980a; Romanovsky, 1984a, b, 1985). Romanovsky assumed that the exponents relating assimilation and respiration to body size are equal. He then calculated threshold food concentrations, taking into account the relative egg size (egg weight:body weight at maturity) and the maximum duration of juvenile development at near-threshold conditions. Because small cladocerans produce relatively large eggs but are able to prolong juvenile development to almost the same degree as larger species, Romanovsky (1984a) concluded that the juveniles of small species have lower threshold concentrations than juveniles of larger species.

The prediction that smaller species should have lower threshold food levels has been supported in two recent laboratory studies with daphnids. When grown together in the same bottles over a wide range of food concentrations (75–2000 μg C/l of *Scenedesmus*), *Daphnia magna*, *D. pulicaria*, and *D. longispina* exhibited different responses to food concentration (Tillmann and Lampert, 1984). The largest species, *D. magna*, failed to reach reproductive maturity within 14 days at the two lowest food concentrations, whereas the smallest, *D. longispina*, matured within 5–6 days under both high and low food levels. At high food concentrations, however, *Daphnia magna* exhibited the highest size-specific growth rates (μg growth μg body wt^{-1} day^{-1}). The intermediate-sized species, *D. pulicaria*, showed an intermediate response at each concentration. Tessier and Goulden (1987) obtained very similar results in a comparison among juveniles of *Ceriodaphnia reticulata*, *D. pulex*, and *D. magna*. The smallest species, *Ceriodaphnia*, grew fastest at low food concentrations, and largest species, *D. magna*, grew fastest when food was abundant, while *D. pulex* was intermediate. A third study, however, produced contrary results. The largest species (*Daphnia pulicaria*) had the lowest threshold for growth and reproduction; the smallest species (*D. cuculata*) had the highest thresholds; and the intermediate-sized species (*D. hyalina*) was intermediate (Gliwicz and Lampert, unpublished manuscript). Because *D. pulicaria* tends to occur in more oligotrophic lakes, and *D. cucullata* occurs in highly eutrophic lakes, these results may represent adaptations to particular environments rather than a trend with body size.

A tendency for threshold food levels to be lower among smaller species appears to hold for comparisons between calanoid copepods and between rotifers. Direct comparisons between freshwater calanoids of different sizes are lacking; however, comparisons between two well-studied genera of marine calanoids, *Calanus* spp. and *Pseudocalanus* spp., provide strong evidence for body-size related responses to food concentration. These genera are the dominant taxa in both the north Atlantic and the north Pacific, and adults differ in body mass by about an order of magnitude. Data reviewed by Frost (1985) show that the smaller genus, *Pseudocalanus*, reaches its maximum rate of egg production at the same food level (about 60 μg C/l as the threshold for egg production in the larger genus, *Calanus* (also see Vidal, 1980). On the other hand, when food is abundant, the duration of juvenile development is about equal in the two genera, indicating that immature stages of the larger genus can grow about four times faster than immature stages of the smaller genus (Frost, 1980). Moreover, at high food levels, adult *Calanus* can produce about 7–8 times more eggs female^{-1} day^{-1} than *Pseudocalanus*.

The very short generation times of rotifers permit direct determinations of the relation between food concentration and population growth rate using semicontinuous cultures (Stemberger and Gilbert, 1985a). Experiments with eight species of rotifers showed that population threshold food concentrations increased linearly over an 18-fold range in body mass (Stemberger and Gilbert, 1985b). Extreme values were exhibited by members of the same genus (*Keratella cochlearis*, 60 μg dry wt/l; *K. crassa*, 1030 μg dry wt/l; Figure 6.1). Food concentrations that barely supported the growth of *K. crassa* actually inhibited the growth of *K. cochlearis*. Although larger rotifers have higher food requirements, they also exhibit considerably higher population growth rates when food is abundant (Figure 6.2). Thus comparisons among rotifers and among calanoid copepods

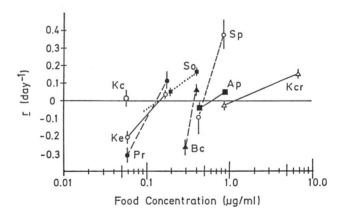

Figure 6.1   Estimates of the threshold food level (food concentration where $r=0$) for rotifers spanning a wide range of body sizes. The species, from smallest to largest, are: Kc, *Keratella cochlearis*; Pr, *Polyarthra remata*; Ke, *Keratella earlinae*; So, *Synchaeta oblonga*; Kcr, *Keratella crassa*; Sp, *Synchaeta pectinata*; Bc, *Brachionus calyciflorus*; and Ap, *Asplanchna priodonta* (after Stemberger and Gilbert, 1985b).

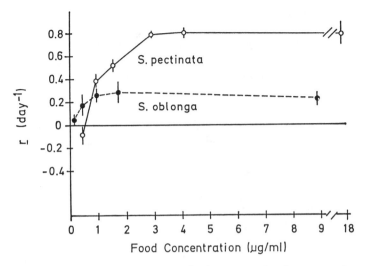

Figure 6.2   Population growth rates (r) vs. food concentration for two species of *Synchaeta*. The smaller species, *S. oblongata*, is favored at low food concentrations (lower threshold for growth) whereas the larger species, *S. pectinata*, is favored at high food concentrations (higher r_{max}; after Stemberger and Gilbert, 1985b).

show trends for smaller species to have lower food thresholds, but for larger species to have higher r_{max} values.

The hypothesis that closely related species of similar size can differentiate along an r–K gradient provides an important contrast with the notion of size-related food thresholds. Hrbáčková and Hrbáček (1978) grew *Daphnia pulex* and *D. pulicaria* with various dilutions and concentrations of natural seston and found that *D. pulicaria* grew faster than *D. pulex* at low food concentrations. *D. pulex* was favored at higher food levels. Previous work had already shown that *D. pulex* was unable to survive with natural summer densities of natural seston from Slapy Reservoir, whereas *D. pulicaria* and the native *D. hyalina* survived and reproduced under the same conditions (Hrbáčková-Esslova, 1963). These results led Hrbáček (1977) to conclude that *D. pulex* and *D. pulicaria*, despite similarities in size and morphology, represent opposite extremes within the genus *Daphnia* in terms of adaptation to high and low food levels. Additional comparisons of the growth rates of species and clones of *Daphnia* from oligotrophic and eutrophic lakes provide further support for the hypothesis that cladocerans may show adaptations to food conditions in their native environments (Hrbáčková and Hrbáček, 1979). Various clones of *Daphnia galeata* from productive and unproductive lakes differed more in their responses to food conditions than did *D. galeata*, *D. pulicaria*, and *D. hyalina* from similar environments. In contrast, a recent chemostat experiment with *D. pulex* and *D. pulicaria* failed to find significant differences in their responses to high and low concentrations of *Scenedesmus* (Taylor, 1985). The low food level, 100 μ C/l, may have been too high to cause strong food limitation in either species.

Some evidence also exists for tradeoffs between efficiency at low food levels and maximal growth rates for calanoid copepods of similar size. For example, Elmore (1983) found that *Diaptomus dorsalis*, a common species in eutrophic lakes in Florida, developed more rapidly at high food levels than *D. floridanus*, despite having smaller eggs. *D. floridanus*, however, a species found in less productive lakes, matured faster and exhibited better survivorship at low food levels than did *D. dorsalis*.

How do threshold concentrations vary between cladocerans, copepods, and rotifers? Muck and Lampert (1984) reported a threshold concentration for individual growth of about 35 μg C/l for *Eudiaptomus gracilis*, whereas *Daphnia* spp. rapidly lost weight at the same food levels. Both freshwater (Williamson et al., 1985; Piyasiri, 1985) and marine (Harris and Paffenhöfer, 1976; Vidal, 1980) calanoids can grow and reproduce at food concentrations $\leq$50 μg C/l, which are values near the threshold for individual growth in daphnids of similar size. Thus calanoids appear to have consistently lower threshold food levels than daphnids. Because copepods mature much more slowly than cladocerans, this result is consistent with the notion that taxa with lower individual growth rates should exhibit lower threshold concentrations (Romanovsky, 1984a). Despite their small sizes, rotifers exhibit a very broad range of threshold food concentrations, which bracket the extremes exhibited by calanoids and cladocerans. The smallest rotifer tested by Stemberger and Gilbert (1985b), *K. cochlearis*, showed positive population growth at food levels similar to the lowest observed for calanoid copepods (60 μg dry wt/l $\simeq$ 25 μg C/l). The four largest species, however, exhibited thresholds of 380–1030 μg dry wt/l; considerably higher than the thresholds for larger cladocerans. Thus, while threshold food concentrations show some general trends with body size among related taxa, these trends break down when comparisons are made across cladocerans, copepods, and rotifers.

Resistance to starvation   When two or more species compete for a single limiting resource under steady state conditions, the species with the lowest threshold concentration should always win (Tilman, 1982). In a fluctuating environment, however, responses to food levels well below and above the threshold level may become important. Such fluctuations in food supply are often caused by oscillations in zooplankton population density (McCauley and Murdoch, 1987). When food levels drop below the threshold level, survival will depend upon an ability to withstand starvation. Threlkeld (1976) developed a model to predict time to starvation based upon size-specific rates of respiration. He calibrated the model by regressing body weight versus respiration rate, using literature data for freshwater and marine crustacean zooplankton. The predicted size-specific survival times were based on four assumptions:

1. Individuals die after losing 50 percent of their body weight.
2. The respiration rate during starvation falls to 50 percent of the rate estimated from literature values.
3. Zooplankton metabolize protein during starvation.
4. Zooplankton do not allocate energy to growth and reproduction during starvation.

Because larger animals have lower weight-specific metabolic rates, the model predicted that larger size should confer greater resistance to starvation. The model, which predicted approximately a doubling of the time to starvation with a twenty-fold increase in weight, tended to overestimate survival times taken from the literature (mainly for daphnids and marine calanoids). Correcting for mortality due to old age, however, appeared to account for much of the difference (Threlkeld, 1976).

Despite the relative ease with which survival in the absence of food can be measured, there have been relatively few estimates of starvation times since Threlkeld's analysis (1976). Although they confirm general body-size trends, these studies point to important effects of the allocation of energy reserves and species-specific differences. Tessier et al. (1983) found that adult daphnids accumulated considerable energy reserves in the form of lipids during the intermolt period and then invested most of this reserve in producing eggs. Continued reproduction during starvation, particularly when starvation began late in the molt cycle, resulted in decreased survival times. Goulden and Hornig (1980) noted a difference in reproductive allocation between two cladocerans; *Daphnia galeata mendotae* continued to reproduce at low food levels, whereas *Bosmina longirostris* exhibited less investment in reproduction under the same conditions. They argued that the more conservative strategy of *Bosmina* was probably important in explaining the persistence of this smaller species during conditions of scarce food.

One important consequence of size-specific starvation rates is that juveniles, especially neonates, should be more vulnerable to starvation than adults. Studies with both cladocerans (e.g., Tessier et al., 1983) and calanoid copepods (e.g., Williamson et al., 1985) suggest that immature stages are more vulnerable to starvation. The energy reserves of cladoceran eggs and neonates, however, vary considerably depending on both the food regime experienced by the mother and on the cladoceran species. Tessier et al. (1983) reported that *D. magna* grown under low food conditions produced neonates with low lipid reserves and with survival times under starvation conditions less than half those of offspring from well-fed mothers. Although well-fed *D. magna* and *D. galeata mendotae* both produced eggs that contained about 20 percent lipid, newborn *D. magna* had considerably higher lipid reserves (Goulden and Henry, 1984; Goulden et al., 1987). Presumably, *D. galeata mendotae* utilized a greater proportion of its energy reserves during embryonic development. The neonates of a third, much smaller cladoceran, *Bosmina longirostris*, particularly those born to low-food mothers, appeared to have much lower lipid reserves than either *Daphnia* species.

Two studies on calanoid copepods further illustrate responses to starvation that cannot be explained solely by body size. Experiments with three genera of marine calanoids of similar size showed that adults of two genera (*Acartia* and *Centropages*) starved within 6 to 10 days, whereas most individuals of a third genus, (*Pseudocalanus*), survived for more than 15 days (Dagg, 1977). Nauplii of both *Pseudocalanus* and *Acartia* survived only about 48 hours without food. Adults of a larger species, *Calanus finmarchicus* exhibited little mortality during 20 days of starvation. In contrast, Burns (1985) found no differences in resistance to starvation among adults of a large species of *Boekella*, which inhabits eutrophic ponds in New Zealand, and two smaller congeners, which inhabit oligotrophic lakes. The nauplii of the smaller species, however, survived about twice as long without food as did the nauplii of the larger species, despite smaller size at birth. The median survivorship of the most resistant nauplii, those of *B. dilatata*, was 16.9 days; longer than the survivorship of the adults under starvation, and more than four times longer than predicted by Threlkeld's (1976) model. Even though the pond species, *B. triarticulata*, produces relatively small eggs, it matures most rapidly when food is abundant (Jamieson and Burns, 1985). This characteristic suggests that starvation tolerance has been sacrificed in favor of fast growth and development.

Only very limited data are available on starvation times of rotifers. *Keratella cochlearis* showed a median starvation time of about 48 hours (Gilbert, 1985), within the range predicted by Threlkeld's model. Thus, despite some interesting exceptions, available evidence points to the importance of body size in determining resistance to starvation among diverse taxa.

Intrinsic rate of increase, r_{max}   Whereas threshold concentrations and resistance to starvation involve responses to low food conditions, r_{max} is the rate of increase when food is abundant. Over a very broad range of body sizes, smaller organisms tend to have higher r_{max} values because of shorter generation times (Smith, 1954). Among zooplankton, however, differences in reproductive mode and the tendency for larger species to produce larger clutches can offset the more rapid maturation of smaller taxa. According to the classic

r–K dichotomy (MacArthur and Wilson, 1967), high r_{max} values are associated with opportunistic species that increase quickly when resources are abundant but fare poorly when resources are scarce. High r_{max} values, however, could also lead to close tracking of resource abundance, consistently strong food limitation, and more rapid competitive displacement (e.g., Huston, 1979; Wiens, 1984). In contrast, slower growing species may be limited to densities well below levels where resource depletion is significant, because of occasional "ecological crunches" (Wiens, 1977). Thus, in a fluctuating environment, species with high r_{max} values have greater potential to reach densities at which interspecific and intraspecific competition become important.

Allan's (1976) review of life history patterns pointed out distinct differences among cladocerans, copepods, and rotifers in their maximum rates of increase. At one extreme, some rotifers are capable of doubling their population size in less than two days; at the other extreme, copepods in some northern lakes produce only one generation per year. Copepods exhibit obligate sexual reproduction and must molt through six naupliar instars and five copepodite instars before reaching sexual maturity. Though lifetime fecundity can be high, it cannot offset the effect of slow development and sexual reproduction in reducing the population rate. Within the calanoids, large species mature at about the same rate as smaller species and often produce larger clutches (Allan and Goulden, 1980). This trend suggests that r_{max} may increase with body size among copepods.

In contrast to copepods, cladocerans and rotifers mature quickly and typically reproduce parthenogenetically. The ability of smaller cladocerans to mature more rapidly tends to offset the larger clutch sizes of larger species. Thus, r_{max} values are similar over a wide span of cladoceran sizes (Allan and Goulden, 1980; Lynch, 1980b), although larger species may tend to exhibit slightly higher rates of increase (Goulden et al., 1978; Romanovsky, 1985). Exceptions to general patterns include chydorids, which produce a maximum of two eggs per clutch and thus exhibit low r_{max} values (Allan and Goulden, 1980); and *Moina*, which matures very quickly despite producing large clutches of small eggs and thus shows unusually high r_{max} values (Romanovsky, 1985).

As mentioned earlier, Stemberger and Gilbert (1985b) have recently shown a strong trend for increasing r_{max} values with increasing body size among rotifers. Large rotifers such as *Brachionus calyciflorus*, *Synchaeta pectinata*, and *Asplanchna priodonta* exhibit r_{max} values much higher than crustacean zooplankton; whereas small rotifers such as *Keratella cochlearis* and *Synchaeta oblongata* exhibit r_{max} values similar to those of cladocerans. One species, *Keratella crassa*, exhibits a high food threshold and a relatively low r_{max} value. Stemberger and Gilbert (1985b) noted that this species has an exceptionally robust lorica, which presumably provides protection from invertebrate predators.

Departures from stable age structure can, under some circumstances, decrease the relevance of life table estimates of r_{max} in fluctuating environments. For example, in Lake Mitchell, Vermont, *Daphnia rosea* and *Bosmina longirostris* hatched synchronously from resting eggs in early May (DeMott and Kerfoot, 1982). The ability of *Bosmina* to reproduce at an earlier age, however, allowed it to increase more rapidly. Similarly, the ability of some copepods to overwinter in late copepodite or adult stages may allow them to increase in numbers during spring blooms much more rapidly than would be predicted from life table estimates of r_{max}.

The resource variance hypothesis   Success when food is limiting may involve two quite distinct factors: the threshold food concentration and resistance to starvation when food levels are below the threshold. Indeed, within each of the three major taxa of freshwater zooplankton, these two factors are often negatively correlated. Smaller species tend to exhibit lower thresholds levels, whereas larger species are more resistant to starvation. Because large rotifers, which are smaller than most cladocerans and copepods, require high threshold levels, the trend breaks down when comparisons are made across rotifers, cladocerans, and calanoids. Nonetheless, size trends may be consistent and strong enough to allow some generalizations about the effects of resource productivity and variability on zooplankton competition, particularly within the major taxonomic groups (see Table 6.1 for a summary).

Romanovsky and Feniova (1985) recently described a new model of body-size-dependent competition among cladocerans. At both high and low rates of resource supply (i.e., high and low primary productivity), resource dynamics are assumed to be controlled by grazers. When primary productivity is low, the smaller species dominates resource dynamics and may be able to exclude the larger species by depressing resources below the larger species' threshold level (Figure 6.3a). When primary productivity is initially high, however, the larger species increases rapidly to very high densities. Because resources are depressed well below the threshold levels of both species, the

Table 6.1   The responses of zooplankton to food abundance*

	Between major taxa	Size trends within major taxa			Overall size trend
		Rotifers	Cladocera	Copepods	
Threshold concentration	Cl>Cop	L>S	L>S	L>S	No
Starvation resistance	Cop≥Cl>R	L>S?	L>S	L>S	Yes
r_{max}	R≥Cl>Cop	L>S	L≥S	L>S?	No

*Summary of general trends with body size (L, large body size; S, small body size) between and within the three major taxonomic categories of freshwater zooplankton (R, rotifers; Cl, cladocerans, Cop, calanoid copepods). The strengths of these trends and specific exceptions are discussed in the text; relationships which have not been directly tested are indicated by a question mark (?).

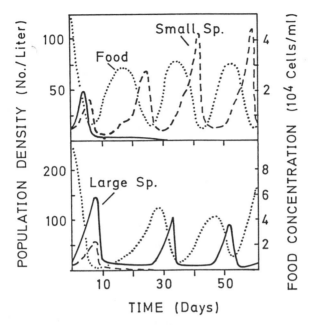

Figure 6.3 Computer simulation of competition between a small zooplankton species with a lower threshold for growth (dashed line) and a larger species with greater resistance to starvation (solid line). When the rate of food supply is low, as in the upper graph, the small species wins by depressing the average food concentration below the threshold value for the larger species. When the rate of food supply is high, lower graph, the larger species depresses the food concentration well below the thresholds for both species and the smaller species starves first (after Romanovsky and Feniova, 1985).

larger species' greater resistance to starvation provides an advantage over the smaller species (Figure 6.3b). The specific formulations of Romanovsky and Feniova (1985) emphasize age class interactions. Adults of the larger species are most resistant to starvation and therefore favored by strong oscillations in resource abundance, whereas juveniles of the larger species are most vulnerable to starvation when resources approach low steady state concentrations. Because controlling resource dynamics is a critical aspect of competitive ability in the model, it also appears that initial species densities and relative r_{max} values could be important in determining the outcome of competition. Presumably, some intermediate range of productivity permits coexistence.

Sommer (1985 and Chapter 3) has demonstrated the effects of fluctuating resources on competition between algae in laboratory chemostat experiments with natural phytoplankton. Pulsed additions of inorganic nutrients increased the number of coexisting species but also resulted in the competitive exclusion of species that dominated under steady-state conditions. Nutrient pulses favor phytoplankton species that efficiently acquire and store temporarily abundant resources for future growth. The zooplankton model differs in two ways:

1. Variation in resources in generated endogenously by oscillations in zooplankton abundance.
2. Species differences in mortality rates as well as growth rates are important.

Nonetheless, both circumstances fall within Levins' (1979) theoretical development of the resource variance hypothesis.

Romanovsky and Feniova (1985) primarily sought to explain how competitive dominance could vary between lakes that differ in productivity. How could the resource variance hypothesis apply to seasonal successions of zooplankton? The most severe oscillations in phytoplankton abundance often occur in mesotrophic and eutrophic temperate lakes during late spring when dense populations of large grazers, mainly *Daphnia*, can severely overgraze phytoplankton causing a "clear water phase" (e.g., Lampert et al., 1986). According to the resource variance hypothesis, the probability of smaller species being competitively excluded would be highest at this time. A tendency for decreased productivity of edible algae during summer, and dampened oscillations in zooplankton abundance could later shift the competitive advantage to smaller species. In their recent review, McCauley and Murdoch (1987) noted that the dynamics of *Daphnia* and edible phytoplankton were characterized by strong cyclical oscillations in some lakes and relatively stable equilibria in others. According to the resource variance hypothesis, the diversity and seasonal dynamics of other coexisting zooplankton would be quite different in different lake types.

6.3   Responses to Food Quality

Dietary breadth and feeding mode   The above analysis of food thresholds, starvation times, and maximum rates of growth is based on the implicit assumption that planktonic grazers exhibit complete overlap in diet. Seasonal succession in the phytoplankton, however, results in significant changes in the size distribution and nutritional value of potential foods. In turn, species differences in feeding modes and diet give rise to a variety of new, often conflicting possibilities. For example, as pointed out by Brooks and Dodson (1965), larger species may gain a competitive advantage if they are able to ingest larger particles while still feeding effectively on small particles. On the other hand, specialization by large grazers on larger particles and by small grazers on smaller particles would decrease the intensity of interspecific competition (Hutchinson, 1959; Wilson, 1975). Finally, species with broad diets and nonselective feeding modes may actually be at a disadvantage when particles that are difficult to ingest, poor in food value, or toxic interfere with feeding on more nutritious particles.

What is the potential for food resource partitioning among suspension-feeding zooplankton? Over a wide range of body sizes, it seems reasonable to expect larger zooplankton to exhibit larger upper size-limits of particles ingested. The often-cited linear relationship between cladoceran body length and the maximum size of plastic bead ingested (Burns, 1968) provides a good starting point. Because most large algae are both nonspherical and breakable, however, experiments with plastic spheres underestimate abilities to ingest large algae. For example, gut analysis shows that *Daphnia* can ingest filamentous green algae several hundred micrometers in length (Nadlin-Hurely and Duncan, 1976), and radiotracer studies indicate substantial feeding on single filaments and small colonies of the cyanobacterium *Aphanizomenon* (Lampert, 1981; Holm et al., 1983). Some large net algae such as *Staurastrum* may occasionally be swallowed by *Daphnia* after being broken in half (Infante, 1973), and cananoid copepods readily bite off parts of large net diatoms (Vanderploeg et al., 1988).

Although limits on the largest particles ingested are related to grazer body size, it is not necessarily the case for lower size limits. Geller and Müller (1981) pointed out that an ability to feed upon very small particles should depend upon the structure of the feeding appendages, and that many larger species have relatively fine filter setules. They hypothesized that the ability of many cladocerans to feed on solitary bacteria could help explain their frequent dominance over calanoid copepods during warmer seasons, when bacteria are often more abundant. Moreover, they hypothesized that mid-summer peaks in *Diaphanosoma brachyurum* and *Chydorus sphaericus*, two cladocerans with very fine filter setules, were due to a greater ability to feed upon bacteria. Several studies using radioactively labeled bacteria and picoplankton (Bogdan and Gilbert, 1984; DeMott, 1985; Brendelberger, 1985; Schoenberg and MacCubbin, 1985) or bacteria-sized microspheres (Gophen and Geller, 1984; Hessen, 1985) have verified that the ability of cladocerans to capture and ingest very small particles is correlated with measurements of the filtering setules and unrelated to body size among the Daphniidae (*Daphnia, Ceriodaphnia,* and *Simocephalus*) and Sididae (*Diaphanosoma, Sida, Holopedium,* and *Pseudosida*). Some small cladocerans, however, including *Chydorus sphaericus* and *Bosmina longirostris,* feed much less effectively on bacteria than would be predicted from their filter mesh sizes (DeMott, 1985). The ability of large *Daphnia* to feed effectively on particles ranging from bacteria to net algae means that the diets of smaller grazers will be a subset of the diet of *Daphnia.* Given this form of inclusive niche overlap, stable coexistence requires that the smaller, specialist species be more adept at collecting their subset of available resources (Miller, 1967).

Arguments for resource partitioning among suspension-feeders have recently been strengthened by evidence that many nondaphnids are adept at selecting between individual food particles (see review by Price, 1988). For example, high speed movies have demonstrated that calanoid copepods can detect and actively capture individual particles (reviewed by Koehl, 1984). The ability of larger calanoids to ingest larger particles combined with a lower size limit for active capture by smaller species could lead to significant particle-size partitioning among coexisting calanoid copepods (Price and Paffenhöfer, 1985).

Direct evidence for particle-size partitioning among coexisting calanoids is, however, very limited. In one example, gut analysis showed that the diet of the larger species, *Diaptomus shoshone*, was restricted to relatively large particles, which were rarely ingested by the smaller species, *Diaptomus coloradensis* (Maly and Maly, 1974). This comparison, however, involves one primarily herbivorous species and one largely predatory species. Vanderploeg et al. (1988) show that *Diaptomus sicilis* handles and ingests large diatoms more effectively than does its smaller congener *D. ashlandi*. There is also some evidence for differing particle-size preferences among calanoid copepods of similar body size. Richman et al. (1980) found that *Eurytemora affinis* tended to specialize on larger particles whereas several, similar-sized coexisting species of *Diaptomus* fed more intensively on smaller particles. On the other hand, Harris (1982) found almost complete overlap in the size distributions of particles ingested by *Pseudocalanus* and *Calanus*, marine copepods that differ by an order of magnitude in body mass.

The ability of suspension-feeding copepods to detect, capture, handle, and ingest large prey such as elongate diatoms (Vanderploeg et al., 1988) and rotifers (Williamson and Butler, 1986) suggests that calanoids can feed more effectively on large food items than can cladocerans of comparable size. In contrast with cladocerans, calanoid copepods show negligible feeding rates on bacteria and very low feeding rates on picoplankton (e.g., Muck and Lampert, 1984; Bogdan and Gilbert, 1987). Thus, cladocerans and copepods differ significantly in their particle-size preferences.

Most studies of copepod feeding have focused on adults, particularly adult females. The lack of work on juveniles represents a serious gap in knowledge because nauplii and copepodites are sensitive to food limitation and show broad overlap in body size between species. Interestingly, the radiotracer studies of Bogdan and Gilbert (1987) showed that nauplii of *Diaptomus minutus* were restricted to feeding on larger particles than adults.

Bogdan and Gilbert (1987) have used principle components analysis to divide a taxonomically diverse group of zooplankton into feeding guilds based on ability to ingest bacteria and small algae (<4 μm diameter). As in comparisons within the Cladocera, ability to feed on small particles was not correlated with body size and was only partially predictable from taxonomy. For example, some rotifers fed effectively on bacteria, small algae, and larger flagellates (e.g., *Keratella cochlearis*), whereas others specialized on medium-size flagellated algae, particularly *Cryptomonas* spp. (*Synchaeta* and *Polyarthra*; also see Gilbert and Bogdan, 1984). The rotifer genera *Keratella* (Gilbert and Bogdan, 1984) and *Brachionus* (Rothhaupt, 1988) provide good examples of particle-size partitioning between small and large congeners.

Although most studies of zooplankton diets have emphasized selection on the basis of particle size, recent studies show that many taxa of freshwater zooplankton can use chemical cues to choose between particles that differ in nutritional value. Whereas daphnids and *Diaphanosoma* did not discriminate between algal-flavored and untreated microspheres, the cladoceran *Bosmina*, calanoid and cyclopoid copepods, and several taxa of rotifers strongly discriminated against untreated microspheres (DeMott, 1986). These differences in diet are associated with fundamental differences in feeding mechanisms. Although studies of the implications of "taste discrimination" for dietary overlap are still in an early stage, it appears that considerable resource partitioning on the basis of food quality may occur among selective and non-discriminating taxa. For example, recent studies with freshwater and marine calanoid copepods (DeMott, 1988a, b; 1989) demonstrate abilities to select between flavored beads and live algae, live and dead algae of the same species, digestible and digestion-resistant algae (gelatinous green algae), and edible and toxic algae (a toxic strain of *Microcystis*). Moreover, as predicted by optimal foraging theory (Lehman, 1976), copepods were most selective when edible algae were abundant but broadened their diets to include more dead and digestion-resistant algae when high-quality food was scarce. Although all four species discriminated between algae and detritus, *Eudiaptomus* and three coexisting marine calanoids (*Arcartia* spp., *Temora longicornis*, and *Pseudocalanus* sp.) exhibited differences in selectivity, which reduced dietary overlap (DeMott 1988a, b).

An ability to use chemical cues to discriminate between individual particles should be especially advantageous when inert or low quality particles are abundant. For example, G.-Toth et al. (1987) reported that two species of *Daphnia* in Lake Balaton, Hungary, fed primarily on suspended mineral particles and contained only 2.4–5.5 cells of algae per gut, whereas the guts of co-occurring *Eudiaptomus* contained 19.6 algae on the average, and must less inert material. Under these circumstances, *Daphnia* may be gaining nutrition from organic chemicals adsorbed on the surface of mineral grains (Arruda et al., 1983), whereas *Eudiaptomus* may obtain most of its nutrition from less numerous living algae.

Vulnerability to interfering particles   Seasonal succession of phytoplankton in eutrophic lakes often leads to blooms of cyanobacteria and dinoflagellates, which are too large to ingest, are low in nutritional value, or are even toxic. These changes in the composition of the phytoplankton are often correlated with the decline of large, generalist feeders, particularly daphnids, and an increase in the relative or absolute abundance of rotifers, copepods, and small cladocerans (e.g., Gliwicz, 1977; Richman and Dodson, 1983; Threlkeld, 1985). Jarvis (1986) showed a significant decline in the in situ filtering rates of *Daphnia pulex* as the abundance of *Microcystis* increased in a South African lake. The filtering rate of *Ceriodaphnia*, which dominated during the period of high *Microcystis* abundance, appeared to be less sensitive to interfering algae. Several laboratory studies have shown that filamentous cyanobacteria lead to reduced feeding rates and increased respiration rates for cladocerans, and that larger species are more strongly affected (Webster and Peters, 1978; Porter and McDonough, 1984; Dawidowicz et al., in press). Recently, Gliwicz and Lampert (unpublished manuscript) have demonstrated that the addition of a filamentous cyanobacterium to laboratory cultures caused a reversal in the rank order of threshold food concentrations, favoring a small species (*Daphnia cucullata*) over a larger congener (*D. pulicaria*).

In experiments with natural seston and plastic beads, Gliwicz (1977) showed that the filtering rates of cladocerans declined during summer blooms of net algae (primarily filamentous cyanobacteria and dinoflagellates), and that the declines were sharpest for larger particles. The filtering rates of two small cladocerans, *Diaphanosoma brachyurum* and *Chydorus sphaericus*, were least affected by interfering algae, and these two species tended to increase in abundance during midsummer cyanobacterium blooms. Decreased filtering rates were hypothesized to be a result of both increased rejection of noningestible particles using the post-abdominal claw and a narrowing of the gap between valves of the carapace. Subsequent laboratory experiments showed that large daphnids exhibit both of these behaviors in response to increased densities of interfering net algae (Gliwicz, 1980; Gliwicz and Siedlar, 1980). More recently, Fulton and Paerl (1987a) found that the filtering rate of *Diaphanosoma brachyurum* was less affected by the presence of colonies of the cyanobacterium *Microcystis* than other cladocerans. Whereas cladocerans readily ingested small to moderately large colonies of *Microcystis*, calanoid copepods ingested very little of the cyanobacterium. Its presence had little or no influence on feeding rates of calanoids on an edible flagellate (Fulton and Paerl, 1987a). This result appears to be another example of the use of taste to select between algae that differ in food value.

In addition to physical interference, some strains of cyanobacteria are toxic, leading to decreased filtering rates and survival rates below those observed in the complete absence of food. For example, Lampert (1981) reported that the ingestion of a single-cell form of *Microcystis aeruginosa* led to reduced filtering rates and increased mortality of *Daphnia pulicaria*. Porter and Orcutt (1980) found that a strain of *Anabaena flos-aquae* caused mortality in *Daphnia magna*, and Fulton and Paerl (1987b) documented that toxicity of a stain of *Microcystis aeruginosa* that they isolated from the Neuse River, North Carolina to several cladoceran species from the same site. Many other species and strains of cyanobacteria are nontoxic but are poor in food value, either because they are poorly assimilated (e.g., Arnold, 1971; Holm et al., 1983) or are nutritionally inadequate (Lampert, 1977a). In addition to physical inhibition and toxicity, some cyanobacteria may exert allelopathic effects on edible algae. Thus, even in carefully designed laboratory experiments, it can be difficult to determine whether the adverse effects of cyanobacteria on zooplankton growth are due to physical inhibition, toxicity, or the scarcity of nutritious algae (Infante and Abella, 1985). Recently, Hanazato and Yasuno (1987a) found that fresh *Microcystis aeruginosa* inhibited the growth of *Ceriodaphnia*, whereas decaying *Microcystis* was a good source of nutrition.

Although cyanobacteria seem to be poor food for crustacean zooplankton, it is not necessarily the case for rotifers. In one interesting example, Starkweather and Kellar (1983) found that *Brachionus calyciflorus* readily grew on a strain of *Anabaena flos-aquae* that is toxic to *Daphnia*. More recently, Fulton and Paerl (1987a) noted that the same rotifer flourished during blooms of the cyanobacterium *Microcystis*. Nonetheless, the growth of *Brachionus* can be extremely sensitive to nutritional quality. Scott (1980) showed that altering the growth rate of the flagellate *Brachiomonas subtilis* altered its proportion of carbohydrates, lipids, and proteins, which in turn influenced the growth rate of the marine rotifer (*Brachionus plicatilis*) that fed upon it.

Differential utilization following ingestion   Even nonselective feeders could be differentially affected by changes in resource composition if they differ in their abilities to digest and assimilate various foods. For example, Infante and Litt (1985) found significant differences between two species of *Daphnia* in their abilities to grow and reproduce on ten species of phytoplankton commonly found in Lake Washington. Because *Daphnia* are relatively nonselective feeders and most algal species were of a readily ingested size, assimilation efficiencies or nutritional sufficiency probably varied between the *Daphnia* species. In a second example, Porter (1975) found differences between two species of *Daphnia* in their ability to break apart and digest colonies of the gelatinous green algae *Sphaerocystis*. DeMott (1983) provided evidence that differing abilities to utilize digestion-resistant green algae influenced the seasonal succession of two species of *Daphnia* in a Vermont lake (see Section 6.7 below).

Little is known about the mechanisms underlying differences in utilization following ingestion. Studies with marine copepods suggest that changes in gut enzymes both between zooplankton species and seasonally within species could be important (e.g., Mayzaud and Poulet, 1978; Mayzaud and Mayzaud, 1985). Laboratory experiments may overestimate nutritional problems if phytoplankton species that are inadequate as the sole source of nutrition contribute to growth when part of a mixed diet (e.g., Taub and Dollar, 1968).

6.4   Effects of Temperature and Other Physicochemical Variables

Species-specific responses to temperature and other environmental variables (e.g., O_2, pH) could drive zooplankton succession. As discussed by Hebert (1978), suboptimal physical conditions probably affect zooplankton fitness by depressing feeding rates or by otherwise adversely influencing the balance between energy intake and metabolic demands. Studies of zooplankton growth illustrate the expected interaction between food levels and temperature. When food is abundant, growth rates increase with increasing temperatures; however, food levels that support growth at low temperatures may be inadequate at higher temperatures (e.g., Neill, 1981a; Orcutt and Porter, 1984). Both Lynch (1977) and Lampert (1977c) pointed out that the metabolic requirements of *Daphnia pulex* increase more rapidly than feeding rates at temperatures above 20°C, and that large individuals are more adversely affected than small individuals. Thus, threshold food concentrations can be quite sensitive to temperature, particularly at extremes. Species differences in temperature optima should not be considered an alternative to competition as an explanation for seasonal succession. Rather, as shown by laboratory experiments with algae (Tilman et al., 1981) and Cladocera (e.g., Loaring and Hebert, 1981; Bengtsson, 1987; also see Section 6.6 below), differing temperature adaptations provide a mechanism for seasonal reversals in competitive ability.

Despite an abundance of studies on the effects of temperature on zooplankton feeding, growth, and development, there are surprisingly few critical comparisons between zooplankton species. In one example, Burns (1969) found that the filtering rates of adult *Daphnia schodleri* and *Daphnia pulex* were similar to each other and relatively constant from 15°C to 25°C. In contrast, two congeners, *D. magna* and *D. galeata mendotae*, appeared to be better adapted to higher temperatures. In these species, filtering rates increased with increasing temperature and at 20°C and 25°C were more than twice the values measured at 15°C. Bengtsson (1986) found several significant interactions between temperature and food level, temperature and species, and food level and species in laboratory life table experiments with *Daphnia magna*, *Daphnia pulex*, and *Daphnia longispina*. Subsequent laboratory competition experiments verified that both temperature and food supply influence the outcome of competition between these species (see Section 6.6 below). In another comparative study, Allan (1977) found that *Bosmina longirostris* was able to reproduce at temperatures from 0–30°C, whereas the reproduction of *Daphnia ambigua* and *Ceriodaphnia quadrangula* was restricted to temperatures above 4° and 8°C, respectively. Seasonal succession at the study site (Frains Lake, Michigan) was in accord with the temperature data; *Bosmina* and *Daphnia* reproduced during winter, whereas *Ceriodaphnia* appeared in late spring and was abundant during summer. Thermal adaptations were less reliable as predictors of seasonal succession in a eutrophic Japanese lake (Hanazato and Yasuno, 1985). Eggs of two common summer species, *Moina micrura* and *Diaphanosoma brachyurum*, developed more rapidly at high temperatures (25°C and 30°C) than did eggs of *Daphnia longispina, Bosmina fatalis,* and *Bosmina longirostris*. Because the two species of *Bosmina* showed virtually identical thermal responses, however, their midsummer succession was hypothesized to be a result of changes in food resources, not temperature (also see Hanazato et al., 1984; Hanazato and Yasuno, 1987b).

Recent experiments have demonstrated extreme thermal differentiation among electrophoretic clones of *Daphnia magna* occupying a shallow lake (Carvalho, 1987; Carvalho and Crisp, 1987). "Summer clones" showed maximal reproductive rates at 20–30°C, whereas "winter clones" were unable to survive at 25°C and above, despite high food levels and opportunities for gradual acclimation. "Winter clones" reproduced from 5–20°C, with a distinct optimum at 15°C, whereas "summer clones" were unable to survive at 5°C. "Spring" and "autumn" clones were intermediate, with reproductive maxima at 20°C. Such high interclonal variability certainly complicates the demonstration of interspecific differences.

Coexisting clones of *Daphnia* also exhibit different tolerances to oxygen stress (Weider and Lampert, 1985). Because dissolved O_2 (e.g., Heisey and Porter, 1977) and pH (Kring and O'Brien, 1976) influence feeding rates, seasonal changes in these variables could also cause reversals in the outcome of competition.

Daphnia populations in shallow lakes and ponds often exhibit high clonal diversity and seasonal clonal successions (e.g., Hebert and Crease, 1980), whereas *Daphnia* in large lakes exhibit low clonal diversity and relatively high stability of clonal frequencies (Mort and Wolf, 1985). Mechanisms that appear to cause successions between clones in ponds may result in successions between pairs of congeners in larger, more stable environments.

6.5 An Overview of Competition Experiments

Competition experiments are the most direct way to examine the role of competition in zooplankton succession. Thus an excellent review of competition experiments among Cladocera (Bengtsson, 1987) provides a good starting point for an examination of the evidence for competitive interactions. Bengtsson's (1987) review includes 20 studies testing competition among 37 species combinations in the laboratory or field. By definition, competition experiments require both single species control treatments and treatments with various combinations of potentially competing species. Because individual animals must be sorted and counted, competition experiments are usually limited to laboratory vessels or small (1–4 liter) field enclosures.

Before considering the results of competition experiments, it is worthwhile to consider the range of taxonomic combinations that have been tested.

More than 50 percent of published competition experiments have tested interactions between species of *Daphnia* or between clones or morphs of the same *Daphnia* species. A majority of the remaining experiments involved competition between *Daphnia* and a second genus, often another daphnid (e.g., *Ceriodaphnia, Simocephalus*), whereas only about 10 percent of the experiments involved two nondaphnids. Although Bengtsson's (1987) review deals solely with competition among Cladocera, it should be pointed out that few competition experiments have been conducted between copepods or between rotifers, and that experiments between Cladocera and other taxa are very limited. The emphasis on *Daphnia* is in part justified by its role as the dominant competitor and grazer in many lakes and ponds (see Chapter 4). The relatively nonselective feeding mode of daphnids, however, ensures a broad diet and results in extremely high dietary overlap among coexisting *Daphnia* species (Kerfoot et al., 1985a). Thus, the emphasis on competition between *Daphnia* species could lead to underestimation of the importance of food resource partitioning among the broader array of coexisting species.

When experiments result in competitive exclusion, "competitive dominance" is easy to define. In experiments not resulting in competitive exclusion, Bengtsson (1987) considered the species that obtained the highest density or greatest biomass to be the "competitive dominant." In many studies, particularly those conducted in the field, it is impossible to determine whether exclusion would have occurred if the experiment had been longer in duration. It is worth noting that all experimental studies of competition among cladocerans, both in the laboratory and in the field, have demonstrated statistically significant differences in population growth and numbers between single species controls and interspecific experimental groups in at least some experimental trials.

The major conclusion of Bengtsson's (1987) review is that no single factor, including body size, r_{max}, or efficiency at low food levels, can account for the patterns of competitive dominance seen among Cladocera (Table 6.2). The larger species was the competitive dominant in a bare majority of cases (60 percent). Because r_{max} tends to be correlated with body size in Cladocera, the species exhibiting the higher r_{max} value was also often the competitive dominant (68 percent support). Bengtsson (1987) followed Romanovsky (1984a, 1985) in assuming that efficiency at low food levels is correlated with small size. Thus, high feeding efficiency was associated with competitive dominance in only 36 percent of cases. Competitive ability and susceptibility to predation were strongly correlated in nine out of ten cases. Again, this result was, in part, because larger species, which were most often dominant, were also more vulnerable to fish predation. Defenses against invertebrate predators, however, were associated with low competitive ability in experiments with species of *Daphnia* (Kerfoot and Pastorok, 1978; Cooper and Smith, 1982) and "morphs" of *Bosmina* (Kerfoot, 1977). Studies in which the outcome of competition varied under different conditions are especially relevant to under-

Table 6.2   Patterns of competitive dominance among Cladocera

	Hypotheses for competitive dominance*				
	SEH	r_{max}	LFE	PS	DEC
Total number of tests	30	25	14	10	21
Number of field tests	10	7	3	6	7
Number of tests with exclusion	15	13	8	7	9
Total number of supporting tests	18	17	5	9	16
Number of supporting field tests	8	5	1	6	7
Supporting tests with exclusion	8	10	3	7	6
Percentage of supporting tests	60	68	36	90	76

Source: Adapted from Bengtsson, 1987.

*SEH, size efficiency hypothesis; r_{max}, r_{max} hypothesis; LFE, low food efficiency hypothesis; PS, hypothesis of trade-off between susceptibility to predation and competitive ability; and DEC, hypothesis that the outcome of competition varies under different environmental conditions.

standing seasonal succession. Of experiments carried out under different environmental conditions, 76 percent showed varying outcomes, including seven of seven field tests. Some circularity may be involved in this perfect agreement in field tests, because changes in environmental conditions are more likely to be considered significant after competitive reversals are observed. Nonetheless, these results indicate how competitive dominance could change seasonally or between habitats (e.g., with depth).

6.6   Reversals in Competitive Advantage in Cladocerans

Until recently, most researchers believed that the competitive abilities of cladocerans were fixed and largely dependent upon body size. Thus, few laboratory experiments have been designed to test how environmental variables influence the outcome of competition. So far, this work has focused on only two variables: temperature and the rate of food supply. Based largely on results showing that *Daphnia magna* was favored by low temperatures in competition with other *Daphnia* species, Bengtsson (1987) concluded that large Cladocera are generally favored at lower temperatures. Although this conclusion is consistent with the observation that some small cladocerans are often most abundant during midsummer (e.g., *Ceriodaphnia, Diaphanosoma, Chydorus*), there does not seem to be any reason to predict that temperature optima should vary systematically with body size. On the contrary, recent studies of clonal succession in *Daphnia magna* demonstrate extreme thermal differentiation among clones of this species (Carvalho and Crisp, 1987; Carvalho, 1987).

The concentration of added food resources is the second variable that has been examined in laboratory competition experiments. All laboratory competition experiments with zooplankton, with the exception of Neill (1975), have involved adding a constant amount of food at 1–3 day intervals. Recent laboratory experiments all support the hypothesis that smaller species are favored by the addition of low concentrations of food resources, and larger species are favored by the addition of high concentrations of food resources (different species of *Daphnia*, Bengtsson, 1987; *Daphnia* vs. *Bosmina*, Goulden et al., 1982; *Daphnia* vs. *Ceriodaphnia* and *Daphnia* vs. *Diaphanosoma*, Romanovsky and Feniova, 1985; *Daphnia* vs. *Diaphanosoma*, Orcutt and Porter, 1985; different species of *Bosmina*, Hanazato and Yasuno, 1987b; Figure 6.4). Because competitive exclusion occurred in only a few trials, these results are also consistent with the notion that variance in resource levels promotes coexistence over a broad range of resource supply rates (Levins, 1979). Thus, agreement with the "resource variance hypothesis" is encouraging.

Several workers have conducted competition experiments in the labo-

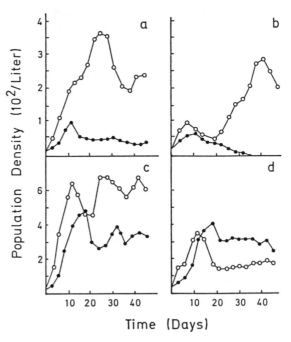

Figure 6.4 The effect of the rate of food supply on the outcome of laboratory competition experiments between *Ceriodaphnia reticulata* (small species, open circles) and *Daphnia pulex* (large species, solid circles). Both species survived in single species cultures at both low (a) and high (c) rates of food supply. When growth together, *Daphnia* was excluded when the rate of food supply was low (b) but was the more abundant species when the rate of food supply was high (d) (after Romanovsky and Feniova, 1985).

ratory with animals and seston collected from the field (e.g., de Bernardi, 1979; Cooper and Smith, 1982; Matveev, 1985a, 1986a; Romanovsky and Feniova, 1985). Such a hybrid approach allows temperature to be controlled and encourages a more detailed analysis of population dynamics than is practical in field competition experiments. In addition, resources can be readily diluted or supplemented, as was done by Romanovsky and Feniova (1985; Figure 6.4). Using this laboratory/field approach with organisms from Lake Glubokoe, Matveev (1986a, 1987) observed that *Diaphanosoma brachyurum* strongly depressed or even excluded *Daphnia hyalina* during the first month of an experiment, but later *D. hyalina* increased and by about 70 days, *D. brachyurum* was excluded. An analysis of demographics revealed that competitive ability was correlated with the relative lengths of time lags between population density and population growth. Presumably, longer time lags are an indication of a species' ability to avoid starvation and to continue growing as food conditions deteriorate (also see Matveev, 1985b).

There is, however, a potential problem with extrapolating from results with semi-continuous cultures to natural systems. In semi-continuous cul-

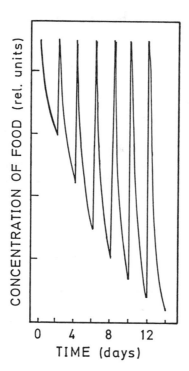

Figure 6.5   Food resource dynamics during laboratory growth experiments with *Daphnia* spp. Food and culture media were exchanged every two days. Individual growth resulted in more rapid depletion of food over time (after Tillmann and Lampert, 1984).

tures, food levels undergo sharp increases and decreases following each food addition (Foran, 1986; Tillmann and Lampert, 1984; Figure 6.5). Although oscillations in food resources appear to be important in many natural systems, such oscillations are driven by oscillations in cladoceran population densities that have periods of several weeks (McCauley and Murdoch, 1987; also see Figure 6.3). Experiments in which resource renewal is divorced from both resource abundance and grazing levels may not be appropriate models for predicting interactions in nature, particularly among genera with differing functional responses to food concentration (e.g., *Daphnia* and *Bosmina*, DeMott, 1982). Experiments with continuous flow cultures (e.g., Lampert, 1976; Williamson et al., 1985) as well as more simulation modeling (e.g., Frost, 1980; Romanovsky and Feniova, 1985) would be helpful. There is also a strong need for manipulating additional resource-related variables in laboratory competition experiments. Despite abundant evidence from feeding experiments and single-species growth experiments, the effects of particle size, particle digestibility, and interfering filaments have not been tested in laboratory competition experiments (but see Rothhaupt, 1988).

6.7   Evidence for Shifts in Competitive Advantage in Nature

When competition experiments are conducted with natural food resources in field enclosures, the results usually vary over time. Given the marked seasonal changes in the abundance and composition of phytoplankton, some variability seems unavoidable, if only because food is more or less limiting at different times. For example, an enclosure experiment of one to two weeks may be too short in duration to reveal strong competition when resources are initially abundant, but may be long enough to demonstrate competition when resources are initially scarce. Thus, it is important to distinguish between seasonal shifts in competitive advantage and seasonal changes in the degree of food limitation.

As in laboratory experiments, the pattern of resource renewal may also be important in field experiments. Some workers have advocated changing the water and food resources within field enclosures at intervals of one to three days (e.g., Dodson, 1974; Smith and Cooper, 1982; Hoenicke and Goldman, 1987). Such a design keeps resources similar to those in the natural lake or pond but could also delay or even prevent divergence between control and competition enclosures due to exploitative competition (Smith et al., 1988). As in laboratory experiments with semi-continuous cultures (e.g., Figure 6.5), intense exploitation would lead to pronounced fluctuations in resource abundance when water is changed. Thus, other workers (e.g., Lynch, 1978; Kerfoot and DeMott, 1980; DeMott, 1983) have incorporated low exchange rates between enclosures and the natural system in their field experiments. In this case, resource dynamics should be sensitive to the effects of grazing mortality and nutrient cycling within the experimental enclosures, but enclosure artifacts would also be more likely. With either design, it is important to show that competition occurs at near-natural population densities.

Seasonal succession of daphnids   Three studies using field enclosures for competition experiments with daphnids provide the best evidence for seasonal reversals in competitive ability in natural communities. One such study involved two series of competition experiments in 3.5-liter enclosures in a fish-free eutrophic pond (Lynch, 1978). *Daphnia pulex* was largely replaced by *Ceriodaphnia reticulata* in the natural community during July. At the same time, *Daphnia pulex* was excluded from all competition enclosures by *Ceriodaphnia reticulata*. As was shown by Neill (1975), the smaller species, *Ceriodaphnia*, was able to outcompete the larger species, *Daphnia*, because juvenile *Daphnia* were very sensitive to resource depression. In a second enclosure experiment, however, begun shortly after the completion of the first, the presence of *Daphnia* strongly depressed the abundance of *Ceriodaphnia* and *Daphnia* actually benefited from the presence of *Ceriodaphnia*. Two possible explanations were proposed for this reversal in competitive advantage. First, water temperature was higher during the first experiment (often above 25°C) and was probably above the optimum for *Daphnia*. Second, both primary productivity and inorganic nutrient levels were lower during the second experiment. It was hypothesized that grazing and nutrient regeneration by *Ceriodaphnia* in the second experiment favored phytoplankton species that were readily utilized by *Daphnia* but not by *Ceriodaphnia*. Whereas the demographic consequences of competition and juvenile mortality matched the predictions of the resource variance hypothesis, the apparent effects of productivity did not. The smaller species, *Ceriodaphnia*, was favored during the period of higher primary productivity. Interpretation of this study is complicated by much higher population densities within small enclosures than were found in the pond.

Smith and Cooper (1982) also reported variable competitive outcomes between *Daphnia pulex*, *Ceriodaphnia* spp., and a third cladoceran, *Moina affinis*, in enclosure experiments in a fishless pond during summer. In the first two of three series of competition experiments, the biomass of intra- and interspecific competitors was a good predictor of the reduction in the growth rate of each of the three cladocerans. This result suggests that each species was more or less equivalent in competitive ability, its effect on resources, and its response to resource depression. In the third series of experiments, conducted when *Daphnia* was increasing in the pond, the presence of *Ceriodaphnia* had no significant effect on *Daphnia*. The inhibiting effect of *Daphnia* on *Ceriodaphnia*, however, was stronger than *Ceriodaphnia*'s effect on itself. Nannoplankton reached their seasonal peak just at the beginning of this experiment. Thus, competitive dominance by the larger species is in agreement with the resource variance hypothesis. By this time, *Moina* had already disappeared from the community, probably as a consequence of interspecific competition.

The competition experiments conducted by Smith and Cooper (1982) lasted only 5 to 8 and one-half days and included a replacement of water in 1-liter glass enclosures with fresh pond water every two days. The short duration of experiments and frequency of resource renewal probably decreased competition and prevented starvation within the experimental enclosures. Thus, despite evidence for strong competition, the results of this study probably understate the level of competition that actually occurred in the pond. Smith and Cooper (1982) used a range of initial densities in both the single-species controls and the competition treatments. Because this design permits a direct evaluation of the relative strengths of inter- and intraspecific competition, it represents an important advance over previous approaches. The rapid seasonal successions that often occur among Cladocera in temporary ponds (e.g., Modlin, 1982; Crosetti and Margaritora, 1987) recommend these habitats for future studies.

In a two-year study of seasonal succession in a small, shallow, mesotrophic lake, DeMott (1983) found that *Daphnia rosea* had an apparent advantage during spring and early summer but was rapidly replaced by a morphologically similar congener, *D. pulicaria*, during late summer and autumn (Figure 6.6b). Competition experiments in both large and small enclosures confirmed that the decline of *D. rosea* was a direct consequence of competition from *D. pulicaria*. Resting eggs, however, allowed the *D. rosea* population to persist until spring, when it was able to increase at higher rate before food became limiting and also was a stronger competitor in experimental enclosures. The reversal in competitive advantage appeared to be caused by a shift in food resource quality. Radiotracer experiments showed that *Daphnia pulicaria* was better able to assimilate the digestion-resistant algae (mainly gelatinous green algae) that predominated during midsummer and autumn. There was no evidence, however, to explain why *D. rosea* was favored during spring and early summer when flagellates and detritus were the dominant food resources. Analysis of resistant algae within *Daphnia* guts revealed extremely high interspecific overlap in diet among individuals collected in the same samples (mean 97 percent, range 92–99 percent overlap; Kerfoot et al., 1985a). Thus, even grazers with identical diets can be differentially affected by seasonal changes in resource competition.

In addition to experiments in large and small enclosures, the Lake Mitchell study of *Daphnia* succession included a detailed analysis of *Daphnia* demographics (DeMott, 1983). In large enclosures as well as in the lake, intraspecific competition caused strong, density-dependent decreases in egg production (Figure 6.7) and density-dependent population growth (also see Kerfoot et al., 1985b). Each *Daphnia* species had both higher reproductive rates and lower mortality rates during its periods of competitive advantage.

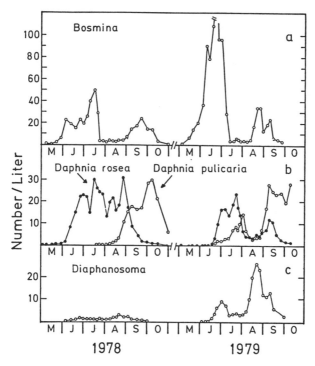

Figure 6.6   Seasonal dynamics of Cladocera in Lake Mitchell, Vermont (after DeMott and Kerfoot, 1982; DeMott, 1983; Kerfoot et al., 1988).

This trend contrasts with the high natality/high mortality, low natality/low mortality pattern that is often used to suggest a balance between competition and predation (e.g., Allan, 1973; Jacobs, 1977b; Tessier, 1986). Thus, a reversal in relative fecundity may be a good indication of a shift in competitive ability in food-limited communities.

Seasonal reversals in relative fecundity have been documented during successions between *Daphnia* species in a wide variety of habitats, ranging from deep reservoirs (Tappa, 1965; Wright, 1965) to fishless ponds (Hebert, 1977). Although these workers did not test for competition or resource limitation experimentally, all three studies showed seasonal reversals in relative fecundity and an inverse relationship between per capita egg production and *Daphnia* abundance. Tappa (1965), Wright (1965), and Hebert (1977) all concluded that shifts in competitive advantage were probably important both in driving seasonal successions and in maintaining species coexistence. Competitive shifts may be important even in lakes with high densities of planktivorous fishes. For example, *Daphnia parvula* and *Daphnia ambigua* frequently coexist in fish ponds and small eutrophic lakes in North America. The success of these small-bodies species often appears dependent upon the removal of

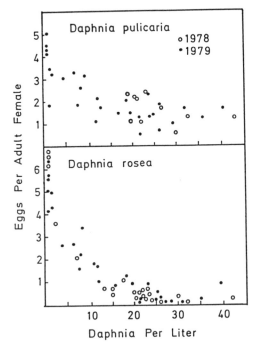

Figure 6.7   The number of eggs per adult female of two *Daphnia* species as a function of the total daphnid density. Data for 1978–79 in Lake Mitchell, Vermont (after DeMott, 1983).

large *Daphnia* by visually feeding fishes (Lynch, 1979; Kerfoot, 1981). Analyses of population dynamics, however, suggest that food limitation and competitive shifts underlie the seasonal succession of *D. ambigua* and *D. parvula* (Allan, 1977; Lei and Armitage, 1980). Unlike the situation in Lake Mitchell, most shifts in reproductive advantage are correlated with temperature and could be caused by either differing temperature optima or seasonal changes in resources. Subsequent research in the Australian pond communities studied by Hebert (1977) shows that the better reproducer during summer, *D. cephalata*, is also less susceptible to predation from *Notonecta* (Grant and Bayly, 1981). Because predation is probably more intense during summer, *D. cephalata* may be favored by both predation and competition during summer.

In some lakes, daphnids fail to show seasonal succession, despite demographic evidence for exploitative competition. For example, analyses of the population dynamics of cladocerans in Lake Glubokoe, a mesotrophic lake near Moscow, provide evidence of density-dependent rates of reproduction, juvenile development, and population growth during June through August (Matveev, 1983, 1986b; Ghilarov, 1984, 1985; Romanovsky, 1984b). Rather than showing a seasonal succession, coexisting populations of *Daphnia*

cucculata and *D. galeata* remained equally abundant and showed highly synchronous fluctuations in population numbers and egg production, which were
correlated with the density of nannoplankton food resources (Ghilarov, 1984).
Other cladocerans (*Bosmina coregoni* and *Diaphanosoma brachyurum*) also
showed peaks in reproduction coincident with peaks of fertility of the two
Daphnia species, but no other species pair demonstrated such high similarity
in patterns of reproduction and population change as the two species of
Daphnia. These two species appear to act as ecological equivalents in Lake
Glubokoe.

To focus exclusively on interactions among coexisting species may in-
correctly give the impression that competitive exclusion is uncommon in zooplankton communities. The low diversity of zooplankton communities is consistent with a strong role for species interactions in community structure
(Pennak, 1957; Hebert, 1982); however, only the introduction of non-native
zooplankton into field enclosures can provide direct evidence of whether
predation, competition, or other environmental factors prevent particular species from colonizing particular habitats. For example, successful introductions
of *Daphnia pulex* into predator-free enclosures in ponds suggest that predation
from a large copepod (Dodson, 1974) and fish (Vanni, 1986) were responsible
for excluding *D. pulex* from the two natural communities. Laboratory and field
experiments, however, showed that food levels, and not predation prevented
Daphnia pulex from colonizing ultra-oligotrophic lakes in British Columbia
(Neill, 1978). Low food levels were a consequence of both low nutrient levels
and grazing by native species, especially *Daphnia rosea* (also see Walters et
al., 1987). When phosphorus and nitrogen were added to large enclosures to
stimulate phytoplankton production, *D. pulex* quickly became the most abundant species. Thus, Neill's (1978) study provides support for the resource
variance hypothesis and is also consistent with Hrbáček's (1977) conclusions
about the resource adaptations of *D. pulex*.

Seasonal succession of *Daphnia* and *Diaphanosoma*   Although experiments should be designed to separate the effects of competition, predation,
various environmental factors, and their interactions, we should not be sur-
prised if species adapt to a suite of seasonal factors. Recent studies show that
cladocerans of the genus *Diaphanosoma* are well adapted to a variety of midsummer conditions. This genus has been described as a warm-water organism
because of its prominence in tropical lakes and its seasonal distribution in
temperate lakes (Hutchinson, 1967). In the temperate zone, *Diaphanosoma*
recruits from resting eggs later in the spring (Herzig, 1984) and consistently
reaches its seasonal peak later in the summer than other cladocerans (Kratz
et al., 1987). Apart from predation, at least four different factors may favor
Diaphanosoma in competition with *Daphnia* during midsummer: blooms of
inedible cyanobacteria, increases in the relative importance of bacteria, in-

creases in water temperature, and decreases in the productivity of shared food resources (i.e., edible nannoplankton).

Midsummer declines in *Daphnia* in eutrophic lakes and subsequent increases in *Diaphanosoma* have been attributed to the greater tolerance of *Diaphanosoma* to interference from filamentous cyanobacteria (Gliwicz, 1977). Recent radiotracer studies confirm that *Diaphanosoma*'s feeding rate on small edible algae is less affected than *Daphnia*'s by the presence of cyanobacteria (Fulton and Paerl, 1987a; Fulton, 1988). Demonstration of *Diaphanosoma*'s lack of vulnerability to cyanobacterium toxins further emphasizes that physical interference rather than toxicity is the key variable (Fulton and Paerl, 1987b). Using in situ life table experiments, Threlkeld (1986) showed that resource-mediated collapses of populations of *Daphnia* and *Ceriodaphnia* coincided with a midsummer bloom of filamentous cyanobacteria. *Diaphanosoma*, however, showed little change in abundance. Because declines in daphnids did not result in reciprocal increases in *Diaphanosoma*, interspecific competition did not seem to be important in limiting *Diaphanosoma*. It seems possible that severe cyanobacterium blooms could prevent cladocerans from reaching densities where exploitative competition for edible algae is significant. Further circumstantial evidence for strong adverse effects of cyanobacteria on cladocerans comes from studies in tropical Lake Valencia (Infante and Riehl, 1984). The complete disappearance of all three native species of cladocerans (*Diaphanosoma*, *Ceriodaphnia*, and *Moina*) for over two years coincided with dense blooms of two filamentous cyanobacteria (*Oscillatoria* and *Lyngbia*). Rotifers and copepods were seemingly unaffected.

Despite correlations, it is often difficult to be sure whether midsummer declines in *Daphnia* are primarily caused by blooms of cyanobacteria, fish predation, or a scarcity of edible algae. Although laboratory experiments show that adding filamentous cyanobacteria could induce *Daphnia* declines (Dawidowicz et al., in press) or cause reversals in competitive ability (Gliwicz and Lampert, unpublished manuscript), evidence from field enclosure experiments is lacking. Field experiments in which the abundance of cyanobacteria and edible algae are directly manipulated would help clarify the mechanisms underlying resource-mediated midsummer declines. On the other hand, several studies show that excluding fish (Lynch and Shapiro, 1981; Schoenberg and Carlson, 1984) or reducing their abundance (e.g., Shapiro and Wright, 1984) can result in increases in *Daphnia* and declines in midsummer cyanobacterium blooms. Thus, cyanobacterium blooms can be a consequence rather than a cause of *Daphnia* declines.

Because *Diaphanosoma* exhibits midsummer peaks in lakes without significant densities of cyanobacteria, greater tolerance to interfering particles cannot be the sole reason for *Diaphanosoma*'s success during midsummer. Other food-related conditions, predators, and water temperature may all be important. For example, Geller and Müller (1981) hypothesized that high efficiency in feeding on bacteria could explain the success of *Diaphanosoma*

during midsummer minima in edible algae. Subsequent studies (DeMott and
Kerfoot, 1982; Bogdan and Gilbert, 1984; DeMott, 1985) have confirmed the
ability of *Diaphanosoma* to feed very effectively on bacteria, but have not
tested the importance of bacteria as a source of nutrition (e.g., Pace et al.,
1983).

The importance of competition from *Daphnia* in limiting the abundance
of *Diaphanosoma* in Lake Mitchell, Vermont has been demonstrated through
field experiments and demographic analysis. *Diaphanosoma* was typically a
minor component of the zooplankton community and its rate of population
growth was strongly dependent on *Daphnia* density (Figure 6.8). Moreover,
a midsummer decline in *Daphnia*, which occurred during an exceptionally
warm period in 1979 (Figure 6.6b), also resulted in a sharp increase in the
abundance of *Diaphanosoma* (Figure 6.6c). The *Daphnia* decline also occurred
in fishless enclosures (Kerfoot et al., 1988) and involved increases in inviable
eggs and decreases in body size (DeMott, 1983), which are symptoms very
similar to those described by Threlkeld (1985, 1986) as being characteristic
of declines induced by cyanobacteria. Removal of *Daphnia* from large enclo-

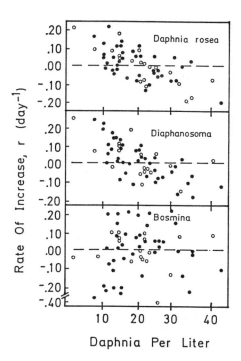

Figure 6.8   Relation between total *Daphnia* density and the population growth rates
of *Daphnia rosea*, *Diaphanosoma brachyurum*, and *Bosmina longirostris*. Open circles,
population in the lake; closed circles, population in the fish exclosures. 1978 data from
Lake Mitchell, Vermont (after DeMott and Kerfoot, 1982).

sures, either by introducing fish (DeMott and Kerfoot, 1982) or by netting (Kerfoot et al., 1988) resulted in rapid, order of magnitude increases in the abundance of *Diaphanosoma*. High temperatures also favor *Diaphanosoma* in laboratory competition experiments with *Daphnia* (J.D. Orcutt, personal communication). Thus, the Lake Mitchell results not only illustrate strong suppression of *Diaphanosoma* by *Daphnia*, but also show that fish predation or temperature-induced shifts in competition can favor *Diaphanosoma* over *Daphnia* (see Chapter 7 for further discussion on selective predation.

In agreement with the resource variance hypothesis, laboratory experiments show that *Diaphanosoma* fares poorly in competition with *Daphnia* at high levels of resource supply (Orcutt and Porter, 1985; Romanovsky and Feniova, 1985). The addition of inorganic nutrients to large enclosures in Lake Mitchell resulted in increases in *Daphnia* and the early disappearance of *Diaphanosoma* (DeMott and Kerfoot, 1982; DeMott, unpublished), providing further support for the hypothesis that competition between these genera is influenced by resource productivity. Thus, in addition to selective fish predation on *Daphnia*, *Diaphanosoma*'s success during midsummer may depend on blooms of cyanobacteria, increases in the relative importance of bacteria, increases in water temperature, and decreases in the productivity of edible algae.

Seasonal succession of *Daphnia* and *Holopedium*   In contrast to *Diaphanosoma*, the other planktonic member of the Sididae, *Holopedium*, is more common in northern latitudes and exhibits a variable seasonal distribution. Studies in two small lakes in which *Holopedium* was the summer dominant provide strong evidence that *Holopedium* was food limited, whereas coexisting *Daphnia* were maintained at low levels by predation from fish and *Chaoborus* (Allan, 1973; Tessier, 1986). In both studies, dense midsummer populations of *Holopedium* exhibited low rates of reproduction and mortality, whereas very low populations of *Daphnia* exhibited high rates of reproduction and mortality. Thus, while intraspecific competition appeared to limit the abundance of *Holopedium* through reduced reproduction, high densities of *Holopedium* did not depress the reproductive rate of *Daphnia* at either site. With enclosure and food addition experiments, Tessier (1986) demonstrated remarkably rapid starvation of *Holopedium* in fish enclosures (99 percent decline in one week) and almost equally rapid increases (>100-fold increases in two weeks) in *Daphnia catawba* during midsummer. Moreover, when fish predation decreased in late summer, *Daphnia catawba* rapidly increased and *Holopedium* subsequently declined. The late summer decline in *Holopedium* coincided with the increase in *Daphnia* during each of three years, providing further evidence that *Daphnia* can strongly depress *Holopedium*.

Demographic analyses from lakes and ponds in which predation appears to be less important further suggest that competition from *Daphnia* can strongly depress or even exclude *Holopedium*. In lakes where *Daphnia* appears to become strongly food limited, *Holopedium* is most abundant during spring and sharply declines or even goes extinct during midsummer. For example, in Mirror Lake, New Hampshire, *Holopedium* was common during spring but became rare as *Daphnia catawba* became abundant in late June (Makarewicz and Likens, 1975). In a second example, only very low densities of strongly food-limited *Holopedium* coexisted with much higher densities of food-limited *Daphnia rosea* in Castle Lake, California (Hoenicke and Goldman, 1987). In a final example, a relatively stable population of *Daphnia rosea* persisted throughout the summer in a small pond in the Colorado Rocky Mountains, whereas populations of *Holopedium* and *Daphnia pulex* v. *minnihaha* went extinct during July and early August (Dodson, 1972). Dodson (1972) argued that predation from *Chaoborus* was important in controlling the abundance of *Daphnia rosea*. In retrospect, however, this hypothesis seems incompatible with more recent evidence that *Holopedium* (O'Brien et al., 1979) and *Daphnia pulex* v. *minnihaha* (Havel and Dodson, 1984) have effective morphological defenses against *Chaoborus*. Low reproductive rates during population declines suggest that *Holopedium* and *D. minihaha* were competitively excluded by *Daphnia rosea*. In general, these results suggest that *Holopedium* is an inferior competitor and that its success during summer requires selective predation on coexisting *Daphnia* species. Walters et al. (1987), however, cite preliminary evidence that *Holopedium* may be favored by low temperatures. Thus, both predation and temperature-dependent competition may control seasonal successions between these two genera.

Seasonal succession between calanoid copepods   Hutchinson's (1951) paper on "copepodology for the ornithologist" focused attention on the idea that competition was important in limiting the number of coexisting copepod species and in determining their seasonal abundance. Noting that calanoid copepods that co-occur in temporary ponds often differ markedly in body size, Hutchinson hypothesized that the size range of particles ingested was related to copepod body size and that competitive exclusion prevented the coexistence of species of similar body size. Differing seasonal patterns in abundance were also hypothesized to promote coexistence in both temporary habitats and lakes, whereas differences in depth distributions were considered important in stratified lakes (see Hutchinson, 1967, for a review of early literature). With only a few exceptions, studies of coexisting calanoids have emphasized differences in body size, seasonal occurrence, and vertical distributions, rather than directly measuring dietary overlap or testing for food limitation and competition.

Because the naupliar and early copepodite stages often suffer high mortality (e.g., Confer and Cooley, 1977), instar analysis can be an important tool in understanding copepod population dynamics regulation (but see Hairston and Twombly, 1985). Unfortunately, the immature stages of calanoid copepods, especially congeners, are difficult to identify to the species level.

Thus, the lack of data on immature instars is a serious limitation in studies of copepod succession.

The obligate sexual mode of reproduction of copepods may influence the nature of species interactions. As pointed out by Watras and Haney (1980), each clutch of freshwater calanoids requires fertilization. In the absence of males, unfertilized eggs disintegrate upon extrusion, leading to a considerable waste of reproductive material. Thus, mate limitation as well as food limitation could depress reproductive rates. Williamson and Butler (1987) have recently developed and applied indices that use the proportion of females with egg sacks and ripened ovaries to separate the effects of food and mate limitation on reproductive rates. Laboratory experiments showed that the interval between clutches was more strongly influenced by food limitation than was the number of eggs per clutch (Williamson and Butler, 1987). Thus, attempts to correlate food levels and reproduction in calanoid copepods need to consider the proportion of mature females bearing egg sacks as well as clutch size and the possibility of mate limitation.

Although pheromones are involved in the location and recognition of mates among calanoids (e.g., Katona, 1973), interspecific couplings are common (DeFrenza et al., 1986). In addition to competition, difficulties in finding conspecific mates could make the invasion of a new species of calanoids difficult in a lake or pond already occupied by a related species. Moreover, selection for reproductive isolation provides an alternative to competition as an explanation for differences in body size and breeding season among co-occurring populations (Chow-Fraser and Maly, 1988). On the other hand, the coexistence of highly diverse (>200 coexisting species) low density assemblages of calanoids in marine systems (Hayward and McGowan, 1979) suggests that the problem of finding mates is not insurmountable.

Although early workers (e.g., Pennak, 1957; Cole, 1961) considered the coexistence of diaptomid copepods unusual, subsequent studies indicate that some combinations of congeners commonly coexist. For example, in a survey of 100 lakes in southern Ontario, the two most common calanoids, *Diaptomus minutus* and *Diaptomus oregonensis*, were found in 70 and 69 lakes, respectively, and coexisted in 45 lakes (Rigler and Langford, 1967). Rather than showing character displacement, as would be predicted if body size were an important niche variable or related to reproductive isolation, coexisting populations of *D. minutus* and *D. oregonensis* were significantly more similar in size than allopatric populations. Comparisons between winter and summer samples often provided evidence for marked seasonal successions. When *D. minutus* and/or *D. oregonensis* occurred alone, they were common during winter. Other species, however, that were rare or limited to the hypolimnion during summer (e.g., *D. sicilis* and *D. birgei*) were frequently winter dominants.

An intensive 4-year study of coexisting *D. minutus* and *D. oregonensis* in Clarke Lake, Ontario documented only modest differences in depth distributions, body size, and seasonal abundance (Sandercock, 1967). Interestingly, between-year differences in water temperature appeared to account for between-year differences in abundance. Although adults of both species were present throughout the year, *D. minutus* was more abundant during late spring while *D. oregonensis* was more abundant during midsummer. More rapid warming of epilimnetic waters during two years favored *D. oregonensis* over *D. minutus*. Perhaps temperature or some factor associated with thermal stratification can shift the relative competitive abilities of these two copepods.

Somewhat surprisingly, the most diverse assemblages of diaptomids have been reported from weakly stratified waters, where opportunities for vertical segregation are minimal. Under these circumstances, seasonal successions are often pronounced. Davis (1961) reported that five species of *Diaptomus* that co-occurred in Lake Erie showed differences in the timing of adult abundance and reproduction. These ranged from *D. sicilis* that reproduced during winter and early spring to *D. oregonensis* that bore eggs from July through November. Data from a broad range of lakes are needed to determine whether periods of reproduction in calanoids are relatively fixed or tend to contract or be displaced in the presence of congeners.

Diaptomids co-occurring in shallow ponds in Saskatchewan, Canada also exhibited very sharp differences in seasonal occurrence (Hammer and Sawchyn, 1968). Two distinct reproductive strategies were noted. The univoltine, vernal species hatched from resting eggs, matured within about three weeks after ponds filled in the spring, and presumably produced only resting eggs. These were followed by summer species that produced several generations before the ponds dried up or were covered with ice. Rapid declines in vernal species, as summer species increased, suggested that competition could be important in the replacement of vernal species by summer species. Perhaps the univoltine species are adapted to very rapid growth under the conditions of high resource abundance, which must occur shortly after the ponds fill with water. Laboratory studies are needed to test whether this strategy of rapid growth compromises the ability to survive and reproduce when food resources are scarce.

Not all coexisting populations of diaptomic copepods provide evidence of ecological differentiation. *Eudiaptomus gracilis* and *Eudiaptomus graciloides* are the dominant calanoids in central and northern Europe and coexist in many lakes. In a study of co-occurring populations in three lakes in northern Germany, Hofmann (1979) found that the temporal and depth overlaps were higher between these two *Eudiaptomus* species than between either one of them and any other zooplankton species. Differences in body length were only about 10 percent or less. Moreover, both species showed synchronous fluctuations in body size and clutch size, which were correlated with the abundance of nannoplankton food. Thus, although food limitation is probably important in determining seasonal dynamics, there is no evidence for differentiation that could stabilize coexistence.

Because many large calanoids are highly predatory, the patterns of co-existing large and small calanoids that Hutchinson (1951) described may often involve significant levels of interspecific predation and only weak potential for interspecific competition. One carefully studied case involves large *Diaptomus shoshone* (1.85–2.47 mm total length) and *Diaptomus coloradensis* (0.97–1.28 mm), which co-occur in high altitude ponds in the Rocky Mountains. The univoltine *D. shoshone* matures first, but differences in body size during the development of the first generation of both species probably preclude both predation and interspecific competition (Maly, 1976). The nauplii and early copepodites of the second generation of *D. coloradensis* are, however, susceptible to predation from *D. shoshone*. Correlations involving a large number of ponds and data from several years for individual ponds revealed that the body sizes and clutch sizes of both species were inversely related to conspecific but not interspecific densities (Maly, 1973). Enclosure experiments provided direct evidence for intraspecific competition in both species (Maly, 1973). Thus food limitation and intraspecific competition influence the demographics and, perhaps, the abundance of these species. The apparent lack of interspecific competition is consistent with the marked differences in diet between these two species (Maly and Maly, 1974).

6.8   Importance of Differences in Feeding Mode and Diet

With the exception of large and small copepods, the experiments and demographic evidence described above involved species with very similar feeding modes and high dietary overlap. Because reversals in competitive advantage can facilitate the coexistence of species with virtually identical diets, it becomes very challenging to demonstrate the effects of specialized feeding modes and food resource partitioning on the intensity and outcome of competition. Clearly, resource partitioning is not a requirement for coexistence in food-limited communities. What is the nature of competition between a generalist, such as *Daphnia*, and more specialized grazers?

Competition between *Daphnia* and *Bosmina*   In contrast to other cladocerans, recent studies show that species in the family Bosminidae are able to select between individual particles that differ in size or quality. *Bosmina* is able to choose algal-flavored spheres over untreated spheres (DeMott, 1986) and, contrary to early perceptions, to feed selectively on larger particles (DeMott, 1985; Bleiwas and Stokes, 1985; Fulton and Paerl, 1987a; Fulton, 1988). Thus, *Bosmina* appears to exhibit food selection capabilities intermediate between nonselective daphnids and even more discriminating groups, such as calanoid copepods and some rotifers.

Competition experiments between *Daphnia* spp. and *Bosmina longirostris* in small-scale enclosures have yielded highly variable results. For example, a total of nine series of small enclosure experiments, each lasting 3–4 weeks, were conducted in Lake Mitchell from June through September of three different years (Kerfoot and DeMott, 1980; DeMott, 1983). No instances of competitive exclusion were observed: in four experiments, competition was not detected; in three experiments, both *Daphnia* spp. and *Bosmina* were depressed in competition treatments; and in two experiments, *Daphnia* was significantly depressed, but *Bosmina* was unaffected by interspecific competition. In enclosure competition experiments in other lakes, Lynch (1978) reported that *Daphnia pulex* strongly depressed but did not exclude *Bosmina longirostris*, whereas Von Ende and Dempsey (1981) found only weak competition between the same species. Kerfoot and DeMott (1980) argued that differences in feeding mode and diet were important in stabilizing coexistence. The argument was based, in part, on the observation that *D. pulex* quickly excluded *Bosmina* when the two species competed for high densities of *Chlamydomonas reinhardi* in laboratory experiments (Kerfoot and DeMott, 1980). This line of reasoning is weakened by new evidence that resource supply rates can influence the outcome of competition between large and small (i.e., *Bosmina*) species (Romanovsky and Feniova, 1985; Goulden et al., 1982).

In Lake Mitchell, as in many other lakes, *Bosmina longirostris* exhibits a bimodal pattern of seasonal abundance, with peaks in early summer and early autumn (Figure 6.6a). In contrast, *Daphnia* spp. remained abundant from June through October (Figure 6.6b). Thus, an attempt to understand this seasonal succession focused on *Bosmina*'s response to changes in resource quality and quantity as well as relationships with *Daphnia* abundance (DeMott and Kerfoot, 1982). Population crashes in *Bosmina* occurred in mid-July in the lake, in several large (12,000 liter) enclosures that excluded fish, as well as in enclosures where fish had strongly depressed the abundance of *Daphnia*. Thus, population crashes could not be attributed to competition from *Daphnia*. In each instance, *Bosmina*'s sharp decline came after a precipitous drop in the abundance of flagellated algae and involved both declines in reproduction and increases in mortality. Conversely, increases in *Bosmina* in spring and autumn followed increases in flagellates and were marked by increasing birth rates and decreasing death rates. These results complement feeding experiments showing that *Bosmina* is very efficient in feeding on flagellates, particularly when these high quality resources are low in abundance (DeMott, 1982; DeMott and Kerfoot, 1982). Thus, *Bosmina*'s ability to feed selectively and efficiently on a subset of available resources reduces the intensity of competitive interactions with *Daphnia* but makes *Bosmina* vulnerable to successional changes in phytoplankton that are not due to *Daphnia* grazing. Subsequent studies in other mesotrophic lakes have also shown that *Bosmina*'s dynamics are sensitive to the abundance of flagellates (Sarnelle, 1986; Vanni, 1987). Complete removal of *Daphnia* in a Lake Mitchell fish enclosure resulted in an order of magnitude increase *Bosmina*'s abundance during the autumn peak (DeMott and Kerfoot, 1982). Thus, although *Daphnia* is unable to exclude *Bosmina*, *Daphnia* can influence the magnitude of abundance peaks.

Based on laboratory competition experiments, Goulden and his colleagues (Goulden and Hornig, 1980; Goulden et al., 1982) suggested that oscillations in *Daphnia* density permit *Bosmina* to persist despite strong competition. The idea behind this formulation of the resource variance hypothesis is that *Bosmina* experiences lower starvation mortality than *Daphnia* during the declining phase of the *Daphnia* oscillation. In Lake Mitchell, however, there was no correlation between natural oscillations in *Daphnia* and *Bosmina*'s growth rate (Figure 6.8), diminishing the merits of this hypothesis for natural populations.

Competition between *Daphnia* and rotifers   As noted in Sections 6.2 and 6.3 above, rotifers differ greatly in their threshold food levels, maximal rates of increase, and food preferences. Although the tendency to lump rotifers into a single group may be justified in part by their small body sizes, it may be more a matter of convenience than a reflection of ecological reality. Laboratory experiments have demonstrated rapid competitive exclusion of *Brachionus calyciflorus* and *Keratella cochlearis* by *Daphnia pulex* (Gilbert, 1985). Because of its exceptionally high food threshold requirements and high maximal growth rate, *B. calyciflorus* is an excellent example of a fugitive species (Hutchinson, 1951). *Brachionus calyciflorus* is largely restricted to ponds and temporary habitats, where it often dominates in the spring following phytoplankton blooms, probably because of its high reproductive rate, and then disappears as *Daphnia* increases (e.g., Daborn and Hayward, 1978). In contrast, *Keratella cochlearis* has a relatively low food threshold. The competitive exclusion of *K. cochlearis* in laboratory experiments was attributed to sensitivity to mechanical damage by large *Daphnia* (Gilbert and Stemberger, 1985) and low resistance to starvation.

Manipulation of *Daphnia*, *Chaoborus*, omnivorous diaptomic copepods, and inorganic nutrients in large enclosures (27,000 liters) clearly demonstrated that competition from *Daphnia* was the primary cause for the low abundance of rotifers in an oligotrophic lake in British Columbia (Neill, 1984). Removal of *Daphnia rosea* stimulated rotifer growth to densities nearly 250 times greater than controls, whereas quadrupling initial *Daphnia* numbers reduced rotifer densities to about one-tenth of control densities. In contrast, removal of the copepods *Diaptomus leptopus* and *Diaptomus kenai* had no detectable effect on rotifer abundance. *Chaoborus* additions benefited rotifers, presumably by causing modest declines in the abundance of *Daphnia* (Neill, 1981b). Nutrient additions had little influence on rotifer abundance unless combined with *Daphnia* removal. Other field experiments in which *Daphnia* was selectively removed from enclosures by fish (e.g., Lynch, 1979) or netting (e.g., Ferguson et al., 1982) have also recorded dramatic increases in rotifer abundance.

Despite dramatic effects on rotifer abundance, Neill's (1984) manipulations of *Daphnia* seemingly had little effect on the relative abundance or spatial/temporal distributions of the three common rotifer species (*Keratella cochlearis, Kellicottia longispina,* and *Polyarthra*). As in a previous descriptive study (Makarewicz and Likens, 1975), coexisting rotifers showed a distinctive succession with very little overlap in time and depth. Apparently the timing of this rotifer succession is independent of the abundance of *Daphnia* as well as invertebrate predators. Presumably, understanding the succession of rotifer species will require insights into the nature of ongoing and evolutionary interactions between rotifer species. Perhaps the apparent niche differentiation among rotifers is related to specialized food preferences or nutritional requirements.

Competition between *Daphnia* and calanoid copepods   Several workers have pointed out that eutrophication tends to favor cladocerans, particularly *Daphnia*, over calanoid copepods. Most hypotheses emphasize the effects of competition mediated through changes in the quantity or quality of food (McNaught, 1975; Muck and Lampert, 1984; Lampert and Muck, 1985), although interactions between food limitation and predation may also be important (Byron et al., 1984). As shown by Richman and Dodson (1983), an abundance of cyanobacteria and other low quality particles could again favor calanoid copepods under highly eutrophic conditions. Comparisons between years in a South African reservoir showed that high concentrations of suspended sediments also favored calanoid copepods over *Daphnia* (Hart, 1986).

Does competition between calanoids and cladocerans influence seasonal dynamics? Neill (1985) reported that two species of *Diaptomus* showed less response to manipulation of *Daphnia* density in large enclosures in an oligotrophic lake than did co-occurring cladocerans and rotifers. A significant increase of only about 20 percent in the abundances of *Diaptomus kenai* and *Diaptomus leptopus* was noted in the treatment involving a 99 percent initial reduction in the abundance of the dominant grazer *Daphnia rosea*. In contrast, the cladoceran *Bosmina longirostris* increased about 10-fold, and two rotifer species increased about 100–300 fold in the same enclosures. In another experimental study, the introduction on *Daphnia pulex* into enclosures in a small eutrophic pond caused a significant decline in the abundance of edible algae and small grazers, but had no detectable impact on the abundance of *Diaptomus pallidus*, one of the dominant native species (Vanni, 1986).

Edmondson (1985) analyzed a 10-year data set for evidence that colonization of Lake Washington by *Daphnia* spp. influenced the seasonal abundance, size, and reproduction of *Diaptomus ashlandi*. In general, comparisons between years showed only a weak tendency for reciprocal changes in abundance, although there were some suggestive correlations. For example, the year in which *D. ashlandi* showed its strongest autumn peak was the only year in which *Daphnia* did not have an autumn peak. Contrary to the competition hypothesis, *Diaptomus* showed higher reproductive rates and larger adult body size after colonization of the lake by *Daphnia*. Higher birth rates in *Diaptomus* were due to a higher percentage of egg-bearing adults and presumably a shorter interclutch period, rather than an increase in clutch size. This report emphasized comparisons between years; a more detailed analysis of seasonal dynamics should prove helpful. For example, *Diaptomus* birth rates appear to be inversely related to *Diaptomus* densities, suggesting that intraspecific competition limited copepod reproduction. The tendency for *D. ashlandi* to reach peak abundance in spring, before *Daphnia* became abundant, and to show poor juvenile survivorship even in years before *Daphnia* colonized the lake, probably reduced the potential for competition between these two genera.

Several characteristics, acting singly or together, may reduce the intensity of competition between *Daphnia* and calanoid copepods. As shown by Lampert and Muck (1985), the very low food thresholds and often greater starvation resistance of calanoid copepods and selective feeding capabilities should reduce the impact of *Daphnia* competition. On the other hand, slower growth rates and narrower diets may prevent copepods from monopolizing resources and strongly depressing *Daphnia* or other cladocerans. In addition, juvenile stages of both *Daphnia* and copepods may act as bottlenecks, thereby decreasing the overall intensity of competition.

In marked contrast to studies in lakes, Soto (1985) demonstrated strong competition between *Daphnia pulex* and *Diaptomus siciloides* in 1000-liter tanks. *Diaptomus* reached high densities (50–100/l) in tanks without *Daphnia* but remained relatively scarce in tanks with *Daphnia*. Introduction of *Daphnia* into tanks with a high abundance of *Diaptomus* resulted in rapid increases in *Daphnia* and sharp declines in *Diaptomus*. Thus, under some conditions, perhaps dependent on the nature of food resources, *Daphnia* can strongly depress the abundance of *Diaptomus*.

6.9   Habitat Partitioning by Depth as a Stabilizing Influence

It is often assumed that zooplankton species that occupy different depth strata will experience reduced competition. Certainly, species restricted to the epilimnion probably interact very weakly with species living in the hypolimnion. Some workers have argued that zooplankton can avoid competition by diurnal migration, even if they occupy the same depth range (e.g., Lane, 1975; Lane et al., 1978). Such reasoning probably applies to terrestrial vertebrates; lizards and birds that forage by day encounter different prey than do their nocturnal counterparts. Zooplankton that feed on the same algae at different times of the day, however, could still experience strong competition.

Although predation from visually-feeding fish is generally considered the most important selective force influencing zooplankton depth distributions (see Chapter 7), there is both circumstantial and experimental evidence that competition could cause vertical segregation. For example, in Aziscoos Lake, Maine a deep reservoir with five coexisting species of *Daphnia*, the smallest species, *D. ambigua*, was found throughout the water column during spring but was restricted to the hypolimnion during summer stratification (Tappa, 1965). This pattern contrasts with that attributed to fish predation in which larger species occupy greater depths. Tappa (1965) speculated that *D. ambigua* was forced into the hypolimnion because of competition from congeners that occupied the epilimnion. Indeed, in shallow lakes, *D. ambigua* is often dominant during winter and spring but is replaced by other daphnids (*D. parvula*, *Ceriodaphnia*) during late spring and summer (e.g., Allan, 1977; Lei and Armitage, 1980). Thus, declines in *D. ambigua* in Aziscoos Lake during late summer, because of declining O_2 concentrations in the hypolimnion, could be blamed on the "ghost of competition past" (Connell, 1980). In another example, hypolimnetic distributions of two *Bosmina* species and the migrational patterns of several other species in Lake Constance seems inconsistent with the predation hypothesis (Müller, 1985; Geller, 1986).

Reciprocal transfer experiments using enclosures at different depths can provide more direct evidence of the role of competition in vertical distributions. Recently, Hoenicke and Goldman (1987) presented evidence that *Daphnia middendorffiana*, which produced large clutches while living in the hypolimnion of oligotrophic Castle Lake, was unable to survive on epilimnetic resources. Two smaller epilimnetic species, *Daphnia rosea* and *Holopedium gibberum*, however, fared better on epilimnetic resources and were negatively affected by *D. middendorffiana* when placed in water containing hypolimnetic seston. Presumably, *D. middendorffiana* was more efficient at utilizing the more abundant but lower quality algae and detritus that predominated in the hypolimnion. In similar experiments in a eutrophic lake, a hypolimnetic species, *Daphnia pulicaria*, fared poorly in epilimnetic incubations during a brief period of high water temperature and low densities of edible algae (Threlkeld, 1980). An epilimnetic congener, *D. galeata mendotae* continued rapid growth during the same period. These results suggest that the epilimnetic species has a more favorable metabolic balance at high temperatures and low food conditions than does the hypolimnetic species. Thus, differing temperature optima or abilities to use specific resources can result in the vertical segregation of potentially competing zooplankton species, thereby stabilizing coexistence.

6.10   Conclusions

Direct experimental evidence that intense, ongoing competition causes seasonal succession is limited to several experimental studies of competition between daphnids (Lynch, 1978; Smith and Cooper, 1982; DeMott, 1983). Similar demographic trends during successions between other closely related cladocerans and calanoid copepods, however, suggest that intense competition and seasonal shifts in competitive ability are frequently important successions between closely related species, as well as clones of parthenogenetically reproducing forms.

The emerging view of the role of competition in zooplankton communities is in good agreement with a model proposed by Hebert (1982). As also discussed by Ghilarov (1984) and Bengtsson (1986), similarities in competitive abilities and resource requirements may increase the probability of coexistence among intensely competing, closely related clones and species (Figure 6.9). Fluctuating environments may favor parallel or even convergent evolution among competing species. The relationship between genetic similarity and the potential intensity of competition is, of course, not a precise one. For example, the similar feeding mode of daphnids and sidids (e.g., *Diaphanosoma*) may lead to intense competition and shifting advantage, even though phylogenetic arguments suggest that these families should be placed in different orders (Fryer, 1987). In at least some lakes, certain congeners of *Daphnia* (Ghilarov, 1984) and *Eudiaptomus* (Hofmann, 1979) show such similar dynamics that they appear to be ecological equivalents. At the other extreme, some congeners have quite different food preferences (e.g., *Diaptomus shoshone* and *D. coloradensis*; Maly, 1973) and a potential for only weak exploitative competition.

This review is based on the premise that the nature and consequences

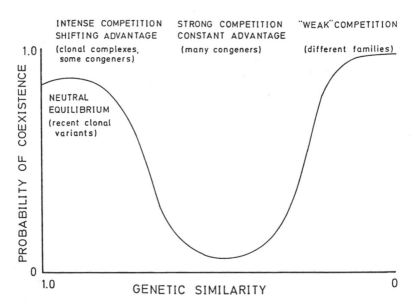

Figure 6.9 Proposed relation between genetic similarity and the probability of coexistence among zooplankton (after Hebert, 1982).

of competition between closely related species (e.g., *Daphnia* congeners) is fundamentally different from competition between distantly related species with different feeding modes and different life histories (e.g., *Daphnia* and rotifers). For example, Neill's (1984) study showed that competition from *Daphnia* caused declines of several orders of magnitude in several rotifer species yet did not seem to influence their seasonal phenologies. Unlike the case with larger organisms (e.g., vertebrates), greatly decreased population density may not substantially increase the probability of local extinction in zooplankton. In general, the same arguments about population size discussed by Hutchinson (1961) in relation to phytoplankton, should also apply to zooplankton. Because zooplankton are relatively small, even species that are near the limits of detectability (0.01 individuals/l) would have large total populations (on the order of 10^7 individuals in a small lake). Asexual reproduction and resting stages should also reduce the change of random extinction.

If, in fact, competitive dominants such as *Daphnia* are unlikely to cause the extinction of more specialized species (rotifers and, perhaps, *Bosmina*), this kind of interaction causes a semantic problem for Hebert's (1982) model. If the probability of competitive exclusion is low, should we call these interactions "weak" even though interspecific competition may be responsible for decreasing the population densities by several orders of magnitude? Lack of understanding of the nature of competition between dissimilar species is one of the principal weaknesses in current knowledge. As in other systems, the persistence of relatively scarce specialized species is difficult to understand, because intraspecific competition is not likely to be a significant stabilizing force when populations densities are very low (Grubb, 1986).

The introduction of this review mentioned resource partitioning and four alternatives:

1. A balance between competition and predation.
2. Shifts in competitive advantage.
3. *r-K* differences combined with alternation between density-independent and density-dependent population growth.
4. Specialized resource tracking.

It seems plausible that niche partitioning and all four nonequilibrium mechanisms might be critical to interactions between different combinations of species in the same zooplankton community. Predation and competition are often considered alternatives. Although size-selective predation by fish can be very effective in eliminating larger zooplankton species, such predation does not necessarily preclude strong competition between small species that can coexist with high densities of planktivorous fish (Benndorf and Horn, 1985; Vanni, 1987).

Evidence reviewed here suggests that competitive outcomes between closely related taxa can be quite sensitive to several seasonally varying factors, including the variance in resource abundance, phytoplankton composition, and temperature. A combination of laboratory experiments, field experiments, and descriptive demographic analyses will probably be needed to demonstrate strong competition and to determine which factor is most important in a particular interaction. The importance of seasonal reversals in competitive ability may help explain the reduced species richness of zooplankton in tropical lakes in comparison to temperate zone lakes (Foran, 1986).

The last two nonequilibrium mechanisms, $r-K$ differences and specialized tracking, appear to be less important as mechanisms for coexistence. Certain rotifers and protozoans (not reviewed here), which dominate during early spring, may be good examples of r-selected fugitive species. Among crustaceans, fugitive species appear to be limited to temporary ponds (e.g., univoltine copepods, Hammer and Sawchyn, 1968; *Moina*, Romanovsky, 1985). Differences along an $r-K$ continuum do not seem to permit the coexistence of congeneric crustaceans in lakes (e.g., *Daphnia pulex* and *D. pulicaria*, Hrbáčková and Hrbáček, 1978; *Diaptomus dorsalis* and *D. floridanus*, Elmore, 1983). Perhaps zooplankton communities approach equilibria with food levels too quickly (e.g., McCauley and Murdoch, 1987; Walters et al., 1987) to allow $r-K$ differences to be significant in promoting coexistence among longer-lived species. In general, zooplankton are less specialized than envisioned in the specialized tracking hypothesis; they eat many different species of phytoplankton. Certainly, *Daphnia*-phytoplankton interactions (McCauley and Murdoch, 1987) are better described by Price's (1984) "steadily renewed" resource category than as "pulsing ephemeral" resources. Nonetheless, it seems likely that population crashes in specialized species, particularly rotifers, are often due to successional changes in the phytoplankton, which are largely independent of grazing.

Resource partitioning is also important in zooplankton communities. Certainly vertical segregation facilitates the coexistence of closely related species, and differences in feeding mode and diet contribute to the coexistence of more distantly related species. As emphasized by Pennak (1957) and Hebert (1978), zooplankton communities exhibit very low species diversity, even in comparison to local species pools. Size constraints imposed by predation are one factor (see Chapter 7). Evidence for strong reciprocal interactions during seasonal succession suggests that competitive exclusion of closely related species is also important. The potential for experimental studies and the range of possible structuring mechanisms make zooplankton communities excellent systems for studying the role of competition in changing environments.

Acknowledgments
I thank Maciej Gliwicz, Joanna Pijanowska, Robert Sterner, Alan Tessier, and Ulrich Sommer for suggesting many improvements in the manuscript. Lara Grubb and Desiree Watson helped with

copy editing. Preparation of this chapter was supported by a summer faculty grant from Indiana-Purdue University at Fort Wayne and NSF grant BSR 8717564.

References

Allan, J.D. 1973. Competition and the relative abundance of two cladocerans. *Ecology* 54: 484–498.

Allan, J.D. 1976. Life history patterns in zooplankton. *American Naturalist* 110: 165–180.

Allan, J.D. 1977. An analysis of seasonal dynamics of a mixed population of *Daphnia* and the associated cladoceran community. *Freshwater Biology* 7: 505–512.

Allan, J.D. and Goulden, C.E. 1980. Some aspects of reproductive variation among freshwater zooplankton, pp. 388–410, in Kerfoot, W.C. (editor), *Ecology and Evolution of Zooplankton Communities*. New England University Press, Hanover, N.H.

Arnold, D.E. 1971. Ingestion, assimilation, survival, and reproduction by *Daphnia pulex* fed seven species of blue-green algae. *Limnology and Oceanography* 16: 906–920.

Arruda, J.A., Marzolf, G.R., and Faulk, R.T. 1983. The role of suspended sediments in the nutrition of zooplankton in turbid reservoirs. *Ecology* 64: 1225–1235.

Bengtsson, J. 1986. Life histories and interspecific competition between three *Daphnia* species in rockpools. *Journal of Animal Ecology* 55: 641–655.

Bengtsson, J. 1987. Competitive dominance among Cladocera: Are single-factor explanations enough? *Hydrobiologia* 145: 19–28.

Benndorf, J. and Horn, W. 1985. Theoretical considerations on the relative importance of food limitation and predation in structuring zooplankton communities. *Archiv für Hydrobiologie Beihefte Ergebnisse der Limnologie* 21: 383–396.

Bleiwas, A.H. and Stokes, P.M. 1985. Collection of large and small particles by *Bosmina*. *Limnology and Oceanography* 30: 1090–1092.

Bogdan, K.G. and Gilbert, J.J. 1984. Body size and food size in freshwater zooplankton. *Proceedings of the National Academy of Sciences USA* 81: 6427–6431.

Bogdan, K.G. and Gilbert, J.J. 1987. Quantitative comparison of food niches in some freshwater zooplankton. A multi-tracer-cell approach. *Oecologia* 72: 331–340.

Brendelberger, H. 1985. Filter mesh-size and retention efficiency for small particles: comparative studies with Cladocera. *Archiv für Hydrobiologie Beihefte Ergebnisse der Limnologie* 21: 135–146.

Brooks, J.L. and Dodson, S.I. 1965. Predation, body size and composition of plankton. *Science* 150: 28–35.

Burns, C.W. 1968. The relationship between body size of filter-feeding Cladocera and the maximum size of particle ingested. *Limnology and Oceanography* 13: 675–678.

Burns, C.W. 1969. Relation between filtering rate, temperature, and body size in four species of *Daphnia*. *Limnology and Oceanography* 14: 693–700.

Burns, C.W. 1985. The effects of starvation on naupliar development and survivorship of three species of *Boeckella* (Copepoda: Calanoida). *Archiv für Hydrobiologie Beihefte Ergebnisse der Limnologie* 21: 297–309.

Byron, E.R., Folt, C.L., and Goldman, C.R. 1984. Copepod and cladoceran success in an oligotrophic lake. *Journal of Plankton Research* 6: 45–65.

Carvalhro, G.R. 1987. The clonal ecology of *Daphnia magna*. II. Thermal differentiation among seasonal clones. *Journal of Animal Ecology* 56: 469–478.

Carvalhro, G.R. and Crisp, D.J. 1987. The clonal ecology of *Daphnia magna* Crustacea: Cladocera. I. Temporal changes in the clonal structure of a natural population. *Journal of Animal Ecology* 56: 453–468.

Chesson, P.L. and Case, T.J. 1986. Overview: Nonequilibrium community theories:

Chance, variability, history, and coexistence, pp. 229–239, in Diamond, J. and Case, T.J. (editors), *Community Ecology*. Harper and Row, New York.

Chow-Fraser, P. and Maly, E.J. 1988. Aspects of mating, reproduction, and co-occurrence in three freshwater calanoid copepods. *Freshwater Biology* 19: 95–108.

Cole, G.A. 1961. Some calanoid copepods from Arizona with notes on congeneric occurrences of *Diaptomus* species. *Limnology and Oceanography* 6: 432–442.

Confer, J.L. and Cooley, J.N. 1977. Copepod instar survival and predation by zooplankton. *Journal of the Fisheries Research Board of Canada* 34: 703–706.

Connell, J.H. 1980. Diversity and the coevolution of competitors, or the ghost of competition past. *Oikos* 35: 131–138.

Cooper, S.D. and Smith, D.W. 1982. Competition, predation, and the relative abundance of two species of *Daphnia*. *Journal of Plankton Research* 4: 859–879.

Crosetti, D. and Margaritora, F.G. 1987. Distribution and life cycles of cladocerans in temporary pools from Central Italy. *Freshwater Biology* 18: 165–176.

Daborn, G. and Hayward, J.A. 1978. Studies on *Daphnia pulex* in sewage oxidation ponds. *Canadian Journal of Zoology* 56: 1392–1401.

Dagg, M. 1977. Some effects of patchy food environments on copepods. *Limnology and Oceanography* 22: 99–107.

Davis, C.C. 1961. Breeding of calanoid copepods in Lake Erie. *Internationale Vereinigung für Theoretische und Angewandte Limnologie: Verhandlungen* 14: 933–942.

Dawidowicz, P., Gliwicz, Z.M., and Gulati, R.D. In press. Can *Daphnia* prevent a blue-green algal bloom in hypereutrophic lakes? *Limnologica*.

DeAngelis, D.L. and Waterhouse, J.C. 1987. Equilibrium and nonequilibrium concepts in ecology. *Ecological Monographs* 57: 1–21.

de Bernardi, R. 1979. An experimental approach to the interspecific competition between two species of *Daphnia: D. hyalina* and *D. pulicaria* (Crustacea). *Vestnik Ceskoslovenske Spolecnosti Zoologicke* 43: 81–93.

DeFrenza, J., Kirner, R.J., Maly, E.J., and Van Leeuwen, H.C. 1986. The relationship of sex size ratio and season to mating intensity in some calanoid copepods. *Limnology and Oceanography* 31: 491–496.

DeMott, W.R. 1982. Feeding selectivities and relative ingestion rates of *Daphnia* and *Bosmina*. *Limnology and Oceanography* 27: 518–527.

DeMott, W.R. 1983. Seasonal succession in a natural *Daphnia* assemblage. *Ecological Monographs* 53: 321–340.

DeMott, W.R. 1985. Relations between filter mesh—size, feeding mode, and capture efficiency for cladocerans feeding on ultrafine particles. *Archiv für Hydrobiologie Beihefte Ergenbnisse der Limnologie* 21: 125–134.

DeMott, W.R. 1986. The role of taste in food selection by freshwater zooplankton. *Oecologia* 69: 334–340.

DeMott, W.R. 1988a. Discrimination between algae and artificial particles by freshwater and marine copepods. *Limnology and Oceanography* 33: 397–408.

DeMott, W.R. 1988b. Discrimination between algae and detritus by freshwater and marine zooplankton. *Bulletin of Marine Science* 43: 486–499.

DeMott, W.R. 1989. Optimal foraging theory as a predictor of chemically mediated selection by suspension-feeding copepods. *Limnology and Oceanography* 1989 34: 140–154.

DeMott, W.R. and Kerfoot, W.C. 1982. Competition among cladocerans: nature of the interaction between *Bosmina* and *Daphnia*. *Ecology* 63: 1949–1966.

Dodson, S.I. 1972. Mortality in a population of *Daphnia rosea*. *Ecology* 53: 1011–1023.

Dodson, S.I. 1974. Zooplankton competition and predation: an experimental test of the size-efficiency hypothesis. *Ecology* 55: 605–613.

Duncan, A. 1985. Body carbon in daphnids as an indicator of the food concentration available in the field. *Archiv für Hydrobiologie Beihefte Ergebnisse der Limnologie* 21: 81–90.

Edmondson, W.T. 1985. Reciprocal changes in abundance of *Diaptomus* and *Daphnia* in Lake Washington. *Archiv für Hydrobiologie Beihefte Ergebnisse der Limnologie* 21: 475–481.

Elmore, J.L. 1983. Factors influencing *Diaptomus* distributions: An experimental study in subtropical Florida. *Limnology and Oceanography* 28: 522–532.

Ferguson, A.J.D., Thompson, J.M., and Reynolds, C.S. 1982. Structure and dynamics of zooplankton communities maintained in closed systems, with special reference to the algal food supply. *Journal of Plankton Research* 4: 523–543.

Foran, J.A. 1986. The relationship between temperature, competition and the potential for colonization of a subtropical pond by *Daphnia magna*. *Hydrobiologia* 134: 103–112.

Frost, B.W. 1980. The inadequacy of body size as an indicator of niches in the zooplankton, pp. 742–753, in Kerfoot, W.C. (editor), *Ecology and Evolution of Zooplankton Communities*. University Press of New England, Hanover, N.H.

Frost, B.W. 1985. Food limitation of the planktonic marine copepods *Calanus pacificus* and *Pseudocalanus* sp. in a temperate fjord. *Archiv für Hydrobiologie Beihefte Ergebnisse der Limnologie* 21: 1–13.

Fryer, G. 1987. Morphology and the classification of the so-called Cladocera. *Hydrobiologia* 145: 19–28.

Fulton, R.S. III. 1989. Grazing on filamentous algae by herbivorous zooplankton. *Freshwater Biology* 20: 263–272.

Fulton, R.S. III and Paerl, H.W. 1987a. Effects of colonial morphology on zooplankton utilization of algal resources during blue-green algal (*Microcystis aeruginosa*) blooms. *Limnology and Oceanography* 32: 634–644.

Fulton, R.S. III and Paerl, H.W. 1987b. Toxic and inhibitory effects of the blue-green alga *Microcystis aeruginosa* on herbivorous zooplankton. *Journal of Plankton Research* 9: 837–855.

Geller, W. 1986. Diurnal vertical migration of zooplankton in a temperate great lake (L. Constance): A starvation avoidance mechanism? *Archiv für Hydrobiologie/Supplement* 74: 1–60.

Geller, W. and Müller, H. 1981. The filtration apparatus of Cladocera: filter mesh–sizes and their implications on food selectivity. *Oecologia* 49: 316–321.

Geller, W. and Müller, H. 1985. Seasonal variability in the relationship between body length and individual dry weight as related to food abundance and clutch size in two coexisting *Daphnia* species. *Journal of Plankton Research* 7: 1–18.

Ghilarov, A.M. 1984. The paradox of the plankton reconsidered; or, why do species coexist. *Oikos* 43: 46–52.

Gilarov, A.M. 1985. Dynamics and structure of cladoceran populations under conditions of food limitation. *Archiv für Hydrobiologie Beihefte Ergebnisse der Limnologie* 21: 323–332.

Gilbert, J.J. 1985. Competition between rotifers and *Daphnia*. *Ecology* 66: 1943–1950.

Gilbert, J.J. and Bodgan, K.G. 1984. Rotifer grazing: in situ studies on selectivity and rates, pp. 97–133, in Meyers, D.G. and Strickler, J.R. (editors), *Trophic Interactions Within Aquatic Ecosystems*, AAAS Symposium Series no. 85. Westview, Boulder, Colorado.

Gilbert, J.J. and Stemberger, R.S. 1985. Control of *Keratella* populations by interference competition from *Daphnia*. *Limnology and Oceanography* 30: 180–188.

Gliwicz, Z.M. 1977. Food size selection and seasonal succession of filter feeding zooplankton in a eutrophic lake. *Ekologia Polska A* 25: 179–225.

Gliwicz, Z.M. 1980. Filtering rates, food size-selection, and feeding rates in cladocerans—Another aspect of interspecific competition in filter-feeding zooplankton, pp. 282–291, in Kerfoot, W.C. (editor), *Ecology and Evolution of Zooplankton Communities*. University of New England Press, Hanover, N.H.

Gliwicz, Z.M., Ghilarov, A., and Pijanowska, J. 1981. Food and predation as major factors limiting two natural populations of *Daphnia cucullata*. *Hydrobiologia* 80: 205–218.

Gliwicz, Z.M. and Lampert, W. In prep. Competitive superiority in three *Daphnia* species and its reversal in the presence of blue-green filaments.

Gliwicz, Z.M. and Siedlar, E. 1980. Food size limitation and algae interfering with food collection in *Daphnia*. *Archiv für Hydrobiologie* 88: 155–177.

Gophen, M. and Geller, W. 1984. Filter mesh size and food particle uptake by *Daphnia*. *Oecologia* 64: 408–412.

Goulden, C.E., Henry, L., and Berrigan, D. 1987. Egg size, postembryonic yolk and survival ability. *Oecologia* 72: 28–31.

Goulden, C.E. and Henry, L.L. 1984. Lipid energy reserves and their role in Cladocera, pp. 167–185, in Meyers, D.G. and Strickler, J.R. (editors), *Trophic Interactions Within Aquatic Ecosystems*, AAAS Selection Symposium 85. Westview Press, Boulder, Colorado.

Goulden, C.E., Henry, L.L., and Tessier, A.J. 1982. Body size, energy reserves and compoetitive ability in three species of Cladocera. *Ecology* 63: 1780–1789.

Goulden, C.E. and Hornig, L.L. 1980. Body size, energy reserves in plankton Cladocera and their consequences to competition. *Proceedings of the National Academy of Sciences USA* 77: 1716–1720.

Goulden, C.E., Hornig, L.L., and Wilson, C. 1978. Why do large zooplankton species dominate? *Internationale Vereinigung für Theoretische und Angewandte Limnologie: Verhandlungen* 20: 2457–2460.

Grant, J.W.G. and Bayly, I.A.E. 1981. Predator induction of crests in morphs of the *Daphnia carinata* King complex. *Limnology and Oceanography* 26: 201–218.

Grubb, P.J. 1986. Problems posed by sparse and patchily distributed species in species-rich plant communities, pp. 207–225, in Diamond, J. and Case, T.J. (editors), *Community Ecology*. Harper and Row, New York.

G.-Toth, L., Zankai, N.P., and Messner, O.M. 1987. Alga consumption of four dominant planktonic crustaceans in Lake Balaton (Hungary). *Hydrobiologia* 145: 323–332.

Hairston, N.G. Jr. and Twombly, S. 1985. Obtaining life table data from cohort analysis: a critique of current methods. *Limnology and Oceanography* 30: 886–892.

Hall, D.J., Threlkeld, S.T., Burns, C.W., and Crowley, P.H. 1976. The size-efficiency hypothesis and the size structure of zooplankton communities. *Annual Review of Ecology and Systematics* 7: 177–208.

Hammer, U.T. and Sawchyn, W.W. 1968. Seasonal succession and congeneric associations of *Diaptomus* spp. (Copepoda) in some Saskatchewan Ponds. *Limnology and Oceanography* 13: 476–484.

Hanazato, T. and Yasuno, M. 1985. Effect of temperature in the laboratory studies on growth, egg development and first parturition of five species of Cladocera. *Japanese Journal of Limnology* 46: 185–191.

Hanazato, T. and Yasuno, M. 1987a. Evaluation of *Microcystis* as a food for zooplankton in a eutrophic lake. *Hydrobiologia* 144: 251–259.

Hanazato, T. and Yasuno, M. 1987b. Experimental studies on competition between *Bosmina longirostris* and *Bosmina fatalis*. *Hydrobiologia* 154: 189–199.

Hanazato, T., Yasuno, M., Iwakuma, T., and Takamura, N. 1984. Seasonal changes in the occurrence of *Bosmina longirostris* and *Bosmina fatalis* in relation to *Microcystis* bloom in Lake Kasumigaura. *Japanese Journal of Limnology* 45: 153–157.

Harris, R.P. 1982. Comparison of the feeding behaviour of *Calanus* and *Pseudocalanus* in two experimentally manipulated enclosed ecosystems. *Journal of the Marine Biological Association of the United Kingdom* 62: 71–91.

Harris, R.P., and Paffenhöfer, G.-A. 1976. The effect of food concentration on cumulative ingestion and growth efficiency of two small marine planktonic copepods. *Journal of the Marine Biological Association of the United Kingdom* 56: 875–888.

Hart, R.D. 1986. Zooplankton abundance, community structure and dynamics in relation to inorganic turbidity, and their implications for a potential fishery in subtropical Lake le Roux, South Africa. *Freshwater Biology* 16: 351–371.

Havel, J.E. and Dodson, S.I. 1984. *Chaoborus* predation on typical and spined morphs on *Daphnia pulex*: Behavioral observations. *Limnology and Oceanography* 29: 487–494.

Hayward, T.L. and McGowan, J.A. 1979. Patterns and structure in an oceanic zooplankton community. *American Zoologist* 19: 1045–1055.

Hebert, P.D.N. 1977. Niche overlap among species in the *Daphnia carinata* complex. *Journal of Animal Ecology* 46: 399–409.

Hebert, P.D.N. 1978. The population biology of *Daphnia*. *Biological Reviews* 53: 387–426.

Hebert, P.D.N. 1982. Competition in zooplankton communities. *Annales Zoologici Fennici* 19: 349–356.

Hebert, P.D.N. and Crease, T.J. 1980. Clonal coexistence in *Daphnia pulex* (Leydig): Another planktonic paradox. *Science* 207: 1363–1365.

Heisey, D. and Porter, K.G. 1977. The effect of ambient oxygen concentration on filtering and respiration rates of *Daphnia galeata mendotae* and *Daphnia magna*. *Limnology and Oceanography* 22: 839–845.

Herzig, A. 1984. Temperature and life cycle strategies of *Diaphanosoma brachyurum*: an experimental study on development, growth, and survival. *Archiv für Hydrobiologie* 101: 143–178.

Hessen, D.O. 1985. Filtering structures and particle size selection in coexisting Cladocera. *Oecologia* 66: 368–372.

Hoenicke, R. and Goldman, C.R. 1987. Resource dynamics and seasonal changes in competitive interactions among three cladoceran species. *Journal of Plankton Research* 9: 397–417.

Hofmann, W. 1979. Characteristics of syntopic populations of *Eudiaptomus gracilis* (Sars) and *E. graciloides* (Lilljeborg) in three lakes with different trophic levels. *Archiv für Hydrobiologie* 86: 1–12.

Holm, N.P., Ganf, G.G., and Shapiro, J. 1983. Feeding and assimilation rates of *Daphnia pulex* fed *Aphanizomenon flos-aquae*. *Limnology and Oceanography* 28: 677–687.

Hrbáček, J. 1977. Competition and predation in relation to species composition of freshwater zooplankton, mainly Cladocera, pp. 305–353, in Cairns, J. (editor), *Aquatic Microbial Communities*. Garland, New York.

Hrbáčková, M. and Hrbáček, J. 1978. The growth rate of *Daphnia pulex* and *Daphnia pulicaria* (Crustacea: Cladocera) at different food levels. *Vestnik Ceskoslovenske Spolecnosti Zoologicke* 42: 115–127.

Hrbáčková, M. and Hrbáček, J. 1979. Rate of the postembryonic development in several populations of the group of species *Daphnia hyalina* at various concentrations of food. *Vestnik Ceskoslovenske Spolecnosti Zoologicke* 43: 253–259.

Hrbáčková-Esslova, M. 1963. The development of three species of *Daphnia* in the surface water of the Slapy Reservoir. *Internationale Revue der gesamten Hydrobiologie* 48: 325–333.

Huston, M. 1979. A general hypothesis of species diversity. *American Naturalist* 113: 81–101.

Hutchinson, G.E. 1951. Copepodology for the ornithologist. *Ecology* 32: 571–577.

Hutchinson, G.E. 1959. Homage to Santa Rosalia, or why are there so many kinds of animals? *American Naturalist* 93: 145–159.

Hutchinson, G.E. 1961. The paradox of the plankton. *American Naturalist* 95: 137–146.

Hutchinson, G.E. 1967. *A Treatise of Limnology. Volume II. Introduction to Lake Biology and the Limnoplankton.* John Wiley and Sons, New York.

Infante, A. 1973. Untersucheungen uber die Ausutzbarkeit verscheidener Algen durch das Zooplankton. *Archiv für Hydrobiologie Supplement* 42: 340–405.

Infante, A. and Abdella, S.E.B. 1985. Inhibition of *Daphnia* by *Oscillatoria* in Lake Washington. *Limnology and Oceanography* 30: 1046–1052.

Infante, A. and Litt, A.H. 1985. Differences between two species of *Daphnia* in the use of 10 species of algae in Lake Washington. *Limnology and Oceanography* 30: 1053–1059.

Infante, A. and Riehl, W. 1984. The effect of Cyanophyta upon zooplankton in a eutrophic tropical lake (Lake Valencia, Venezuela). *Hydrobiologia* 113: 293–298.

Jacobs, J. 1977a. Coexistence of similar zooplankton species by differential adaptation to reproduction and escape in an environment with fluctuating food and enemy densities. I. A model. *Oecologia* 29: 233–247.

Jacobs, J. 1977b. Coexistence of similar zooplankton species by differential adaptation to reproduction and escape in an environment with fluctuating food and enemy densities. II. Field analysis of *Daphnia*. *Oecologia* 30: 313–329.

Jamieson, C.D. and Burns, C.W. 1985. Copepod distribution patterns: Life history adaptations to food and temperature. *Internationale Vereinigung für Theoretische und Angewandte Limnologie: Verhandlungen* 22: 3169.

Jarvis, A.C. 1986. Zooplankton community grazing in a hypertrophic lake (Hartbeespoort Dam, South Africa). *Journal of Plankton Research* 8: 1065–1078.

Katona, S.K. 1973. Evidence for sex pheromones in planktonic copepods. *Limnology and Oceanography* 18: 574–583.

Kerfoot, W.C. 1977. Competition in cladoceran communities: The cost of evolving defenses against copepod predation. *Ecology* 58: 303–313.

Kerfoot, W.C. 1981. Long-term replacement cycles in cladoceran communities. A history of predation. *Ecology* 62: 216–233.

Kerfoot, W.C. and DeMott, W.R. 1988. Foundations for evaluating community interactions: the use of enclosures to investigate the coexistence of *Daphnia* and *Bosmina*, pp. 725–741, in Kerfoot, W.C. (editor), *Ecology and Evolution of Zooplankton Communities*. University Press of New England, Hanover, N.H.

Kerfoot, W.C., DeMott, W.R., and DeAngelis, D.L. 1985a. Interactions among cladocerans: Food limitation and exploitative competition. *Archiv für Hydrobiologie Beihefte Ergebnisse der Limnologie* 21: 431–452.

Kerfoot, W.C., DeMott, W.R., and Levitan, C. 1985b. Nonlinearities in competitive interactions: Component variable or system response? *Ecology* 66: 959–965.

Kerfoot, W.C., Levitan, C., and DeMott, W.R. 1988. *Daphnia*-phytoplankton interactions: density dependent shifts in resource quality. *Ecology* 69: 1806–1825.

Kersting, K. 1983. Direct determination of the "threshold food concentration" for *Daphnia magna*. *Archiv für Hydrobiologie* 96: 510–514.

Koch, A.L. 1974. Coexistence resulting from an alteration of density dependent and density independent growth. *Journal of Theoretical Biology* 44: 373–386.

Koehl, M.A.R. 1984. Mechanisms of particle capture by copepods at low Reynolds numbers: Possible modes of selective feeding, pp. 135–166, in Meyers, D.G. and Strickler, J.R. (editors), *Trophic Interactions Within Aquatic Ecosystems*, AAAS Selected Symposium 85. Westview Press, Boulder, Colorado.

Kratz, T.K., Frost, T.M., and Magnuson, J.J. 1987. Inferences from spatial and temporal variability in ecosystems: long-term zooplankton data from lakes. *American Naturalist* 129: 830–846.

Kring, R.L. and O'Brien, W.J. 1976. Accommodation of *Daphnia pulex* to altered pH conditions as measured by feeding rate. *Limnology and Oceanography* 21: 313–315.

Lampert, W. 1976. A directly coupled, artificial two-step food chain for long-term experiments with filter-feeders at constant food concentrations. *Marine Biology* 37: 349–355.

Lampert, W. 1977a. Studies on the carbon balance of *Daphnia pulex* De Geer as related to environmental conditions. I. Methodological problems of the use of ^{14}C for the measurement of carbon assimilation. *Archiv für Hydrobiologie, Supplement* 48: 287–309.

Lampert, W. 1977b. Studies on the carbon balance of *Daphnia pulex* De Geer as related to environmental conditions. II. The dependence of carbon assimilation on animal size, temperature, food concentration, and diet species. *Archiv für Hydrobiologie, Supplement* 48: 310–335.

Lampert, W. 1977c. Studies on the carbon balance of *Daphnia pulex* De Geer as related to environmental conditions. IV. Determination of the "threshold" concentration as a factor controlling the abundance of zooplankton species. *Archiv für Hydrobiologie, Supplement* 48: 361–368.

Lampert, W. 1978. A field study on the dependence of the fecundity of *Daphnia* on food concentration. *Oecologia* 36: 363–369.

Lampert, W. 1981. Inhibitory and toxic effects of blue-green algae on *Daphnia*. *Internationale Revue der gesamten Hydrobiologie* 66: 285–298.

Lampert, W., Fleckner, W., Rai, H., and Taylor, B.E. 1986. Phytoplankton control by grazing zooplankton: A study on the spring clear-water phase. *Limnology and Oceanography* 31: 478–490.

Lampert, W. and Muck, P. 1985. Multiple aspects of food limitation in zooplankton communities: The *Daphnia-Eudiaptomus* example. *Archiv für Hydrobiologie Beihefte Ergebnisse der Limnologie* 21: 311–322.

Lampert, W. and Schober, U. 1980. The importance of "threshold" food concentrations, pp. 264–267, in Kerfoot, W.C. (editor), *Ecology and Evolution of Zooplankton Communities*. Press of New England, Hanover, N.H.

Lane, P.A. 1975. The dynamics of aquatic systems: A comparative study of the structure of four zooplankton communities. *Ecological Monographs* 45: 307–336.

Lane, P.A., Makarewicz, J.C., and Likens, G.E. 1978. Zooplankton niches and the community structure controversy. *Science* 200: 458–463.

Lehman, J.T. 1976. The filter feeder as an optimal forager, and the predicted shape of feeding curves. *Limnology and Oceanography* 21: 501–516.

Lei, C.H. and Armitage, K.B. 1980. Population dynamics and production of *Daphnia ambigua* in a fish pond, Kansas. *University of Kansas Science Bulletin* 25: 687–715.

Levins, R. 1979. Coexistence in a variable environment. *American Naturalist* 113: 765–783.

Levitan, C. 1987. Formal stability analysis of a planktonic freshwater community, pp. 71–100, in Kerfoot, W.C. and Sih, A. (editors), *Predation: Direct and Indirect Impacts on Aquatic Communities*. University Press of New England, Hanover.

Loaring, J.M. and Hebert, P.D.N. 1981. Ecological differences among clones of *Daphnia pulex*. *Oecologia* 51: 162–168.

Lynch, M. 1977. Fitness and optimal size in zooplankton populations. *Ecology* 58: 763–774.

Lynch, M. 1978. Complex interactions between natural coexploiters *Daphnia* and *Ceriodaphnia*. *Ecology* 59: 552–564.

Lynch, M. 1979. Predation, competition, and zooplankton community structure: An experimental study. *Limnology and Oceanography* 24253–272.

Lynch, M. 1980a. Predation, enrichment, and the evolution of cladoceran life histories:

a theoretical approach, pp. 367–376, in Kerfoot, W.C. (editor), *Ecology and Evolution of Zooplankton Communities*. University Press of New England, Hanover, N.H.

Lynch, M. 1980b. The evolution of cladoceran life histories. *Quarterly Review of Biology* 55: 23–42.

Lynch, M. and Shapiro, J. 1981. Predation, enrichment, and phytoplankton community structure. *Limnology and Oceanography* 26: 86–102.

MacArthur, R.H. and Wilson, E.O. 1967. *The Theory of Island Biogeography*. Princeton University Press, Princeton, N.J.

Makarewicz, J.C. and Likens, G.E. 1975. Niche analysis of a zooplankton community. *Science* 190: 100–103.

Maly, E.J. 1973. Density, size and clutch of two high altitude diaptomid copepods. *Limnology and Oceanography* 18: 840–848.

Maly, E.J. 1976. Resource overlap between co-occurring copepods: effects of predation and environmental fluctuation. *Canadian Journal of Zoology* 54: 933–940.

Maly, E.J. and Maly, M.P. 1974. Dietary differences between two co-occurring calanoid copepod species. *Oecologia* 17: 325–333.

Matveev, V.M. 1983. Estimating competition in cladocerans using data on dynamics of clutch size and population density. *Internationale Revue der gesamten Hydrobiologie* 68: 785–798.

Matveev, V.M. 1985a. Delayed density dependence and competitive ability in two cladocerans. *Archiv für Hydrobiologie Beihefte Ergebnisse der Limnologie* 21: 453–459.

Matveev, V.M. 1985b. Competition and population time lags in *Bosmina* (Cladocera, Crustacea). *Internationale Revue der gesamten Hydrobiologie* 70: 491–508.

Matveev, V.F. 1986a. Long-term changes in the community of planktonic crustaceans in Lake Glubokoe in relation to predation and competition. *Hydrobiologia* 141: 33–43.

Matveev, V.F. 1986b. History of the community of planktonic Cladocera in Lake Glubokoeo (Moscow Region). *Hydrobiologia* 141: 145–152.

Matveev, V.F. 1987. Effect of competition on the demography of planktonic cladocerans *Daphnia* and *Diaphanosoma*. *Oecologia* 74: 468–477.

Mayzaud, P. and Mayzaud, O. 1985. The influence of food limitation on the nutritional adaptation of marine zooplankton. *Archiv für Hydrobiologie Beihefte Ergebnisse der Limnologie* 21: 223–233.

Mayzaud, P. and Poulet, S.A. 1978. The importance of the time factor in the response of zooplankton to varying concentrations of naturally occurring particulate matter. *Limnology and Oceanography* 23: 1144–1154.

McCauley, E. and Murdoch, W.W. 1987. Cyclic and stable populations: plankton as paradigm. *American Naturalist* 129: 97–121.

McNaught, D.C. 1975. A hypothesis to explain the succession from calanoids to cladocerans during eutrophication. *Internationale Vereinigung für Theoretische und Angewandte Limnologie: Verhandlungen* 19: 724–731.

Miller, R.S. 1967. Pattern and process in competition. *Advances in Ecological Research* 4: 1–74.

Modlin, R.F. 1982. Successional changes, variations in population densities, and reproductive strategies of Cladocera in two temporary ponds in North Alabama. *Journal of Freshwater Ecology* 1: 589–598.

Mort, M.A. and Wolf, H.G. 1985. Enzyme variability in large-lake *Daphnia*. *Heredity* 55: 27–37.

Muck, P. and Lampert, W. 1984. An experimental study on the importance of food conditions for the relative abundance of calanoid copepods and cladocerans. I. Comparative feeding studies with *Eudiaptomus gracilis* and *Daphnia longispina*. *Archiv für Hydrobiologie, Supplement* 66: 157–179.

Müller, H. 1985. The niches of *Bosmina coregoni* and *Bosmina longirostris* in the ecosystem of Lake Constance. *Internationale Vereinigung für Theoretische und Ange-wandte Limnologie: Verhandlungen* 22: 3137–3143.

Murdoch, W.W. and McCauley, E. 1985. Three distinct types of dynamic behavior shown by a single planktonic system. *Nature* 316: 628–630.

Nadlin-Hurley, C.M. and Duncan, A. 1976. A comparison of *Daphnia* gut particles with sestonic particles present in two Thames Valley reservoirs throughout 1970 and 1971. *Freshwater Biology* 6: 109–123.

Neill, W.E. 1975. Experimental studies of microcrustacean composition and efficiency of resource utilization. *Ecology* 56: 809–826.

Neill, W.E. 1978. Experimental studies on factors limiting colonization by *Daphnia pulex* Leydig of coastal montane lakes in British Columbia. *Canadian Journal of Zoology* 56: 2498–2507.

Neill, W.E. 1981. Developmental responses of juvenile *Daphnia rosea* to experimental alteration of temperature and natural seston concentration. *Canadian Journal of Zoology* 38: 1357–1362.

Neill, W.E. 1981b. Impact of *Chaoborus* predation upon the structure and dynamics of a crustacean zooplankton community. *Oecologia* 48: 164–177.

Neill, W.E. 1984. Regulation of rotifer densities by crustacean zooplankton in an oligotrophic montane lake in British Columbia. *Oecologia* 61: 175–181.

Neill, W.E. 1985. The effects of herbivore competition upon the dynamics of *Chaoborus* predation. *Archiv für Hydrobiologie Beihefte Ergebnisse der Limnologie* 21: 483–491.

O'Brien, W.J., Kettle, D., and Riessen, H. 1979. Helmets and invisible armor: structures reducing predation from tactile and visual planktivores. *Ecology* 60: 287–294.

Orcutt, J.R. Jr. and Porter, K.G. 1984. The synergistic effects of temperature and food concentration on life history parameters of *Daphnia. Oecologia* 63: 300–306.

Orcutt, J.D. and Porter, K.G. 1985. Food level effects on the competitive interactions of two co-occurring cladoceran zooplankton: *Diaphanosoma brachyurum* and *Daphnia ambigua. Archiv für Hydrobiologie Beihefte Ergebnisse der Limnologie* 21: 465–474.

Pace, M.L., Porter, K.G., and Feig, Y.S. 1983. Species- and age-specific differences in bacterial resource utilization by two co-occurring cladocerans. *Ecology* 64: 1145–1156.

Pennak, R.W. 1957. Species composition of limnetic zooplankton communities. *Limnology and Oceanography* 2: 222–232.

Piyasiri, S. 1985. Methodological aspects of defining food dependence and food thresholds in freshwater calanoids. *Archiv für Hydrobiologie Beihefte Ergebnisse der Limnologie* 21: 277–284.

Porter, K.G. 1975. Viable gut passage of gelatinous green algae ingested by *Daphnia. Internationale Vereinigung für Theoretische und Angewandte Limnologie: Verhandlungen* 19: 2840–2850.

Porter, K.G. and Mcdonough, R. 1984. The energetic cost of response to blue-green algal filaments by cladocerans. *Limnology and Oceanography* 29: 365–369.

Porter, K.G. and Orcutt, J.D. Jr. 1980. Nutritional adequacy, manageability, and toxicity as factors that determine food quality of green and blue-green algae for zooplankton, pp. 538–554, in Kerfoot, W.C. (editor), *Ecology and Evolution of Zooplankton Communities.* University Press of New England, Hanover, N.H.

Price, H.J. 1988. Feeding mechanisms in marine and freshwater zooplankton. *Bulletin of Marine Science* 43: 327–343.

Price, H.J. and Paffenhöfer, G.-A. 1985. Perception of food availability by calanoid copepods. *Archiv für Hydrobiologie Beihefte Ergebnisse der Limnologie* 21: 115–124.

Price, P.W. 1985. Alternative paradigms in community ecology, pp. 353–386, in Price, P.W., Slobodchikoff, C.N., and Gaud, W.S. (editors), *A New Ecology: Novel Approaches to Interactive Systems.* John Wiley and Sons, New York, N.Y.

Richman, S., Bohon, S.A., and Robbins, S.E. 1980. Grazing interactions among freshwater calanoid copepods, pp. 219–240, in Kerfoot, W.C. (editor), *Ecology and Evolution of Zooplankton Communities*. University Press of New England, Hanover, N.H.

Richman, S. and Dodson, S.I. 1983. The effect of food quality on feeding and respiration by *Daphnia* and *Diaptomus*. *Limnology and Oceanography* 28: 948–956.

Rigler, F.H. and Langford, R.R. 1967. Congeneric occurrences of species of *Diaptomus* in southern Ontario lakes. *Canadian Journal of Zoology* 45: 81–90.

Romanovsky, Y.E. 1984a. Individual growth rate as a measure of competitive advantages of cladoceran crustaceans. *Internationale Revue der gesamten Hydrobiologie* 69: 613–632.

Romanovsky, Y.E. 1984b. Prolongation of postembryonic development in experimental and natural cladoceran populations. *Internationale Revue der gesamten Hydrobiologie* 69: 149–157.

Romanovsky, Y.E. 1985. Food limitation and life history strategies in cladoceran crustaceans. *Archiv für Hydrobiologie Beihefte Ergebnisse der Limnologie* 21: 363–372.

Romanovsky, Y.E. and Feniova, I.Y. 1985. Competition among Cladocera: Effects of different levels of food supply. *Oikos* 44: 243–252.

Rothkaupt, K.-O. 1988. Mechanistic resource competition theory applied to experiments with zooplankton. *Nature* 333: 660–662.

Sandercock, G.A. 1967. A study of selected mechanisms for the coexistence of *Diaptomus* spp. in Clarke Lake, Ontario. *Limnology and Oceanography* 12: 97–112.

Sarnelle, O. 1986. Field assessment of the quality of phytoplanktonic food available to *Daphnia* and *Bosmina*. *Hydrobiologia* 131: 47–56.

Schoenberg, S.A. and Carlson, R.E. 1984. Direct and indirect effects of zooplankton grazing on phytoplankton in a hypereutrophic lake. *Oikos* 42: 291–302.

Schoenberg, S.A. and MacCubbin, A.E. 1985. Relative feeding rates on free and particle-bound bacteria by freshwater macrozooplankton. *Limnology and Oceanography* 30: 1084–1089.

Schoener, T.W. 1982. The controversy over interspecific competition. *American Scientist* 70: 586–595.

Scott, J.M. 1980. Effect of growth rate of the food alga on the growth/ingestion efficiency of a marine herbivore. *Journal of the Marine Biological Association of the United Kingdom* 60: 691–702.

Shapiro, J. and Wright, D.I. 1984. Lake Restoration by biomanipulation: Round Lake, Minnesota, the first two years. *Freshwater Biology* 14: 371–383.

Smith, D.W. and Cooper, S.D. 1982. Competition among Cladocera. *Ecology* 63: 1004–1015.

Smith, D.W., Cooper, S.D., and Sarnelle, O. 1988. Curvilinear density dependence and the design of field experiments on zooplankton competition. *Ecology* 69: 868–869.

Smith, F.E. 1954. Quantitative aspects of population growth, pp. 277–294, in Boel, E.J. (editor), *Dynamics of Growth Processes, Growth Symposium II*. Princeton University Press, Princeton, N.J.

Sommer, U. 1985. Comparison between steady state and non-steady state competition: Experiments with natural phytoplankton. *Limnology and Oceanography* 30: 335–346.

Sommer, U., Gliwicz, Z.M., Lampert, W., and Duncan, W. 1986. The PEG-model of seasonal succession of planktonic events in fresh waters. *Archiv für Hydrobiologie* 106: 433–471.

Soto, D. 1985. Experimental evaluation of copepod interactions. *Internationale Vereinigung für Theoretische und Angewandte Limnologie: Verhandlungen* 22: 3199–3204.

Starkweather, P.L. and Kellar, P.E. 1983. Utilization of cyanobacteria by *Brachionus calyciflorus: Anabaena flos-aquae* (NRC-44-1) as a sole or complementary food source. *Hydrobiologie* 104: 373–377.

Stemberger, R.S. and Gilbert, J.J. 1985a. Assessment of threshold food levels and population growth in planktonic rotifers. *Archiv für Hydrobiologie Beihefte Ergebnisse der Limnologie* 31: 269–276.

Stemberger, R.S. and Gilbert, J.J. 1985b. Body size, food concentration and population growth in planktonic rotifers. *Ecology* 66: 1151–1159.

Strong, D.R. 1986. Density vagueness: Abiding the variance in the demography of real populations, pp. 257–268, in Diamond, J. and Case, T.J. (editors), *Community Ecology*, Harper and Row, New York.

Tappa, D.W. 1965. The dynamics of the association of six limnetic species of *Daphnia* in Aziscoos Lake, Maine. *Ecological Monographs* 35: 395–423.

Taub, F.B. and Dollar, A.M. 1968. The nutritional inadequacy of *Chlorella* and *Chlamydomonas* as food for *Daphnia pulex*. *Limnology and Oceanography* 13: 607–617.

Taylor, B.E. 1985. Effects of food limitation on growth and reproduction of *Daphnia*. *Archiv für Hydrobiologie Beihefte Ergebnisse der Limnologie* 21: 285–296.

Tessier, A.J. 1986. Comparative population regulation of two planktonic Cladocera (*Holopedium gibberum* and *Daphnia catawba*). *Ecology* 67: 285–302.

Tessier, A.J. and Goulden, C.E. 1982. Estimating food limitation in cladoceran populations. *Limnology and Oceanography* 27: 707–717.

Tessier, A.J. and Goulden, C.E. 1987. Cladoceran juvenile growth. *Limnology and Oceanography* 32: 680–686.

Tessier, A.J., Henry, L.L., Goulden, C.E., and Durand, M.W. 1983. Starvation in *Daphnia*: energy reserves and reproductive allocation. *Limnology and Oceanography* 28: 667–676.

Threlkeld, S.T. 1976. Starvation and the size structure of zooplankton communities. *Freshwater Biology* 6: 489–496.

Threlkeld, S.T. 1979. Estimating cladoceran birth rates: The importance of egg mortality and the egg age distribution. *Limnology and Oceanography* 24: 601–612.

Threlkeld, S.T. 1980. Habitat selection and population growth of two cladocerans in seasonal environments, pp. 346–357, in Kerfoot, W.C. (editor), *Ecology and Evolution of Zooplankton Communities*. University Press of New England, Hanover, N.H.

Threlkeld, S.T. 1985. Resource variation and the initiation of midsummer declines of cladoceran populations. *Archiv für Hydrobiologie Beihefte Ergebnisse der Limnologie* 32: 333–340.

Threlkeld, S.T. 1986. Resource-mediated demographic variation during the midsummer succession of a cladoceran community. *Freshwater Biology* 16: 673–684.

Tillmann, U. and Lampert, W. 1984. Competitive ability of differently sized *Daphnia* species: an experimental test. *Journal of Freshwater Ecology* 2: 311–323.

Tilman, D. 1982. *Resource Competition and Community Structure*. Princeton University Press, Princeton, New Jersey.

Tilman, D., Mattson, M., and Langer, S. 1981. Competition and nutrient kinetics along a temperature gradient: An experimental test of a mechanistic approach to niche theory. *Limnology and Oceanography* 26: 1020–1033.

Vanderploeg, H.A., Paffenhöfer, G.-A., and Liebig, J.R. 1988. *Diaptomus* vs. net phytoplankton: Roles of algal size and morphology in grazing avoidance. *Bulletin of Marine Science* 43: 377–394.

Vanni, M.J. 1986. Competition in zooplankton communities: Suppression of small species by *Daphnia pulex*. *Limnology and Oceanography* 31: 1039–1056.

Vanni, M.J. 1987. Food availability, fish predation, and the dynamics of a zooplankton community coexisting with planktivorous fish. *Ecological Monographs* 57: 61–88.

Vidal, J. 1980. Physioecology of zooplankton. I. Effects of phytoplankton concentration, temperature, and body size on the growth rate of *Calanus pacificus* and *Pseudocalanus* sp. *Marine Biology* 56: 111–134.

Von Ende, C.N. and Dempsey, D.O. 1981. Apparent exclusion of the cladoceran *Bosmina longirostris* by the invertebrate predator *Chaoborus americanus*. *American Midland Naturalist* 105: 240–248.

Walters, C.J., Krause, E., Neill, W.E., and Northcote, T.G. 1987. Equilibrium models for seasonal dynamics of plankton biomass in four oligotrophic lakes. *Canadian Journal of Fisheries and Aquatic Sciences* 44: 1002–1017.

Watras, C.J. and Haney, J.F. 1980. Oscillations in the reproduction of *Diaptomus leptopus* (Copepoda: Calanoida) and their relation to rates of egg-clutch production. *Oecologia* 45: 94–103.

Webster, K.E. and Peters, R.H. 1978. Some size-dependent inhibitions of larger cladoceran filterers in filamentous suspensions. *Limnology and Oceanography* 23: 1238–1245.

Weider, L.J. and Lampert, W. 1985. Differential response of *Daphnia* genotypes to oxygen stress: respiration rates, haemoglobin content and low-oxygen tolerance. *Oecologia* 65: 487–491.

Wiens, J.A. 1977. On competition and variable environments. *American Scientist* 65: 590–597.

Wiens, J.A. 1984. Resource systems, populations, and communities, pp. 397–436 in Price, P.W., Slobodchikoff, C.N., and Gaud, W.S. (editors), *A New Ecology: Novel Approaches to Interactive Systems*. John Wiley and Sons, New York, N.Y.

Williamson, C.E. and Butler, N.M. 1986. Predation on rotifers by the suspension-feeding copepod *Diaptomus pallidus*. *Limnology and Oceanography* 31: 393–402.

Williamson, C.E. and Butler, N.M. 1987. Temperature, food, and mate limitation of copepod reproductive rates: separating the effects of multiple hypotheses. *Journal of Plankton Research* 9: 821–836.

Williamson, C.E., Butler, N.M., and Forcina, L. 1985. Food limitation in naupliar and adult *Diaptomus pallidus*. *Limnology and Oceanography* 1283–1290.

Wilson, D.S. 1975. The adequacy of body size as a niche difference. *American Naturalist* 109: 769–784.

Wright, J.C. 1965. The population dynamics and production of *Daphnia* in Canyon Ferry Reservoir, Montana. *Limnology and Oceanography* 10: 583–590.

Zaret, T.M. 1980. *Predation and Freshwater Communities*. Yale University Press, New Haven, Connecticut.

7

The Role of Predation in Zooplankton Succession

Z. Maciej Gliwicz and Joanna Pijanowska

Department of Hydrobiology
University of Warsaw
Warsaw, Poland

7.1 Introduction

For the last two decades predation has been considered to be a major driving force in shaping zooplankton communities and in determining their density and structure. In early studies, species succession was usually considered to be the result of differences in ecological tolerance to various abiotic environmental factors, such as light intensity and water density or viscosity (Hutchinson, 1967). By the early 1960s, the time of worldwide "productivity" research, the competition for resources was usually considered to be the primary driving force. Later, after the pioneering papers by Hrbáček (Hrbáček, 1962; Hrbáček et al., 1961) were cited by Brooks and Dodson (1965) and Hall et al. (1976), predation was considered to be the major factor responsible for successional events.

There was, however, still an argument against the predation hypothesis, as follows: Species replacement can be caused by resource limitation alone, but it does not seem likely to result from predation alone. This is because the predator may be able to shift to another food source when its preferred prey becomes scarce. Now a compromise has prevailed and both factors, competition and predation, are believed to interact to cause species replacements in seasonal succession. At the same time, the role of abiotic factors can not be ignored because they are well synchronized with changes in the magnitude of food limitation and predation intensity. Therefore, these factors can be commonly used by prey as cues for changes in prey behavior, physiology, and morphology that can weaken the impact of predation and competition.

In populations of planktonic animals mortality is mainly caused by predation rather than for other reasons, such as senescence, starvation, parasitism, or abiotic factors (e.g., increased flushing rates, extremely high temperatures, or anoxia). Due to the physical structure of offshore habitat, zooplankton are more exposed to predation than the prey in most other environments, and this is the reason why predation seems to be the most important cause of mortality. However, there are many reasons why the effects of predation may be reduced:

1. Most planktivorous predators can easily switch from one prey species to another.

2. The density of the initial prey may quickly be reestablished after a predator has diverted its attention to another species of prey. Prey populations can quickly recover due to their short generation times and high reproduction rates.

3. The flexibility of the life-history traits of prey allows it to withstand intense predation since features which determine vulnerability to predation (e.g., size at maturity) can be easily adjusted.

4. Prey can easily detect a predator presence in a viscous environment and adjust their behavior and morphology to become less susceptible.

5. Because many predators have long lifespans, the responses of predators to changes in prey density may be merely behavioral, not numerical.

Moreover, the importance of predation in seasonal succession is not easily seen since changes in predator abundance and activity are frequently synchronized with seasonal changes in both abiotic and biotic factors. The impact of predation on a prey population is often transitory and, therefore, it may not be easily noticed in the field. Although transitory, effects of predation may be meaningful. This is why predation is frequently considered to primarily produce an evolutionary impact and only secondarily a contemporary demographic effect (Thorp, 1986). Nevertheless, only demographic effects can really contribute to the events occurring in the course of seasonal succession in zooplankton communities.

The general patterns of zooplankton seasonal succession are summarized in the PEG-model (Sommer et al., 1986, and Section 1.2). The mechanisms that underlie competition and food limitation were analyzed in Chapter 6. The emphasis of this chapter is on those mechanisms underlying seasonal patterns which can be either expected or proven to result from predation. Before the role of predation in seasonal succession is evaluated (see Section 7.10), it is necessary to discuss what predators are most commonly involved (Section 7.2) and to list the characteristics which underlie prey vulnerability (Section 7.3). We then can examine the magnitude of the predators' impact (Sections 7.4 to 7.7), compare the importance of vertebrate and invertebrate predation (Section 7.8), and distinguish direct and indirect effects of predation (Section 7.9). Finally, following the assessment of predation in seasonal succession, we will discuss the mechanisms of species replacement that are directly caused (Section 7.11) or initiated (Section 7.12) by predation, and we will consider the predictability of predation-induced events in seasonal succession (Section 7.13).

7.2   The Nature of the Predators

Many species of fish, salamanders, waterfowl, watermites, insects, crustaceans, and rotifers are known to be obligatory or facultative planktivores. Planktivorous fishes are better understood and have been more thoroughly studied than the other planktivores. They are considered to be the most important predators in aquatic systems because each fish species is planktivorous in juvenile stages, at least. Moreover, their importance as obligatory planktivores also stems from:

1. Their large feeding capacities (voracity).
2. Their broad spectra of food size that render nearly all zooplankton species endangered.
3. Their behavioral flexibility, i.e., their ability to shift from one prey type to another, and from one habitat to another.
4. Interspecific similarities in their size-selective effects on prey species.

Other planktivores, which are mostly invertebrates, are more restricted in prey size. Moreover, they differ significantly in their selectivities, thus exerting a very broad range of effects of plankton prey assemblages. Obviously, the basic pattern of prey selection by any planktivore is the result of both prey density and prey vulnerability. Therefore, the diet of a predator is composed either of the most abundant prey species when predation is nonselective, or of the most vulnerable prey species when predation is highly selective. The percent composition of prey species in the predator diet would be the same as that of the prey present in the former case, or an overrepresentation of some prey species in the diet in the latter case. Neither of these two extremes could be observed in the field since a highly selective predator, if an optimal forager, would shift to a less preferred but more abundant prey species as soon as the density of its preferred prey was greatly reduced.

The majority of the planktivores are highly selective and can be classified into groups on the basis of their patterns of prey selection. Such a classification was originally proposed by Zaret (1978, 1980) and is still generally accepted. As prey body size and visibility increases, prey becomes more attractive to some predators. These are mostly visually oriented animals and include fish, salamanders, aquatic birds, some insects, and crustaceans (Figure 7.1). Preference for smaller prey is observed only in the youngest stages when predators have a small mouth. Such prey-size selection imposed by fish gape limitation is well documented (e.g., Blaxter, 1966; Rosenthal and Hempel, 1970; Wong and Ward, 1972). The duration of gape limitation, however, is brief and fish larvae feed on the largest prey by three to four weeks after hatching (Rosenthal and Hempel, 1970; Hunter, 1979).

Prey selection as a function of prey body size is common for size-dependent predators (exclusively invertebrates), which rely mainly on tactile information rather than vision for detecting and locating prey. Their prey size preference drops at a certain point as larger prey become too difficult to capture and ingest (see Figure 7.1). The available data on food selection suggests that visual predators prefer prey larger than 1 mm while size-dependent tactile predators prefer prey smaller than 1 mm (Dodson, 1974a). There are, however, some invertebrates which can consume prey species of a broader size range, e.g., *Mysis* (Murtaugh, 1981) and some *Chaoborus* species (Swift and Fedorenko, 1975; Swift, in preparation).

In addition to prey body size, shape, and visibility, prey motility can also be an important factor (Zaret, 1980). Recently Greene (1985, 1986) proposed a new functional classification of predators that is based on the tactics used by predators to locate and subdue their prey. Freshwater vertebrate predators are classified as either nonvisual, filter feeders or visual, particulate feeders, while freshwater invertebrate predators are

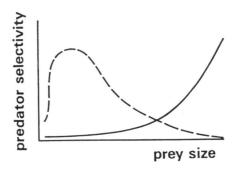

Figure 7.1   The selectivity of vertebrate (————) and invertebrate (– – – – –) predators as a function of the size of their prey (after Zaret, 1978).

classified as either cruising or ambush raptorial feeders. Visual, particulate feeders rely on body size, pigmentation, or motion to detect prey and they remove the largest and most conspicuous prey. Nonvisual, filter feeders select their prey more or less randomly with respect to size. Invertebrate predators are restricted to small-prey body size regardless of their raptorial tactics.

Many predators change both their habitat and their behavior as they progress from early juvenile to late adult stages. A drastic change is often observed in feeding grounds and feeding mode. This may result in a shift in the food spectrum toward larger prey items. This shift is best known in many fish species, in copepods, and in midge larvae (for a review see Werner and Gilliam, 1984). In planktivorous fish it is associated with substantial growth-related morphological changes (Hartmann, 1983, 1986). In invertebrate predators it is related to instar-to-instar increase in body size (Fedorenko, 1975; Wilson, 1975; Hare and Carter, 1987; Moore and Gilbert, 1987; Swift, in preparation). Concomitant changes occur in social behavior (e.g., a shift from schooling in juvenile fish to individual foraging in adults) and in habitats used (e.g., from littoral to offshore in fish, Baumann and Kitchell, 1974; Threlkeld, 1979; and invertebrates, Kajak et al., 1978).

The magnitude of the impact of predation and its seasonal timing are determined by two factors. First, they depend on seasonal changes in predator abundance. Second, they depend on predator feeding rates, which are a function of body size and metabolic activity. The abundance of larval and juvenile fish is affected mostly by abiotic factors that determine the rate of hatching success and the number of juveniles in the lake. In planktivorous fishes that hatch in May (e.g., perch, yellow perch, and smelt), or June to July (bleak and roach), mortality becomes extremely high later in the season. Thus, juvenile fish of the new cohort (which we will call $0+$) usually remain abundant and increase in biomass until midsummer when predation by piscivores causes their decline (yellow perch, Mills and Forney, 1983; smelt, Jachner, in preparation). Despite their small body size, $0+$ planktivorous fishes are probably much more important in predation than older-year classes since survivorship from $0+$ to $1+$ is usually lower than 1 percent (e.g., 0.01–0.15 percent in European perch, Nyberg, 1976; Viljanen and Holopainen, 1982). This high mortality begins within the first 100 days of postembryonic life (Werner and Gilliam, 1984).

In contrast to fish, seasonal changes in abundance cannot be as easily predicted for invertebrate predators. Although the rate of hatching of resting eggs is also temperature-dependent and its timing may be predicted in many species, high variability in the mortality rate of the earliest instars may result in either high or low densities of older individuals (Neill and Peacock, 1980). However, their short life cycles allow invertebrates to take advantage of other peaks in prey abundance that are

again controlled by highly unpredictable biotic factors, such as food availability and predation rates. Feeding rates of both vertebrate and invertebrate predators seem easier to predict seasonally since both groups of
animals have temperature-dependent metabolic rates and individual feeding rates which increase as a function of body size. Also, in both groups,
total daily food ration decreases with increasing body size; younger individuals are certainly more voracious. For instance, juvenile (0+) yellow
perch are reported to eat an amount equal to 21 percent of their body
weight per day (Mills and Forney, 1983), which is nearly five times as
much as older yellow perch consume (Nakashima and Leggett, 1978).
Also, midge larvae were observed to consume up to 100 percent of their
body weight per day soon after hatching, but only 20 percent during the
rest of the summer as older instars (Kajak and Rybak, 1979). If prey are
abundant, food ration drastically increases with increasing temperature
according to Krogh's curve. This is why the highest rations are always
observed in midsummer in both planktivorous fish (e.g., from 1 percent
in the beginning of June to 30 percent at the end of July in cisco; Enderlin,
1981) and invertebrate predators (e.g., from 1 percent in winter to 20
percent in midsummer in midge larvae; Kajak and Rybak, 1979).

The late spring or the beginning of summer seems to be the time
when predation on zooplankton has the greatest impact in most lakes in
the temperate zone. This is the time of high temperature that results in
high feeding intensity of planktivores and the time of highest density of
juvenile fish that have already attained a large-enough gape size to cover
the whole range of prey species. Hatching seems to be timed to permit
cropping of zooplankton in its spring phase of increase. A predator with
a long lifespan does not respond numerically to rapid changes in the prey
density. It can, however, respond by shaping its life history according to
predictable peaks in abundance of its major prey species (Riessen, 1985).
Predators' food selectivity combined with their density and phenology
should be seasonally reflected in those characteristics of prey individuals,
clones, and species which are most susceptible to predation.

7.3   Prey Vulnerability

Whereas food demands have to be satisfied throughout an individual's lifespan, predation occurs only during restricted periods and particular life stages
of the prey. The duration of the periods of vulnerability and immunity of
individuals to predation varies not only from one species to another but also
from year to year and from lake to lake. Vulnerability to predation is determined by each of the prey characteristics that are important to predators at
each step of the predation sequence, from prey location to its ingestion, through

its pursuit, attack, capture, and handling (Holling, 1959; Greene, 1983; Pijanowska, 1985; O'Brien, 1987). Planktivores differ in their capabilities at each step, and prey differ in their vulnerability to be located, attacked, and ingested. Moreover, the prey may efficiently reduce the danger of predation by evolving defense strategies for each step of the predation sequence (Kerfoot et al., 1980; Greene, 1983; Pijanowska, 1985; O'Brien, 1987). Prey population density determines the probability of predator-prey encounter and the profitability of prey pursuit. Prey vulnerability originates from a variety of morphological, physiological, and behavioral features such as size, shape, and pigmentation (Zaret, 1972a, 1972b; Zaret and Kerfoot, 1975; Kerfoot, 1980), palatability (Kerfoot et al., 1980; Kerfoot, 1982), motion (Ware, 1973; Wright and O'Brien, 1982, 1984), and escape abilities (Szlauer, 1965, 1968; Zaret and Suffern, 1976).

When visual, gape-limited predators are dominant, prey species with indetectable motion, small size, high transparency, and hence, low general visibility are less endangered (Figure 7.2, left). When tactile, size-dependent predators are dominant, prey with large body size, complicated morphology, and escape responses are less affected (Figure 7.2, right) (Zaret, 1978, 1980). When both types of predators are present, those least endangered are the species that can develop exoskeleton protuberances to inhibit capture by invertebrates and which, at the same time, can reduce body size and visibility to escape predation by visually oriented planktivores (Zaret, 1972a, 1972b; Dodson, 1974b; O'Brien et al., 1979). Prey vulnerability combined with the predators' preferences would decide which zooplankton species is preferred and hence suffers the highest mortality. The most affected species might eventually be drastically reduced in numbers.

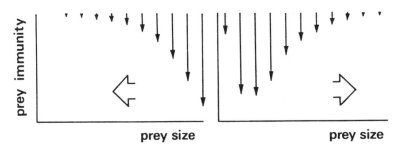

Figure 7.2   Size distribution of the immunity of prey exposed to predation by vertebrates (left) or by invertebrates (right), based on Zaret's (1978) assumption as shown in Figure 7.1. The danger is diminished either by a decrease or an increase in body size depending on which kind of predator is present.

7.4   Evaluating the Impact of Predators

Although there is a common belief that predation can be a major factor responsible for species replacements, very little evidence has been collected to prove that seasonal succession of zooplankton is driven by predation. Nevertheless, in the vast body of literature (for reviews see Hall et al., 1976; Zaret, 1980; O'Brien, 1987; Lazzaro, 1987), there is certainly enough relevant information to suggest which successional events might primarily be due to predation. This topic will be discussed in Sections 7.10 to 7.12. In the three following sections (7.5 to 7.7), we will focus on more general information on the role of predation in zooplankton community structure.

There are two different approaches that can be used to analyze the importance of predation on zooplankton succession. The first approach is to compare the feeding abilities of the predators present (ingestion rates) with the abilities of the prey to recover (clutch sizes combined with generation times). The second approach is to compare zooplankton communities and populations in any of the following ways:

1. In various years in one lake in which predation level has been either manipulated (introductions, removals) or changed in a natural way (fish winter kills, infectious diseases).
2. In different lakes with known different predation levels.
3. In enclosures.
4. In closed laboratory systems with various inocula of predators.

Each of these four may indicate what could be expected in the course of seasonal events in a particular location. Evidence emerging from the enclosure studies seems to be of particular importance since predation can be kept at desired levels while the enclosed community of prey species may remain more or less the same as in nature. This is not the case when the evidence comes from closed laboratory systems or microcosms where one is limited to an unnatural combination of selected species inocula. On the other hand, strong evidence may also emerge from natural or human-imposed, whole-lake experiments when a foreign predator is introduced into a system where it has previously not coexisted with local prey. Equally strong effects may be expected when an autochtonous predator is eliminated from the system.

7.5   The Impact of Predators at the Community Level

Predator feeding rate and prey ability to recover   The most general and the most frequently demonstrated effect of predation is that on the abundance of zooplankton and the structure of its community. Extreme examples show that:

1. Up to 100 percent of the daily net production of crustaceans may be consumed in summer by a planktivorous fish (*Lepomis*) in a shallow North American lake (Hall et al., 1970).

2. One-third of the total zooplankton standing crop can be removed by the freshwater sardine *Limnothrissa* overnight in an African reservoir (Gliwicz, 1986a).

3. Up to 100 percent of the daily net production can be eaten daily by *Chaoborus* in a eutrophic European lake (Kajak and Rybak, 1979).

4. Up to 13 percent of the total zooplankton standing crop can be ingested daily by *Mesocyclops edax* in an Ontario lake (Brandl and Fernando, 1979).

Whereas removing the entire daily net production of prey can merely bring the rate of growth of the prey populations to a standstill, eliminating a large fraction of the prey standing crop can result in a substantial prey density decline. This becomes more likely as the turnover time of prey standing crop and the generation time increases, and the prey clutch size decreases. Prey that are highly fecund may show little or no numerical response to predation despite the fact that they experience high mortality.

Year-to-year variation   Numerous studies have demonstrated that the impact of planktivorous fish on zooplankton communities results in shifts in the size distribution of zooplankton communities (Figure 7.3). Such shifts are observed either from large to small-sized species, such as rotifers, small cla-

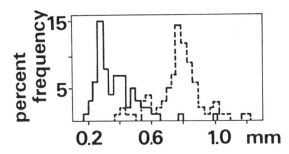

Figure 7.3   Body-size distribution of zooplankton in Crystal Lake, Connecticut, 10 years before the introduction of the planktivorous fish *Alosa aestivalis* (- - - - -) and 10 years after their introduction (———) (after Brooks and Dodson, 1965).

docerans, and copepods when fish were introduced, or from small zooplakton to larger cladocerans and copepods after fish removal (Hrbáček et al., 1961; Hrbáček, 1962; Hrbáček and Novotna-Dvorakova, 1965; Brooks and Dodson, 1965; Gliwicz, 1963; Galbraith, 1967; Kajak et al., 1972; Lynch 1979; Weglenska et al., 1979; Scavia et al., 1986). Similar shifts may result from year-to-year variations in fish abundance in the same lake. An elegant example has recently been published (Cryer et al., 1986) of a naturally generated biannual cycle in juvenile roach abundance which results in striking differences in the size distribution of the zooplankton community in summer depending on whether the roach are either abundant or rare (Figure 7.4). These effects are always best pronounced in fish ponds, where either a biannual or a three-year cycle in fish rearing leads to striking differences in zooplankton size distribution. For instance Fott et al. (1980a, 1980b) report an overwhelming dominance of large-bodied *Daphnia pulicaria* in the years of low fish standing crop, and of zooplankton composed of small crustaceans and rotifers in the years of high fish densities.

Lake-to-lake comparisons   Differences of similar character are also commonly observed when comparing lakes with different fish stocks (Figure 7.5). When planktivorous fish are abundant, communities of zooplankton in lakes and pods are composed of small-bodied species, whereas, in the absence of fish, zooplankton is dominated by large species (Hall et al., 1976; Hutchinson, 1971; Nillson and Pejler, 1973; Grygierek et al., 1966; Stenson et al., 1978; Hurlbert and Mulla, 1981; Langeland, 1982; Shapiro and Wright, 1984; Post and McQueen, 1987). Shifts in zooplankton species composition are also observed under the pressure of omnivorous, filter-feeding fish which, in contrast to particulate feeders, do not rely on vision and act as escape-selective predators. For instance, zooplankton is usually composed of evasive calanoid copepods in those ponds where gizzard shad or blue tilapia are abundant and

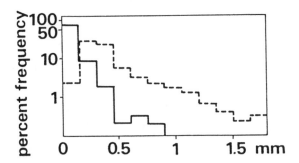

Figure 7.4   Body-size distribution of zooplankton in Alderfen Broad, England, in summer 1981 when planktivorous fry of *Rutilus rutilus* was abundant (———) and in summer 1982 when the fry was nonabundant (-----) (after Cryer et al., 1986).

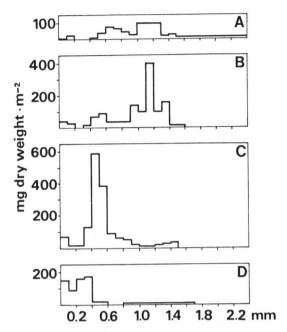

Figure 7.5   Body-size distribution of zooplankton in four Norwegian lakes with different planktivorous fish stocks: A, in fishless Sandtjønna; B, in Nedra Stavatjønn containing a scarce *Salmo trutta* population; C, in Øvre Stavatjønn containing *Salvelinus alpinus* population; D, in Langvatn containing populations of *S. alpinus* and of *Gasterosteus aculeatus* (after Langeland, 1982).

of nonevasive rotifers and cyclopoid copepods in those ponds where these filter feeders are rare or absent (Drenner et al., 1982, 1984).

Predation also affects zooplankton species diversity. In the presence of predators, the separation of prey species in space and time is often much less distinct than it is in the absence of predators (Miracle, 1974; Dawidowicz and Pijanowska, 1984). For the last 20 years evidence has accumulated that diversity is increased when predators selectively remove individuals of the most abundant species, thus facilitating the coexistence of potential competitors by reducing the uptake of limited resources according to the mechanism originally proposed by Paine (1966) and recently confirmed by Glasser (1978, 1979).

Enclosure studies   The mechanism mentioned above has been also confirmed by manipulating predator levels in in situ enclosures and microcosms. This is further discussed in Section 7.9 and in Chapter 6. The impact of predators in enclosures is usually much stronger than in the field due to both high predators' inocula and the lack of refugia for prey. Therefore, shifts in zooplankton species composition and in size distribution are faster and more distinct when invertebrate (Neill, 1981) or vertebrate (Anderson et al., 1978; Goad, 1984) predation is applied. Short-term numerical responses may easily lead to extinctions of large-bodied species when enclosures are stocked with high densities of planktivorous fish (Lynch and Shapiro, 1981). This might seem to contradict the role of predation in promoting high species diversity. However, predation can rarely be blamed for species extinctions in the field where predator density is never as high as it is in an enclosure and prey can escape the danger of predation more easily.

7.6   The Impact of Predators at the Population Level

Many examples show that nearly half of prey population may be eliminated daily by a vertebrate (Gliwicz, 1986a) or an invertebrate (Allan, 1973) predator. Such a high mortality rate induced by predation seems likely to cause prey population extinction if two conditions are fulfilled. First, the high predation impact is persistent. Second, the losses are not compensated by increased juvenile survival and reproduction in prey population despite improved food conditions associated with elevated predation. Such extinctions are, therefore, most likely to occur under the condition of severe food limitation. For instance, large *Daphnia* is eliminated within a few years in high altitude lakes in Poland after planktivorous fish are introduced (Gliwicz, 1985). When drastically reduced, but not completely eliminated, large *Daphnia* may reappear in high density when predation weakens as a result of a winter kill (Weglenska et al., 1979), a parasite disease of fish (De Bernardi and Giussani, 1978), or a reduced planktivore abundance caused by piscivores (Shapiro and Wright, 1984; Scavia et al., 1986). Hall et al. (1976) doubted that invertebrate predation alone can cause the extinction of zooplankton populations. In some lakes, however, invertebrates were shown to contribute to elimination of small zooplankton species. For instance, freshwater mysids are believed to be partly responsible for extinctions of cladoceran populations in Lake Tahoe, California (Richards et al., 1976; Goldman et al., 1979). Elimination of *Daphnia pulicaria*, *D. rosea*, and *Bosmina longirostris* was related to *Mysis* and kokanee predation (Morgan et al., 1978). *Bosmina longirostris* reappeared, however, in plankton of Lake Tahoe after three years, perhaps as the result of a 15-fold reduction in *Mysis relicta* density (Threlkeld, 1981). Another possible example of the impact of predation is the increase in *Daphnia* that followed the decrease in *Neomysis* abundance in Lake Washington (Edmondson and Litt, 1982).

Predation may also alter the size (age) distribution by selective removal of animals of one particular size. Many examples are known that illustrate this effect. For instance, the shift in *Daphnia pulicaria* toward smaller body size after brook charr were introduced to Tatra lakes in Poland is known from paleolimnological evidence (Stasiak, 1981). Conversely, *Daphnia* body size increased after *Alburnus alburnus* disappeared from a subalpine lake (De Ber-

nardi and Giussani, 1978). Shifts in the mean size of *Daphnia hyalina* (Vijverberg and Densen, 1984), and of *Daphnia hyalina* and *Ceriodaphnia quadrangula*
(Cryer et al., 1986) occurred with year-to-year variation in the fish stock.
There were also seasonal changes in body size of *Daphnia cucullata* with
seasonally changing fish and invertebrate pressure (Gliwicz et al., 1981), and
a decrease in mean body size in *D. rosea* and *Bosmina longirostris* in lakes
with higher predation rates (Northcote and Claroto, 1975; Northcote et al.,
1978).

Predation, especially by vertebrates, can influence fecundity in a prey
population in three different ways:

1. By selective removal of egg-bearing females (Mellors, 1975; De
    Bernardi, 1979; Gliwicz, 1981) or removal of eggs (e.g., ripping
    off copepod egg-sacks by *Chaoborus*, Swift and Forward, 1981).

2. By selective removal of females with larger clutch sizes (Sand-
    strom, 1980; Gliwicz, 1981; Dawidowicz and Gliwicz, 1983;
    Vuorinen et al., 1983).

3. By selective removal of large individuals which results in a de-
    crease in average body size in the population and, in turn, in a
    smaller average clutch size.

The last of the three cannot be observed when clutch size tend to
increase as a result of improved food conditions associated with elevated
predation, a phenomenon observed by Neill (1975), Lynch (1979), DeMott
and Kerfoot (1982), and Vanni (1986, 1987). Predation may also alter sex
ratios in a population by selective removal of females or males (Hairston
et al., 1983).

7.7   Impact on Life-history Traits, Morphology, and Behavior

Predation has a general effect on prey life-history tactics, life cycles, morphology, and behavior, since it may promote the coexistence of genotypes
(or phenotypes) that otherwise would have been competitively excluded.
Genotypes which are less vulnerable to predation have lower reproductive
ability. Individuals which have not evolved any antipredator defense mechanisms or which do not manifest any behavioral or structural responses to
increased predation pressure are the first to be eliminated from an exploited
population.

Under pressure from visual predators that selectively remove large individuals, it is better for the zooplankton to invest all its energy into reproduction as early as possible, mature early, and stay small. In the case of

increased mortality of small individuals, it is better to postpone maturity and to invest all the energy into somatic growth first, grow to a large size, and then have larger number of offspring (Stearns, 1976; Lynch, 1977; Gabriel, 1982; Dorazzio and Lehman, 1983). These are two ways in which the life cycle of prey has evolved to reduce predation.

Avoidance of temporal overlap with a predator can occur if the prey can enter diapause, as is seen in copepods (Strickler and Twombly, 1975; Nilssen, 1977; Elgmork, 1980; Hairston and Munns, 1984; Hairston, 1987), or by production of resting eggs as in cladocerans (Stross, 1973; Threlkeld, 1979). These are two effective ways for prey to bide time when predators are particularly dense or active. The copepod life cycle can differ drastically between neighboring lakes with and without planktivorous fish (Papinska, 1988). The most vulnerable stages of copepods are usually timed to appear in the lake during the periods of lowest planktivore activity (Nilssen, 1977).

Many prey species may reduce body size and pigmentation to decrease the probaiblity of being detected by visual predators (Brooks, 1965; Green, 1967; Zaret, 1972a, 1972b; Dodson, 1974b; Zaret and Kerfoot, 1975) or periodically develop elaborate protuberances (gelatinous sheaths, horns, spines, and helmets) to foil invertebrate predators (Gilbert, 1967; Gilbert and Waage, 1967; Allan, 1973; O'Brien and Vinyard, 1978; O'Brien et al., 1979; Stenson, 1987). Some of these responses are triggered by the presence of specific predators, while others are induced by such environmental cues as temperature or photoperiod. In addition to morphological adaptations many prey species have evolved direct escape responses (Szlauer, 1965, 1968; Gilbert, 1985), or habits of vertical migrations to prevent predator encounters in space and time (Zaret and Suffern, 1976; Stich and Lampert, 1981, 1984; Gliwicz, 1986b). All these responses are considered to be the result of either genetic variability (clonal replacements) or phenotypic plasticity of a genotype. Both components are probably involved in each response, regardless of whether it is a change in morphology (e.g., cyclomorphosis, Manning et al., 1978; Brock, 1980; Black, 1980) or behavior (e.g., diel vertical migrations, Ringelberg, 1980; Weider, 1984). Since the costs of evolving defensive mechanisms can be high, well-protected individuals are often poorer competitors. Therefore, defenses are found only in those instars that are more susceptible to predation and are not maintained in the seasons when predation pressure weakens or in lakes where predation rates are low (Green, 1967; Dodson, 1974b, Stich and Lampert, 1981; Riessen, 1984).

Variations in life-history tactics, life cycles, morphology, and behavior are probably complex adaptations to changing predation regimes. However, care must be taken to avoid attributing all those variations to predation alone since another important factor, resource limitation, may also be involved. Moreover, one should remember how difficult it is to distinguish numerical and evolutionary effects of predation in the field. The same effect can be caused either by a selective elimination of individuals or by the evolution of antipredator responses. For example, decreased body size at first reproduction can be caused either by selective removal of large individuals (numerical response) or by younger individuals entering reproduction (evolutionary response to selective pressure upon large, ovigerous females).

7.8 Relative Importance of Vertebrates and Invertebrate Predators

Unlike fish predation, the importance of which in structuring planktonic communities is very well documented, invertebrate predation plays a more ambiguous role (Figure 7.6). From many enclosure studies (e.g., Neill and Peacock, 1980; Neill, 1981) it is apparent that invertebrate predators are capable

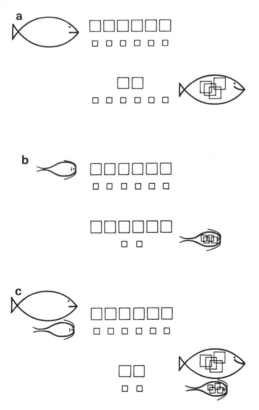

Figure 7.6 General effects of (a) a vertebrate and (b) an invertebrate predator on the body-size distribution of a population or a community of zooplankton herbivores. The squares represent zooplankton of two different sizes. The combined effect (c) of vertebrate and of invertebrate predation can be obscure if both kinds of predation are equally important.

of exerting important effects on prey assemblages, especially in the absence of vertebrate predators. These effects are usually muffled when combined with the vertebrate's impact. Also, the effects that result from the differential impact of fish and invertebrates are often difficult to observe in situ when both types of predators coexist and exert simultaneous impacts on zooplankton communities and populations (Figure 7.6C). Invertebrate predators may dilute the effects of vertebrate selectivity by affecting the opposite end of the prey size spectrum. On the other hand, however, they may also divert the attention of vertebrate predators from their regular herbivorous prey allowing for an increase in prey (Figure 7.7).

A temporary shift in prey size distribution toward larger individuals may also occur. This, however, cannot be a long-lasting effect since as soon as the invertebrate predator becomes less abundant, the vertebrate predator will switch back to herbivorous prey and the zooplankton size distribution will shift again toward smaller forms, as shown in Figure 7.6A. The combination of the two effects of vertebrate predators on zooplankton size distribution— removal of larger prey (a direct effect) and decreasing predation on smaller prey by eliminating invertebrate predators (an indirect effect) (see Figure 7.7E)—

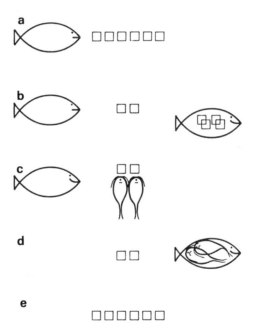

Figure 7.7 The effect of a vertebrate predator, which is strong when no invertebrate predators are present (a and b); can be weakened when an invertebrate predator appears, and diverts the vertebrate's attention from the major herbivore prey (c and d); which then can temporarily increase its density (e) before drawing the vertebrate's attention again.

is probably most important in causing shifts in zooplankton species composition (Stenson, 1978; Lynch et al., 1981).

The overwhelming effects of predators on zooplankton communities and populations that are so obvious from the changes induced by predator introductions and removals or from the changes in predation level, must also play an important role in seasonal changes in zooplankton in a more or less direct way.

7.9   Direct and Indirect Effects of Predation

Mortality in prey population seems to be the only direct effect of predation. If not compensated by reproduction, it results in prey numerical decline (Figure 7.8A). If selective, it further results in a shift in the age-size distribution of the prey population. It may also alter the sex ratio (Figure 7.8B). Furthermore, if ovigerous females are selected by the predator, especially those with larger clutches (see Section 7.6), the predation may also result in a decrease in prey fecundity (see Figure 7.8B). Therefore, although the final direct effect is a lower density, it may be caused by changes in both natality and mortality. Therefore, predation can still be very important in causing a prey population decline, even where there is no apparent increase in death rate.

With no measurable increase in mortality, but a dramatic decrease in

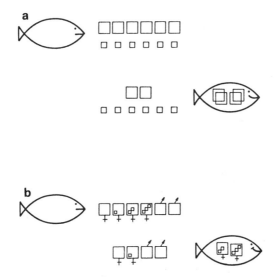

Figure 7.8   Direct effects of a size-selective vertebrate predator on a prey population: on its density and body size distribution (a), and on its sex ratio and mean clutch size (b).

reproduction rate, the population density may suddenly drop. A density decrease of this type could easily be attributed to food limitation instead of a predator-caused decrease in reproduction. No evidence is known that would make it possible to judge the importance of this mechanism in the field. Such evidence seems difficult to obtain, since this effect would be weakened by an overall increase in prey reproduction rate due to increased food levels resulting in an indirect way from the impact of predators. This has been shown by manipulations of predation levels in many enclosure studies. For instance, Vanni (1986) demonstrated elevated survivorship and fecundity of cladoceran individuals that managed to avoid predation in enclosures stocked with high densities of planktivorous sunfish. If predation is nonselective and so intense that densities of coexisting populations are kept well below the threshold resource level necessary to produce food limitation, species with the highest intrinsic rates of increase would be favored (see Chapter 6). Such a situation, however, is not common in the field because usually one species can always escape predation and attain densities high enough to cause food limitation again.

All other effects of predation on the course of seasonal succession in a zooplankton community can be categorized as indirect effects since they are produced only as a consequence of a decrease in population density of the species most affected. Improved food conditions that result from a decrease in one species, could cause an increase in reproduction in surviving individuals of that species and also in individuals of other species that utilize the same food resources (Figure 7.9). The shift from one species to the other in the course of seasonal succession could be, therefore, only initiated by predation, while the final result is the result of the change in the level of resources. Species replacement can be initiated in this way only when food is a limiting factor. This effect occurs through weakening exploitive competition at both the population and interspecific level. It seems obvious, however, that the same effect may also occur in the absence of resource alteration if predation reduced interference competition by decreasing the abundance of superior interfering competitors.

Intuitively, one would assume that there are no other ways that predators can cause a species replacement. It is possible, however, that a successional event can also be initiated in a more direct way. This possibility stems from the fact that a predator's feeding mode changes throughout its lifespan (Figure 7.10). For example, it is well known that juvenile fishes shift their size preference from very small prey when they are still gape limited, to large prey when mature (see Section 7.2). In many fish species, this shift occurs within a period of a few weeks. If young, yellow perch or smelt migrate offshore as a synchronized cohort at the beginning of summer, a change in mortality would be expected for various zooplankton species. For instance, such a shift would occur if growing predators switched first from a smaller rotifer to *Bosmina* and then to *Daphnia*, regardless of which of the three prey species

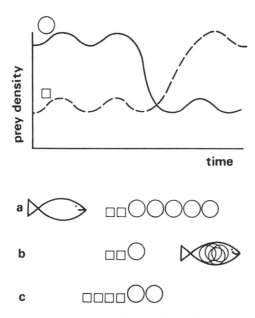

Figure 7.9 Species' succession as an indirect effect of size-selective predation by a vertebrate predator (top). Only a decrease in population density of the larger prey (a to b) results directly from predation, whereas a subsequent increase in the smaller prey population (b to c) is the result of the increase in resources that results from the decrease in number of the larger species.

was competitively superior. In this way—even without any indirect effect of interspecific competition—predation can directly cause species' replacement (see Section 7.11). This effect seems more predictable than the indirect effect of increased resources caused by a decline in number of the dominant competitor (see Section 7.13). In order to determine how predators actually control the course of seasonal succession in zooplankton, we will first examine those successional events that may be intuitively attributed to predation.

7.10   Successional Events Attributed to Predation

Species diversity   One important effect of predation on zooplankton seasonal succession is a drastic alteration of the course of succession when the predation intensity has been changed in either an artificial (Grygierek et al., 1966; Hillbricht-Ilkowska and Weglenska, 1973; Hurlbert and Mulla, 1981; Langeland, 1982) or a natural way (De Bernardi and Giussani, 1978; Edmondson and Litt, 1982; Cryer et al., 1986). Pourriot (1983) observed that in lakes where fish predation was intense there were more peaks in density in rotifer and crustacean populations throughout the seasons. Dawidowicz and

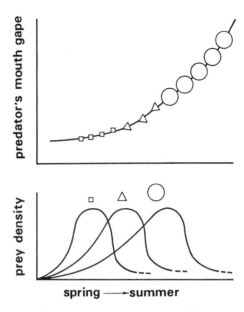

Figure 7.10   A theoretical succession of species as an effect of size-selective predation by a vertebrate predator (bottom). If the predators appear as a synchronized cohort of individuals growing in size, the first most-affected prey species would be a small-bodied herbivore that probably was the first to increase in abundance because of its short generation time. Later, when the predator's mouth gape gets larger (top), bigger herbivores would be eaten and their densities would therefore also decrease. (Upper graph adapted from Hartmann, 1983.)

Pijanowska (1984) reported that fish predation allowed for simultaneous peaks of abundance of many herbivorous cladocerans, which were temporally separated in a neighboring fish-free lake. This is probably the reason why the species diversity of zooplankton is often greatest in midsummer when both vertebrate and invertebrate predators are abundant. This high summer diversity is reported from many lakes (Sommer et al., 1986; Morin, 1987) and can be seen in others from data on the seasonal dynamics of zooplankton. Species diversity is always lower in the early spring when only fast-growing populations of early species are abundant, as well as later in the spring when resources are often monopolized by superior competitors at a time well before most planktivorous fish hatch and also before invertebrate predators multiply.

Spring-to-summer shifts   The other common phenomenon that can be attributed to predation is a clear spring-to-summer shift in the size distribution of zooplankton from large-bodied to small-bodied species and from large to small individuals within particular crustacean populations. An example of this phenomenon is given in Figure 7.11. Also, the spring-to-summer shifts

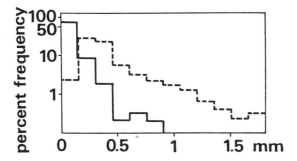

Figure 7.11 The spring-to-summer shift in the body-size distribution of zooplankton: in a year of good recruitment of *Rutilus rutilus* in Alderfen Broad, England, the large-bodied zooplankton with high densities of *Daphnia* and *Ceriodaphnia* that is present in May (-----), is replaced in August by a community with high densities of rotifers and small *Bosmina* (———), after the period of intense feeding by the fry of roach (from Cryer et al., 1986).

in relative abundance of rotifers, cladocerans, and cyclopoid and calanoid copepods are frequently attributed to an increase in predation. These shifts are common and most predictable at the beginning of the season when the usual sequence is a series of peaks in abundance from a ciliate to a rotifer species, then to a small-bodied cladoceran, and, finally, to a larger cladoceran. Figure 7.12 shows such a sequence in a Swedish lake. This sequence has been observed in the spring in many lakes by Patalas (1963), Nauwerck (1963), De Bernardi and Canali (1975), Eloranta (1982), DeMott (1983), Lehtovaara and Sarvala (1984), and Kerfoot (1987). It is also evident from a set of data from a large number of lakes that is summarized by Sommer et al. (1986).

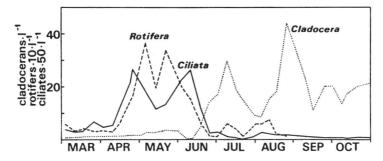

Figure 7.12 The spring-to-summer shift from small-bodied, herbivorous zooplankton—ciliates (———) and rotifers (-----)—to cladocerans in Lake Erken, Sweden (-----). The species seen are, first, small *Bosmina*, next, larger *Daphnia*, *Diaphanosoma*, and *Ceriodaphnia* (compare with Figure 7.13) (from Nauwerck, 1963).

Midsummer species replacements   Shifts in zooplankton composition are also frequently observed in midsummer. For instance, a cladoceran assemblage of *Daphnia-Bosmina* is replaced in part by a combination of *Ceriodaphnia, Diaphanosoma,* and *Chydorus* (e.g., Berg and Nygaard, 1929; Nauwerck, 1963; Patalas, 1963; De Bernardi and Canali, 1975; Gliwicz, 1977; Beattie et al., 1978; Sommer et al., 1986; Mills et al., 1987). Figure 7.13 shows the species

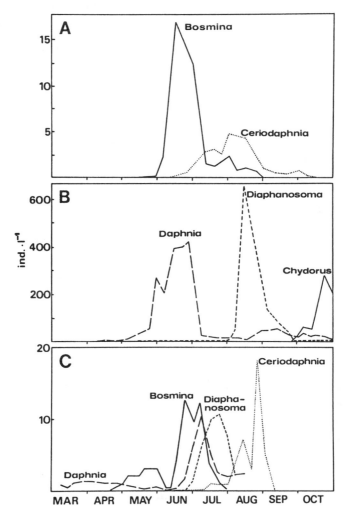

Figure 7.13   Examples of the midsummer succession of herbivorous, cladoceran zooplankton in 3 different lakes: A, in Vassikkalampi Pond, Finland (Eloranta, 1982); B, in Lago Maggiore, Italy (De Bernardi and Canali, 1975); C, in Lake Erken, Sweden (Nauwerck, 1963).

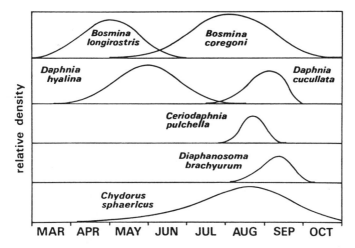

Figure 7.14   A generalized sequence of cladoceran species based on data from 12 lakes of the Friesian Lake District, The Netherlands (Beattie et al., 1978). The congeneric species are probably not as widely separated in time as it appears in this graph, and more overlap is usually observed (compare with Figures 7.16 and 7.17).

succession for three different lakes and Figure 7.14 shows the sequence based on data from 12 Dutch lakes. In other lakes summer replacement of *Daphnia* by *Holopedium* has been reported (Tessier, 1986).

"Summer genera" are usually less vulnerable to fish predation than spring species, because they are either very small (*Chydorus*), extremely transparent (*Holopedium*), or have high escape abilities (*Diaphanosoma*). Most of the papers cited above are purely descriptive, but some supply evidence that species replacement results from competition alone (see Chapter 6). In some of the papers, however, predation is shown to be one of the two major factors (Tessier, 1986) or the principal factor (Mills et al., 1987) responsible for species replacement.

Tessier (1986) presents convincing evidence that such replacement results from a combination of two specific characteristics: *Daphnia* are more affected by food limitation and higher vulnerability to fish predation than *Holopedium*. Mills et al. (1987) show that the predation by fish is responsible for the replacement of large *Daphnia pulex* by *Bosmina*, *Chydorus*, and two small *Daphnia* species. Similar replacements were also observed in rotifers in the form of shifts from soft-bodied and nonevasive *Synchaeta* and *Notholca* to spiny *Keratella* and *Kellicotia* and to an evasive-swimming *Polyarthra* (e.g., Carlin, 1943, and other authors reviewed in Hutchinson, 1967). A similar replacement sequence (Figure 7.15) was attributed by Stemberger and Evans (1984) to predation by *Cyclops bicuspidatus*.

Sometimes, species replacements are observed within single genera. An

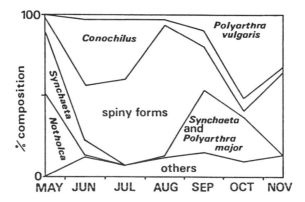

Figure 7.15   An example of the shift in composition of rotifer zooplankton from soft-bodied, vulnerable species (*Notholem, Synchaeta, Conochilus*) to spiny (*Keratella, Kel-licotia*) and evasive-swimming (*Polyarthra*) forms, Lake Michigan (from Stemberger and Evans, 1984).

example of this is the midsummer replacement of *Bosmina longirotris* by *B. coregoni* shown in Figure 7.16 (e.g., Patalas, 1963; Beattie et al., 1978; Culver, 1980). *Daphnia galeata* and *D. hyalina (longispina)* are often replaced by *D. cucullata* in central European lakes (Figure 7.17A; Patalas, 1963; Gliwicz,

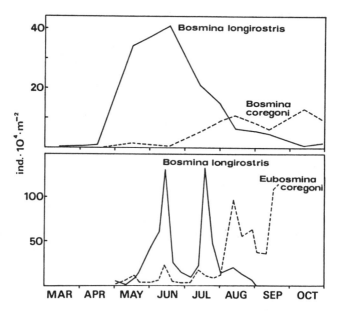

Figure 7.16   Two examples of the succession of *Bosmina* species: top, in Zabinki Lake, Poland (from Patalas, 1963); and bottom, in Lake Ontario, Canada (from Culver, 1980).

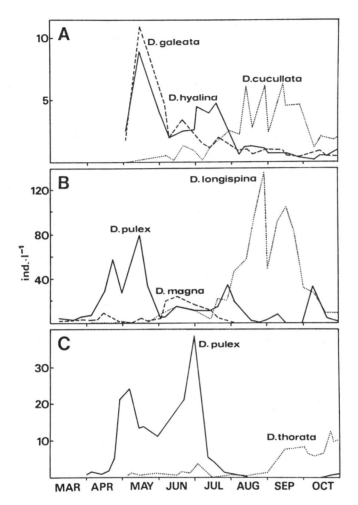

Figure 7.17   Three examples of the succession of *Daphnia* species from large-bodied to small-bodied ones: A, in Klostersee, Germany (from Seitz, 1977); B, in Lake Bysjon, Sweden (from Coveney et al., 1977); C, in Lake Washington, USA (from Edmondson and Litt, 1982).

1977; Seitz, 1977), whereas other *Daphnia* species are found to replace each other in Scandinavian lakes (Figure 7.17B; Coveney et al., 1977) and in North American lakes (Figure 7.17C; Birge, 1898; Edmondson and Litt, 1982; DeMott, 1983).

These midsummer shifts in congeneric species are frequently attributed to fish predation, especially when a large or less evasive species is replaced by a smaller and less conspicuous one. This is true in the case of the replacement of *D. galeata* and *D. hyalina* by *D. cucullata*, the replacement of *D. magna* and *D. pulex* by *D. longispina* (Coveney et al., 1977), *D. galeata mendotae* by *D. retrocurva* (Birge, 1898; Hutchinson, 1967), and *D. pulicaria* by *D. thorata* (Edmondson and Litt, 1982). This is, however, not always the case since sometimes a shift is observed from a less conspicuous species to a more conspicuous one or no clear difference can be detected between the two species. For instance, DeMott (1983) reports a midsummer shift from *D. rosea* to *D. pulicaria*, that is replicable from year to year in a small lake. Neither congeneric species seems to be more susceptible to fish predation, which is generally weak in the lake, so that one of the two large *Daphnia* may be abundant throughout the summer. This example may imply that species' replacement can be caused by exploitive competition alone (see Chapter 6).

Another example that at first glance seems equally difficult to explain by size-selective predation, a consecutive summer replacement of *Daphnia galeata* by *D. hyalina* in a large central European lake, is reported by Geller (1986). He suggests that temperature-related change in an interspecific advantage is responsible for the replacement. Predation, however, may play an important role here, too. Although the two species are similar in size and pigmentation, and their conspicuousness should not differ, their vulnerability does differ since *D. galeata* remains in the epilimnion, whereas *D. hyalina* migrates dielly into the hypolimnion where the light intensity is too low for visual predation. In other lakes, *D. hyalina* remains in the hypolimnion day and night (Pijanowska and Dawidowicz, 1987; Gliwicz and Pijanowska, 1988).

The ability to find a daytime or a permanent refuge in the hypolimnion may be a reason why not every midsummer replacement is from a more to a less conspicuous form. For example, *Ceriodaphnia*, which increases after declines in *Daphnia* and *Bosmina* densities (see Figures 7.13 and 7.14), is neither smaller than *Bosmina* nor less transparent than *Daphnia*. It avoids predation by remaining in the hypolimnion (or the metalimnion if the hypolimnion is anoxic) where it is frequently found in midsummer (e.g., Nauwerck, 1963; Eloranta, 1982). The ability of a species to survive the summer period of high predator abundance by remaining in deep water although food resources are low, seems to be as advantageous as other antipredator defenses, such as the escape ability of *Diaphanosoma* (Drenner and McComas, 1980) and *Polyarthra* (Gilbert, 1985), the small body size of *Chydorus*, and the spininess of *Keratella* (Stemberger and Evans, 1984). All these defenses are seen in midsummer species of zooplankton communities in eutrophic lakes. In this way, the vulnerability of a susceptible species can be reduced by an appropriate change in behavior.

Predation seems to be the most likely explanation for many cases of zooplankton species replacements. This statement is based on the fact that most observations are in perfect agreement with the results of an overview of the general impact of planktivorous fish and invertebrate predators (see Sections 7.5 and 7.6). Rarely, however, have species replacements been shown to be the result of direct and indirect effects of predation. Most papers on zooplankton seasonal succession merely describe changes in populations without directly investigating or even mentioning mechanisms. Whereas some successional replacements have persuasively been shown to result from interspecific competition (see Chapter 6), very few replacement phenomena have been proved in a compelling way to be directly caused (see Section 7.11) or initiated (see Section 7.12) by predation.

7.11 Successional Events Directly Caused by Predation

Sometimes a sequence of peaks in population densities of various zooplankton species can be mistakenly ascribed to competition. Such a sequence is commonplace in late spring or early summer in eutrophic lakes when a peak in a ciliate species is followed by a peak in a rotifer species, and then a peak of *Bosmina* is followed in turn by a peak in *Daphnia* (see Section 7.10, and Figures 7.12 and 7.13). Each species, in turn, has a longer generation time (although not necessarily a smaller clutch size) and this may be the reason why the smallest species is first and the largest is last reaching its maximum density in the spring when food resources increase. Each subsequent species can then be assumed to be a more effective competitor for resources than the one before it since competitive advantage is believed to increase with increasing body size (see Chapter 6). A conclusion that follows is that each subsequent species outcompetes the preceding one. In other words, the sequence can be understood as a typical successional replacement of populations caused by resource limitation or interspecific competition alone.

The possibility cannot be excluded, however, that other forces are involved. First, the sequence may be produced by temperature-cued successive hatching of resting eggs. Second, it may be caused by predation. The time of this common phenomenon, the end of April, May, or the beginning of June, is the time when planktivorous fish hatch because of an increase in water temperature. Juvenile fish appear as a synchronized cohort of gape-limited predators that grow and switch their feeding mode from smaller to larger prey (see Section 7.2). This may cause the observed sequence from small to large zooplankton species. One example can be cited for juvenile roach, which appear in high densities every second spring due to good recruitment in a small English lake (Cryer et al., 1986). As each of the three dominant cladoceran species appears in the young roach diet, its density is observed to decline. This process begins with the smallest *Bosmina* followed by larger *Ceriodaphnia* and still larger *Daphnia*. The same sequence has been observed in Polish lake where juvenile smelt shift their diet from rotifers to *Bosmina* and then to *Daphnia*. The associated sequence of population declines progresses from rotifers to *Daphnia* (Jachner, in preparation).

Such sequential population declines could be attributed to predation in other lakes when more information on feeding biology, phenology, and abundance of planktivorous fish are collected. Similar sequences of population decline may also stem from ontogenetic shifts in size-selective predation by invertebrates such as midge larvae (e.g., Hare and Carter, 1987; Moore and Gilbert, 1987) and cyclopoid copepods (e.g., Wilson, 1975). The effect should be expected to be more distinct later in the season since not only the size of the gape but also the feeding rates of predators increase due to both increased body sizes and increased water temperature (see Section 7.2). The phenomenon of the spring sequence of density peaks in species of increasing body size is evolutionary "perceived" by predators that can make a good use of the seasonality in zooplankton. Riessen (1985) called this phenomen "exploitation of the prey seasonality by the predator" when describing the timing of hatching in a pelagic water mite that results in a peak in its nymphs' density at the precise time of the peak in its cladoceran prey.

The role of predation in shaping the sequence of successional events in zooplankton is probably restricted to those periods of time when, because of climatic factors such as light intensity, photoperiod, and temperature, a sudden synchronous change in predation intensity occurs. Although interspecific competition must simultaneously be occurring, its effects may be blocked or overshadowed by the impact of sudden increases of predation affecting one prey population after another. In most cases, however, seasonal shifts in species composition and abundance seem only to be initiated by predation that causes population declines in more vulnerable competitors. Once the population density of the most exploited prey species has been reduced by predation below a critical level, the initial step of successional replacement has occurred and food limitation will become a governing factor at this stage (see Chapter 6). Thus, predation seems to be a major force merely in initiating the process of species replacement.

7.12   The Causes of Prey Density Reduction

A high predation-induced mortality in a prey population results in a decrease in population density and, consequently, in an increase in the ability of competing species to sequester resources in an extended ecological space (see Section 7.9). Since each decline in numbers implies individual losses from a population, it is very tempting to attribute these declines to increased mortality caused by predation even though no information is available on predator density and feeding. However, if it is not due to senescence, the decline in density can also be caused by other factors, such as starvation, increased flushing rates, and drastic changes in abiotic factors. Moreover, a decline can also be the result of a decreased natality with mortality remaining unaltered and thus be attributed to food limitation as in the case of *Daphnia rosea* in Lake Mitchell (DeMott, 1983).

In most cases where predation is the most likely explanation, it is derived

from the coincidence of high predator density and feeding rates with declines
in prey populations (e.g., Hall, 1964; Wright, 1965; Mills and Forney, 1983;
Stemberger and Evans, 1984). An example of this is shown in Figure 7.18.
The reason for the population decline will not be known unequivocally, however, until prey death rates and predator diet data are known. These two
requirements have rarely been fulfilled. For instance, maximum death rates
in *Daphnia* species are reported to coincide with *Epischura* abundance in Lake
Washington (Edmondson and Litt, 1982). Although the remains of *Daphnia*
were found in fecal pellets of the copepod, it is still unknown whether the
peaks of *Epischura* impact were really synchronized with *Daphnia* declines.

High death rates in *Keratella cochlearis* were reported at the same time
as the highest densities of *Asplanchna* (Bosch and Ringelberg, 1985) and in
Daphnia at the time of high densities of *Leptodora* and *Bythotrephes* (De Bernardi and Giussani, 1975; Figure 7.19) but in neither case was the predator's
diet composition known. In other papers, information on changes in feeding
rate and in diet composition are known and correspond with prey population
declines but no death rates are estimated to demonstrate that the declines
were actually caused by increased mortality. For instance, there was a dramatic
decline in *Daphnia, Ceriodaphnia,* and *Bosmina* at the same time that each
species was reported to enter the diet of roach (Cryer et al., 1986). Similarly,
a sudden decrease in *Cyclops* density was demonstrated at the time when
copepods first entered the charr diet (Dawidowicz and Gliwicz, 1983).

More compelling is evidence that increased mortality rates coincide with
the time of an increase in the proportion of the predator's diet made up by
the prey species. An elegant paper by Threlkeld (1979) comes close to fulfilling
these requirements and, in addition, it provides an insight into the prey population structure. A drastic midsummer decline in population densities of two
Daphnia species in Wintergreen Lake apparently resulted from increased mor-

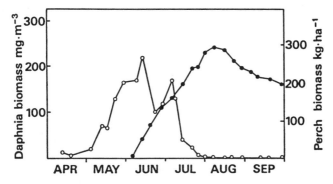

Figure 7.18   Seasonal changes from April to September in *Daphnia pulex* population
density expressed as its biomass (○), plotted against changes in the biomass of juvenile
(0+) *Perca flavescens* (●) in Oneida Lake, New York (from Mills and Forney, 1983).

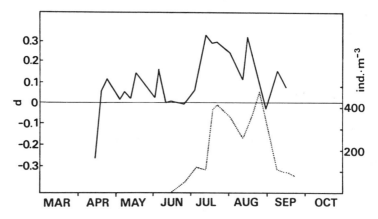

Figure 7.19   Seasonal changes in death rate in *Daphnia hyalina* population (d, ——)
against changes in the indicated density of *Leptodora* and *Bythotrephes* (-----) in Lago
Maggiore (from De Bernardi and Giussani, 1974).

tality caused by bluegill sunfish moving offshore from the littoral. Although
the birth rates in both populations were found to decrease during the decline,
food limitation seemed to be unlikely since the clutch size did not decrease
in the smallest egg-bearing females which should have been the first affected
if a food shortage was involved. This is shown in Figure 7.20. Unfortunately,
other sets of data in which peaks in prey population death rates were observed
at the same time that the prey increased in a predator's diet, are not relevant
to seasonality since they either come from tropical lakes (Gliwicz, 1986a),
follow a sudden change of habitat by planktivorous fish (Fairchild, 1982,
1983), a fish introduction (e.g., Coveney et al., 1977), or various manipulations
of predators (data from fish ponds and enclosures).

The effect of predation may be further magnified because not only death
rates but birth rates may also be affected by predators (see Section 7.6).
Although birth rates are frequently ignored with respect to predation and are
often attributed to resources limitation, a simultaneous decrease in prey birth
rate may contribute to a dramatic decline in prey density.

From a review of information on seasonal changes in zooplankton density, it appears that predation plays an important role in initiating two distinct
seasonal events. Both are quite replicable from lake to lake. The first one is
the drastic decline in cladoceran populations at the end of the spring clear-
water phase. The second is the frequently observed decline in the density of
a dominant *Daphnia* species in midsummer. The first event has been seen in
Lake Constance (Lampert and Schober, 1978), Lake Biel (Tschumi et al., 1982),
and Lake Schöhsee (Lampert, 1988), and probably occurs in many more. The
second has been observed in Lake Mendota (Birge, 1898), in Base Lane Lake

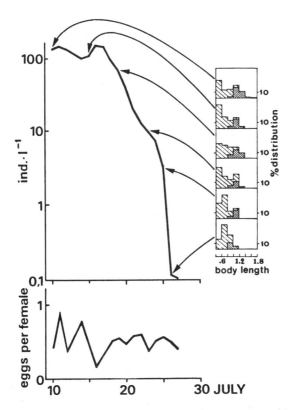

Figure 7.20   Top, the midsummer decline in the population density of *Daphnia galeata mendotae*. The shaded portions in each size fraction in the body length distribution are given for six dates during the period of decline. This decline is observed at the same time that the clutch size of the smallest individuals (1.0–1.2 mm body length) remains unaltered, although (bottom) the overall fecundity of the population (mean number of eggs per female) decreases (after Threlkeld, 1979).

(Hall, 1964), in Canyon Ferry Reservoir (Wright, 1965), in Sunfish Lake (Clark and Carter, 1974), in Lago Maggiore (De Bernardi, 1974), in Lake Kvernavatn (Larsson et al., 1985), and in many others. Although these declines can seldom be confidently attributed to predation, in most of them predation probably plays a principal role, not necessarily by causing prey extinction, but at least preventing prey recovery. Food limitation resulting from extremely low phytoplankton biomass (clear-water phase) or from a large abundance of interfering algae (summer declines), may also affect prey density at these times. It is not known whether predation-induced mortality or starvation is responsible for rapid population declines at the end of the clear-water phase or during summer blooms of interfering algae.

The role of predation in initiating species replacements can also be deduced from the observations of sudden increases in densities of prey species.

A species that is heavily exploited for a time sometimes recovers when predation pressure is subsequently relaxed. This has been observed in Lake Constance where *Daphnia* began to increase exponentially when predation decreased as the old generation of carnivorous *Cyclops vicinus* retired from the lake pelagial at the end of May (Lampert and Schober, 1978). It has also been reported in other lakes where *Daphnia* density greatly increased at a time when either planktivorous fish were decimated by a piscivore (Mills and Forney, 1983) or the attention of planktivorous fish was diverted by the sudden appearance of *Leptodora*, a more attractive prey that had just hatched from resting eggs (Gliwicz et al., 1981).

Besides purely demographic effects of predators on prey populations there are also other effects that may be equally important in initiating a successional event in zooplankton. First, prey may temporally withdraw into dormant stages. Second, it may withdraw to a predator-free space, for instance to the hypolimnion, either permanently or only transiently (diel migrations). Each of the two effects results in an expanded ecological space for other competing species. Other changes in a prey population, which can be induced by a selective predator, may result in a decline in the competitive advantage of certain individuals as a result of restructured morphology or altered behavior that makes other species competitively superior in resources utilization (see Section 7.7). This, however, may not result in a species replacement since predators readily switch to other more vulnerable prey. Despite the fact that they may lose their competitive superiority, species that are well protected against predation do not cede "lebensraum" easily. The phenomenon of the predation-caused decline in a dominant zooplankton species must be commonplace, and it is certainly more or less predictable in various lakes and various seasons.

7.13   Predictability of Predation Effects

The successional shifts induced by predation should be more predictable than those that result from food limitation, because the phenology of predation is more predictable than the phenology of food limitation, at least in lakes of the temperate zone. There are two major reasons for this. First, planktivorous predators have lifespans many times longer than their herbivorous zooplankton prey which, in turn, have much longer lifespans than their algal food. Predictable patterns of seasonal change are most likely to be created by organisms that have fixed life cycles but slow demographic responses to changes in resources (Lampert, 1987). Seasonal changes in the magnitude of fish predation impact are more closely related to the change in water temperature than to density changes in zooplankton prey populations. This is due to the fact that feeding activity is temperature-dependent and juvenile hatching is cued by an increase in the temperature of lake water. Also, drastic changes in photoperiod make predation by visually oriented planktivores more or less intense and this is again a highly predictable seasonal change (see Section 7.2).

Second, in contrast to predation, food limitation is rarely caused by climatic factors. The usual situation in which food resources become highly limiting is created by herbivorous zooplankton, regardless of whether it is a spring, clear-water phase with extremely scarce phytoplankton or a phase of a summer bloom of inedible and interfering algae (see Sommer et al., 1986). The seasonal change in the magnitude of predation seems to be less dependent on the food resources of the predator, i.e., on planktonic prey density. The predator's response would be functional rather than numerical. When a preferred prey is depleted, a predator switches to alternate resources. If prey density becomes so low that predators go hungry, they usually do not die of starvation. They can survive a period of low food because of their body reserves, and they are still there when the prey population recovers.

Although these two reasons seem to be more relevant to vertebrate predators, both are also valid for invertebrate predators. Some invertebrates may show a rapid numerical response to changing prey densities (e.g., parthenogenetic cladocerans). But, on the other hand, the life cycle of an invertebrate (e.g., a midge), its emergence and the timing of egg laying, is probably as fixed as is the life cycle of smelt, perch, or bass. They all hatch at fixed water temperatures, thus appearing in a lake as a synchronized cohort at a particular time of the year, and shift from one to another prey species or from one to another habitat as they grow in size. This may also explain why the following three consistent patterns of zooplankton seasonal succession should be associated with predation rather than with food limitation.

1. The early spring sequence of the peaks in densities from small to large zooplankton species.
2. The drastic declines in cladoceran densities at the end of the spring clear-water phase.
3. The midsummer declines in *Daphnia* populations.

These three observations are the only ones that seem to be replicable from year to year and from lake to lake in the temperate zone.

Even that limited predictability is not seen in tropical lakes where neither the water temperature nor the photoperiod changes much throughout the year. Life cycles of aquatic predators cannot, therefore, be climatically cued. Thus seasonal changes in predator feeding activity and population density are less predictable. The change in the magnitude of predation by visually oriented planktivores in a tropical lake may, however, be related to other aspects of light regime rather than to the photoperiod or the annual distribution of radiation. For instance, the predation rate in tropical lakes is a function of the lunar periodicity. The highest impact of planktivorous fish on zooplankton has been observed during the period of the full moon and a few days after, when a "moon trap" is set for migratory zooplankton (Gliwicz, 1986a). This phenomenon is highly predictable in tropical lakes, especially in arid regions where the cloud cover does not disrupt the regularity of the moon trap, but its predictability is related to the lunar cycle rather than to seasonality. If the phenomenon of lunar periodicity in predation intensity is not restricted to tropical lakes, it may seriously confound the seasonal sequence of events in temperate lakes and weaken the predictability of seasonal effects of predation. Furthermore, "light traps" are probably also set by a combination of clouds on windy nights or days when the moon or the sun may suddenly emerge from the clouds and illuminate planktonic animals that ascended near the surface when the light intensity was low.

There are also other phenomena in temperate lakes that may decrease the predictability of predation effects on zooplankton seasonal succession.

1.   Predators may change habitats or may switch from one prey to another. This may stem from the diet flexibility of predators and may be related to optimal foraging behavior and to prey patchiness.

2.   Although an overall seasonal change in water temperature is highly predictable, the temperature at a given depth may be determined by highly unpredictable changes in the wind strength and direction. The wind velocity would determine the depth of the mixed layer in a lake and thus also decide which fraction of resting eggs will hatch from those deposited in bottom sediments at various depths. Therefore, a sequence of discrete windy periods may bring about a sequence of discrete cohorts of an invertebrate predator (e.g., *Leptodora*). The sequence would appear in the lake as when subsequent portions of eggs are activated in the sediments by the temperature increase that results from each consecutive increase in the mixed layer depth. With each cohort the rate of predation on small-bodied zooplankton would be increased temporarily while predation on the large-bodied zooplankton would decrease as fish focus their attention on the invertebrate predator (see Sections 7.5 and 7.12).

3.   The seasonal sequence of predation intensity may be severely altered by nonseasonal events leading to an overall change in the relative abundance of vertebrate and invertebrate predators. This may either be a "winter kill" by winter anoxia in a lake or a parasitic disease in a fish population, each resulting in a fishless situation for next season. This may also be an accidental intrusion

of a new predator to the system. Finally, this may be a result of variations from year to year in the reproductive success of a plantivorous fish or an invertebrate predator.

Despite the high seasonal predictability that is the result of the predator life cycle, the seasonality of predation effects is diluted because of the variability of prey responses and the nature of predator-prey interactions (see Section 7.1). This is the reason why only a few effects of predation on the course of seasonal succession can be successfully predicted.

7.14 Concluding Remarks

Few successional events are consistent from year to year and from lake to lake. When they are consistent, they are more probably attributed to the effects of predation than to those of food limitation alone, because the seasonal changes in predation intensity are easier to predict from the seasonality of the climate than the seasonal changes in the abundance and availability of resources. The most important role of predation in the course of seasonal succession is probably one of merely initiating the sequence of species replacement, which may be completed only when competition for resources is involved. This is so unless predation is so intense that the densities of competing populations are kept well below the threshold level necessary to produce food limitation.

Today, the effects of predation are still only asserted rather than demonstrated in initiating successional events. The lack of direct evidence from the field has forced us to infer its importance from our general knowledge of the impact of predation on zooplankton communities. Therefore, more experimental work is needed in lakes where predation is thought to be important. The role of predation in species replacements in the course of seasonal succession may become clearer as a result of field-enclosure experiments in which both excluding and adding predators provide solid evidence to show whether they lead to patterns of succession that are similar or different from those observed in the lake itself.

Acknowledgments
We are grateful to Bill DeMott, Uli Sommer, and Bob Sterner for their critical reviews of the first draft of the manuscript and for the linguistic improvements they made. For discussions and useful suggestions we are also grateful to Joanna Gliwicz and Piotr Dawidowicz. Our special thanks go to Mike Swift who improved our wording. For technical assistance we are grateful to Borys Wasiuk and Danka Kozera.

References

Allan, J.D. 1973. Competition and the relative abundance of two cladocerans. *Ecology* 54: 484–498.

Anderson, G., Berggren, H., Cronberg, G., and Gelin, C. 1978. Effects of planktivorous and bentivorous fish on organisms and water chemistry in eutrophic lakes. *Hydrobiologia* 59: 9–15.

Baumann, P.C. and Kitchell, J.F. 1974. Diel patterns of distribution and feeding of bluegill (*Lepomis macrochirus*) in Lake Wingra, Wisconsin. *Transactions of the American Fisheries Society* 103: 255–260.

Beattie, D.M., Golterman, H.L., and Vijverberg, J. 1978. Introduction to the limnology of Friesian lakes. *Hydrobiologia* 58: 49–64.

Berg, K. and Nygaard, G. 1929. Studies of the plankton in the Lake Frederiksborg Castle. *Memoires de l'Academie Royale des sciences et des lettres de Danemark, Section des Sciences, serie 9*, 1: 223–316.

Birge, E.A. 1898. Plankton studies on Lake Mendota. II. The *Crustacea* of the plankton from July 1894, to December 1896. *Transactions of the Wisconsin Academy of Sciences, Arts and Letters* 11: 274–451.

Black, R.W. 1980. The genetic component of cyclomorphosis in *Bosmina*, pp. 456–469, in Kerfoot, W.C. (editor), *Evolution and Ecology of Zooplankton Communities*. University Press of New England, Hanover, New Hampshire.

Blaxter, J.H.S. 1966. The behavior and physiology of herring and other clupeids., pp. 261–393, in Russel, F.S. (editor), *Advances in Marine Biology*. Academic Press, New York.

Bosch, F.V.D. and Ringelberg, J. 1985. Seasonal succession and population dynamics of *Keratella cochlearis* (Ehrb.) and *Kellicotia longispina* (Kellicot) in Lake Maarsseveen I (Netherlands). *Archiv für Hydrobiologie* 103: 273–290.

Brandl, Z. and Fernando, C.H. 1979. The impact of predation by the copepod *Mesocyclops edax* (Forbes) on zooplankton in three lakes in Ontario, Canada. *Canadian Journal of Zoology* 57: 940–942.

Brock, D.A. 1980. Genotypic succession in the cyclomorphosis of *Bosmina longirostris* (Cladocera). *Freshwater Biology* 10: 239–250.

Brooks, J.L. 1965. Predation and relative helmet size in cyclomorphic *Daphnia*. *Proceedings of the National Academy of Sciences, USA* 53: 119–126.

Brooks, J.L. and Dodson, S.I. 1965. Predation, body size and composition of plankton. *Science* 150: 28–35.

Carlin, B. 1943. Die Planktonrotatorien des Notalostrom: zur Taxonomie und Ökologie der Planktonrotatorien. *Meddelanden från Lunds Universitets Limnologiska Institution* 5: 1–255.

Clark, A.S. and Carter, J.C.H. 1974. Population dynamics of cladocerans in Sunfish Lake, Ontario. *Canadian Journal of Zoology* 52: 1235–1242.

Coveney, M.F., Cronberg, G., Enell, M., Larsson, K., and Oloffson, L. 1977. Phytoplankton, zooplankton and bacteria—standing crop and production relationships in a eutrophic lake. *Oikos* 29: 5–21.

Cryer, M., Peirson, G., and Townsend, C.R. 1986. Reciprocal interactions between roach, *Rutilus rutilus*, and zooplankton in a small lake: Prey dynamics and fish growth and recruitment. *Limnology and Oceanography* 31: 1022–1038.

Culver, D. 1980. Seasonal variation in the sizes at birth and first reproduction in Cladocera, pp. 358–366, in Kerfoot, W.C. (editor), *Evolution and Ecology of Zooplankton Communities*. University Press of New England, Hanover, New Hampshire.

Dawidowicz, P. and Gliwicz, Z.M. 1983. Food of brook charr in extreme oligotrophic conditions. *Environmental Biology of Fishes* 8: 55–60.

Dawidowicz, P. and Pijanowska, J. 1984. Population dynamics in cladoceran zooplankton in the presence and absence of fishes. *Journal of Plankton Research* 6: 953–959.

De Bernardi, R. 1974. The dynamics of a population of *Daphnia hyalina* (Leydig) in Lago Maggiore, Northern Italy. *Memorie del Istituto italiano di idrobiologia* 31: 221–243.

De Bernardi, R. 1979. Some problems in the study of population dynamics of zooplankton. *Bolletino di zoologia* 46: 179–189.

De Bernardi, R. and Canali, S. 1975. Population dynamics of pelagic cladocerans in Lago Maggiore. *Memorie del Istituto italiano di idrobiologia* 32: 365–392.

De Bernardi, R. and Giussani, G. 1975. Population dynamics of three cladocerans of Lago Maggiore related to predation pressure by a planktophagous fish. *Internationale Vereinigung für theoretische und angewandte Limnologie, Verhandlungen* 19: 2908–2912.

De Bernardi, R. and Giussani, G. 1978. Effect of mass fish mortality on zooplankton structure and dynamics in a small Italian lake (Lago de Annone). *Internationale Vereinigung für theoretische und angewandte Limnologie, Verhandlungen* 20: 1045–1048.

DeMott, W.R. 1983. Seasonal succession in a natural *Daphnia* assemblage. *Ecological Monographs* 53: 321–340.

DeMott, W.R. and Kerfoot, W.C. 1982. Competition among cladocerans: nature of the interaction between *Bosmina* and *Daphnia*. *Ecology* 63: 1949–1966.

Dodson, S.I. 1974a. Zooplankton competition and predation: an experimental test of size-efficiency hypothesis. *Ecology* 55: 605–613.

Dodson, S.I. 1974b. Adaptive change in plankton morphology in response to size-selective predation: A new hypothesis of cyclomorphosis. *Limnology and Oceanography* 19: 721–729.

Dorazzio, R.M. and Lehman, J.T. 1983. Optimal reproductive strategies in age-structured populations of zooplankton. *Freshwater Biology* 13: 157–175.

Drenner, R.W. and McComas, S.R. 1980. The roles of zooplankter escape ability and fish size selectivity in the selective feeding and impact of planktivorous fish, pp. 587–593, in Kerfoot, W.C. (editor), *Evolution and Ecology of Zooplankton Communities*. University Press of New England, Hanover, New Hampshire.

Drenner, R.W., de Noyelles, F. Jr., and Kettle, D. 1982. Selective impact of filter-feeding gizzard shad on zooplankton community structure. *Limnology and Oceanography* 27: 965–968.

Drenner, R.W., Taylor, S.B., Lazzaro, X., and Kettle, D. 1984. Particle-grazing and plankton community impact of an omnivorous cichlid. *Transactions of the American Fisheries Society* 113: 397–402.

Edmondson, W.T. and Litt, A.H. 1982. *Daphnia* in Lake Washington. *Limnology and Oceanography* 27: 272–293.

Elgmork, K. 1980. Evolutionary aspects of diapause in freshwater copepods, pp. 411–417, in Kerfoot, W.C. (editor), *Evolution and Ecology of Zooplankton Communities*. University Press of New England, Hanover, New Hampshire.

Eloranta, P.V. 1982. Zooplankton in the Vassikalampi pond, a warm water effluent recipient in Central Finland. *Journal of Plankton Research* 4: 813–837.

Enderlin, O. 1981. When, where, what and how much does the adult cisco, *Coregonus albula* (L.) eat in the Bothnian Bay during the ice-free season. *Report of the Institute of Freshwater Research, Drottningholm* 59: 21–32.

Fairchild, G.W. 1982. Population responses of plant-associated invertebrates to foraging by largemouth bass fry *Micropterus salmoides*. *Hydrobiologia* 96: 169–176.

Fairchild, G.W. 1983. Birth and death rates of a littoral filter feeding microcrustacean, *Sida crystallina* (Cladocera) in Cohran Lake, Michigan. *Internationale Revue der gesamten Hydrobiologie* 68: 339–350.

Fedorenko, A.Y. 1975. Instar and species specific diets in two species of *Chaoborus*. *Limnology and Oceanography* 20: 238–249.

Fott, J., Desortova, D., and Hrbacek, J. 1980a. A comparison of the growth of flagellates under heavy grazing stress with a continuous culture, pp. 395–401, in Sikyta, B., Fencl, Z., and Polacek, V. (editors), *Continuous Cultivations of Microorganisms*. Institute of Microbiology, Czechoslovakian Academy of Sciences, Prague.

Fott, J., Pechar, L., and Prazakova, M. 1980. Fish as a factor controlling water quality in ponds. *Developments in Hydrobiology* 2: 255–261.

Gabriel, W. 1982. Modelling reproductive strategies of *Daphnia*. *Archiv für Hydrobiologie* 95: 69–80.

Galbraith, M.G. Jr. 1967. Size-selective predation on *Daphnia* by rainbow trout and yellow perch. *Transactions of the American Fisheries Society* 96: 1–10.

Geller, W. 1986. Diurnal vertical migration of zooplankton in a temperate great lake (L. Constance): A starvation avoidance mechanism? *Archiv für Hydrobiologie* Suppl. 74: 1–60.

Gilbert, J.J. 1967. *Asplanchna* and postero-lateral spine production in *Brachionus calyciflorus*. *Archiv für Hydrobiologie* 64: 1–62.

Gilbert, J.J. 1985. Escape response of the rotifer *Polyarthra*: a high speed cinematographic analysis. *Oecologia* (Berlin) 66: 322–331.

Gilbert, J.J. and Waage, J.K. 1967. *Asplanchna, Asplanchna*-substance and posterolateral spine length variation of the rotifer *Brachionus calciflorus* in a natural environment. *Ecology* 48: 1027–1031.

Glasser, J.W. 1978. The effect of predation on prey resource utilization. *Ecology* 59: 724–732.

Glasser, J.W. 1979. The role of predation in shaping and maintaining the structure of communities. *American Naturalist* 113: 31–41.

Gliwicz, Z.M. 1963. The influence of the stocking of the Tatra lakes on their biocenosis. *Chronmy Przyrode Ojczysta* 5: 27–35.

Gliwicz, Z.M. 1977. Food size selection and seasonal succession of filter feeding zooplankton in a eutrophic lake. *Ekologia polska* 25: 175–225.

Gliwicz, Z.M. 1981. Food and predation in limiting clutch size of cladocerans. *Internationale Vereinigung für theoretische und angewandte Limnologie, Verhandlungen* 21: 1562–1566.

Gliwicz, Z.M. 1985. Predation or food limitation: an ultimate reason for extinction of planktonic cladoceran species. *Archiv für Hydrobiologie, Beihefte Ergebnisse der Limnologie* 21: 419–430.

Gliwicz, Z.M. 1986a. A lunar cycle in zooplankton. *Ecology* 67: 882–897.

Gliwicz, Z.M. 1986b. Predation and the evolution of vertical migration in zooplankton. *Nature* 320: 746–748.

Gliwicz, Z.M., Ghilarov, A.M., and Pijanowska, J. 1981. Food and predation as major factors limiting two natural populations of *Daphnia cucullata* Sars. *Hydrobiologia* 80: 205–218.

Gliwicz, Z.M. and Pijanowska, J. 1988. Predation and resource distribution in shaping behavior of vertical migration in zooplankton. *Bulletin of Marine Science* 43: 951–965.

Goad, J. 1984. A biomanipulation experiment in Green Lake, Seattle, Washington. *Archiv für Hydrobiologie* 102: 137–153.

Goldman, C.R., Morgan, M.D., Threlkeld, S.T., and Angeli, N. 1979. Population dynamics analysis of the cladoceran disappearance from Lake Tahoe, California-Nevada. *Limnology and Oceanography* 24: 289–297.

Green, J. 1967. The distribution and variation of *Daphnia lumnholtzi* (*Crustacea: Cladocera*) in relation to fish predation in Lake Albert, East Africa. *Journal of Zoology* 151: 181–197.

Greene, C.H. 1983. Selective predation in freshwater zooplankton communities. *Internationale Revue der gesamten Hydrobiologie* 68: 297–315.

Greene, C.H. 1985. Planktivore functional groups and patterns of prey selection in pelagic communities. *Journal of Plankton Research* 7: 35–40.

Greene, C.H. 1986. Patterns of prey selection: implications of predator foraging tactics. *American Naturalist* 128: 824–839.

Grygierek, E., Hillbricht-Ilkowska, A., and Spodniewska, I. 1966. The effect of fish on plankton community in ponds. *Internationale Vereinigung für theoretische und angewandte Limnologie, Verhandlungen* 16: 1359–1366.

Hairston, N.G. Jr. 1987. Diapause as a predator-avoidance adaptation, pp. 287–290, in Kerfoot, W.C. and Sih, A. (editors), *Predation, Direct and Indirect Impacts on Aquatic Communities*. University Press of New England, Hanover, New Hampshire.

Hairston, N.G. Jr. and Munns, W.R. Jr. 1984. The timing of copepod diapause as an evolutionary stable strategy. *American Naturalist* 123: 733–751.

Hairston, N.G. Jr., Walton, W.E., and Li, K.T. 1983. The causes and the consequences of sex-specific mortality in a freshwater copepod. *Limnology and Oceanography* 28: 935–947.

Hall, D.J. 1964. An experimental approach to the dynamics of a natural population of *Daphnia galeata mendotae*. *Ecology* 45: 94–112.

Hall, D.J., Cooper, W.E., and Werner, E.E. 1970. An experimental approach to the production dynamics and structure of freshwater animal communities. *Limnology and Oceanography* 15: 839–928.

Hall, D.J., Threlkeld, S.T., Burns, C.W., and Crowley, P.H. 1976. The size-efficiency hypothesis and the size structure of zooplankton communities. *Annual Revue of Ecology and Systematics* 7: 177–208.

Hare, L. and Carter, J.C.H. 1987. Zooplankton populations and the diets of three *Chaoborus* species (*Diptera, Chaoboridae*) in a tropical lake. *Freshwater Biology* 17: 275–290.

Hartmann, J. 1983. Two feeding strategies of young fishes. *Archiv für Hydrobiologie* 96: 496–509.

Hartmann, J. 1986. Interspecific predictors of selected prey of young fishes. *Archiv für Hydrobiologie, Beihefte Ergebnisse der Limnologie* 22: 373–386.

Hillbricht-Ilkowska, A. and Weglenska, T. 1973. Experimentally increased fish stock in the pond type Lake Warniak. VII. Numbers, biomass and production of zooplankton. *Ekologia polska* 21: 533–552.

Holling, C.S. 1959. The components of predation as revealed by a study of small mammal predation of the European pine sawfly. *Canadian Journal of Entomology* 91: 293–320.

Hrbáček, J. 1962. Species composition and the amount of zooplankton in relation to the fish stock. *Rozpravy Ceskoslovenske Akademie Ved, Rada mathematicko-prirodovedecka* 72 (10): 1–114.

Hrbáček, J. and Novotna-Dvorakova, M. 1965. Plankton of four backwaters related to their size and fish stock. *Rozpravy Ceskoslovenske Akademie Ved, Rada mathematicko-prirodovedecka* 75 (13): 1–65.

Hrbáček, J., Dvorakova, M., Korinek, V., and Prochazkova, L. 1961. Demonstration of the effect of the fish stock on the species composition of zooplankton and the intensity of metabolism of the whole plankton association. *Internationale Vereinigung für theoretische und angewandte Limnologie, Verhandlungen* 14: 192–195.

Hunter, J.R. 1979. The feeding behavior and ecology of marine fish larvae, in Bardach, J.E. (editor), *The Physiological and Behavioral Manipulation of Food Fish as Production and Management Tools*.

Hurlbert, W.H. and Mulla, M.S. 1981. Impacts of mosquito fish (*Gambusia affinis*) predation on plankton communities. *Hydrobiologia* 83: 125–151.

Hutchinson, B.P. 1971. The effect of fish predation on the zooplankton on ten Adirondack Lakes, with particular reference to the alewife, *Alosa pseudoharengus*. *Transactions of the American Fisheries Society* 100: 325–335.

Hutchinson, G.E. 1967. *A Treatise on Limnology*. Vol. II. *Introduction to Lake Biology and the Limnoplankton*. John Wiley and Sons, New York, 1115 pp.

Kajak, Z., Dusoge, K., Hillbricht-Ilkowska, A., Pieczynski, E., Prejs, A., Spodniewska, I., and Weglenska, T. 1972. Influence of the artificially increased fish stock on the lake biocenosis. *Internationale Vereinigung für theoretische und angewandte Limnologie, Verhandlungen* 18: 228–235.

Kajak, Z. and Rybak, J. 1979. The feeding of *Chaoborus flavicans* Wiegen (*Diptera, Chaoboridae*) and its predation on a lake zooplankton. *Internationale Revue der gesamten Hydrobiologie* 64: 361–378.

Kajak, Z., Rybak, J., and Ranke-Rybicka, B. 1978. Fluctuations in numbers and changes in the distribution of *Chaoborus flavicans* (Meigen) (*Diptera, Chaoboridae*) in the eutrophic Mikolajskie Lake and dystrophic Lake Flosek. *Ekologia polska* 26: 259–272.

Kerfoot, W.C. 1980. Commentary: transparency, body size, and prey conspicuousness, pp. 609–617, in Kerfoot, W.C. (editor), *Evolution and Ecology of Zooplankton Communities*. University Press of New England, Hanover, New Hampshire.

Kerfoot, W.C. 1982. A question of taste: crypsis and warning coloration in freshwater zooplankton communities. *Ecology* 63: 538–554.

Kerfoot, W.C. 1987. Cascading effects and indirect pathways, pp. 57–70, in Kerfoot, W.C. and Sih, A. (editors), *Predation, Direct and Indirect Impacts on Aquatic Communities*. University Press of New England, Hanover, New Hampshire.

Kerfoot, W.C., Kellog, D.L. Jr., and Strickler, J.R. 1980. Visual observations of live zooplankters: Evasion, escape and chemical defenses, pp. 10–27, in Kerfoot, W.C. (editor), *Evolution and Ecology of Zooplankton Communities*. University Press of New England, Hanover, New Hampshire.

Lampert, W. 1987. Predictability in lake ecosystems: the role of biotic interactions. *Ecological Studies* 61: 333–346.

Lampert, W. 1988. The relative importance of food limitation and predation in the seasonal cycle of two *Daphnia* species. *Internationale Vereinigung für theoretische und angewandte Limnologie, Verhandlungen* 23:713–718

Lampert, W. and Schober, U. 1978. Das regelmassige Auftreten von Fruhjahrs Algenmaximum und "Klarwasserstadium" im Bodensee als Folge von climatischen Bedingungen und Wechselwirkungen zwischen Phytound Zooplankton. *Archiv für Hydrobiologie* 82: 364–386.

Langeland, A. 1982. Interactions between zooplankton and fish in a fertile lake. *Holarctic Ecology* 5: 273–310.

Larsson, P., Johnsen, G., and Steigen, A.L. 1985. An experimental study of the summer decline in a *Daphnia* population. *Internationale Vereinigung für theoretische und angewandte Limnologie, Verhandlungen* 22: 3131–3136.

Lazzaro, X. 1987. A review of planktivorous fishes: Their evolution, feeding behaviors, selectivities and impacts. *Hydrobiologia* 146: 97–167.

Lehtovaara, A. and Sarvala, J. 1984. Seasonal dynamics of total biomass and species composition of zooplankton in the littoral of an oligotrophic lake. *Internationale Vereinigung für theoretische und angewandte Limnologie, Verhandlungen* 22: 805–810.

Lynch, M. 1977. Fitness and optimal body size in zooplankton populations. *Ecology* 58: 763–774.

Lynch, M. 1979. Predation, competition, and zooplankton community structure. *Limnology and Oceanography* 24: 253–272.

Lynch, M., Monson, B., Sandheinrich, M., and Weider, L. 1981. Size specific mortality rates in zooplankton populations. *Internationale Vereinigung für theoretische und angewandte Limnologie, Verhandlungen* 21: 363–368.

Lynch, M. and Shapiro, J. 1981. Predation, enrichment and phytoplankton community structure. *Limnology and Oceanography* 26: 86–102.

Manning, B.J., Kerfoot, W.C., and Berger, E.M. 1978. Phenotypes and genotypes in cladoceran populations. *Evolution* 32: 365–374.

Mellors, W.K. 1975. Selective predation on ephippial *Daphnia* and the resistance of ephippial eggs to digestion. *Ecology* 56: 974–980.

Mills, E.L. and Forney, J.L. 1983. Impact on *Daphnia pulex* of predation by yellow perch in Oneida Lake, New York. *Transactions of the American Fisheries Society* 112: 154–161.

Mills, E.L., Forney, J.L., and Wagner, K.J. 1987. Fish predation and its cascading effect on the Oneida Lake food chain, pp. 118–131, in Kerfoot, W.C. and Sih, A. (editors), *Predation, Direct and Indirect Impacts on Aquatic Communities*. University Press of New England, Hanover, New Hampshire.

Miracle, M.R. 1974. Niche structure in freshwater zooplankton: a principal components approach. *Ecology* 55: 1306–1316.

Moore, M. and Gilbert, J.J. 1987. Age-specific *Chaoborus* predation on rotifer prey. *Freshwater Biology* 17: 223–236.

Morgan, M.D., Threlkeld, S.T., and Goldman, C.R. 1978. Impact of the introduction of kokanee (*Oncorhynchus nerka*) and oppossum shrimp (*Mysis relicta*) on a sub-alpine lake. *Journal of Fisheries Research Board of Canada* 35: 1572–1579.

Morin, P.J. 1987. Salamander predation, prey facilitation, and seasonal succession in microcrustacean communities, pp. 174–187, in Kerfoot, W.C. and Sih, A. (editors), *Predation, Direct and Indirect Impacts on Aquatic Communities*. University Press of New England, Hanover, New Hampshire.

Murtaugh, P.A. 1981. Size-selective predation on *Daphnia* by *Neomysis mercedis*. *Ecology* 62: 894–900.

Nakashima, B.S. and Leggett, W.C. 1978. Daily ration of yellow perch (*Perca flavescens*) from Lake Memphremagog, Quebec-Vermont, with a comparison of methods for in situ determinations. *Journal of Fisheries Research Board of Canada* 35: 1597–1603.

Nauwerck, A. 1963. Die Beziehungen zwischen Zooplankton und Phytoplakton im See Erken. *Symbolae Botanicae Upsalienses* 17 (5): 1–163.

Neill, W.E. 1975. Experimental studies of microcrustacean competition, community composition and efficiency of resource utilization. *Ecology* 56: 809–826.

Neill, W.E. 1981. Impact of *Chaoborus* predation upon the structure and dynamics of a crustacean zooplankton community. *Oecologia* (Berlin) 48: 164–177.

Neill, W.E. and Peacock, A. 1980. Breaking the bottleneck: interactions of invertebrate predators and nutrients in oligotrophic lakes, pp. 715–724, in Kerfoot, W.C. (editor), *Evolution and Ecology of Zooplankton Communities*. University Press of New England, Hanover, New Hampshire.

Nilssen, J.P. 1977. Cryptic predation and the demographic strategy of two limnetic cyclopoid copepods. *Memorie del Istituto italiano di idrobiologia* 34: 187–196.

Nilsson, N.A. and Pejler, B. 1973. On the relation between fish, fauna and zooplankton composition in North Swedish lakes. *Report of the Institute of Freshwater Research, Drottningholm* 53: 51–76.

Northcote, T.G. and Clarotto, J. 1975. Limnetic macrozooplankton and fish predation in some coastal British Columbia lakes. *Internationale Vereinigung für theoretische und angewandte Limnologie, Verhandlungen* 19: 2378–2393.

Northcote, T.G., Walters, J., and Hume, J.M.B. 1978. Initial impacts of experimental fish introductions onthe macrozooplankton of small oligotrophic lakes. *Internationale Vereinigung für theoretische und angewandte Limnologie, Verhandlungen* 20: 2003–2012.

Nyberg, P. 1976. Production and food consumption of perch in two Swedish forest lakes. *Scripta Limnologica Upsaliensia 421, Klotenprojekted Raport* 6: 1–97.

O'Brien, W.J. 1987. Planktivory by freshwater fish: thrust and parry in the pelagia, pp. 3–16, in Kerfoot, W.C. and Sih, A. (editors), *Predation, Direct and Indirect Impacts on Aquatic Communities*. University Press of New England, Hanover, New Hampshire.

O'Brien, W.J., Kettle, D., and Riessen, H. 1979. Helmets and invisible armor: Structures reducing predation from tactile and visual planktivores. *Ecology* 60: 287–294.

O'Brien, W.J. and Vinyard, G.L. 1978. Polymorphism and predation: The effect of invertebrate predation on the distribution of two varieties of *Daphnia carinata* in South India ponds. *Limnology and Oceanography* 23: 452–460.

Paine, R.T. 1966. Food web complexity and species diversity. *American Naturalist* 100: 65–75.

Papinska, K. 1988. The effect of fish predation on *Cyclops* life cycle. *Hydrobiologia* 167/168: 449–453.

Patalas, K. 1963. Seasonal changes in pelagic crustacean plankton in six lakes of Wegorzewo district. *Roczniki Nauk Rolniczych* 82: 209–234.

Pijanowska, J. 1985. Antipredator defense mechanisms in zooplankton. *Wiadomosci ekologiczne* 31: 123–172 (in Polish with English summary).

Pijanowska, J. and Dawidowicz, P. 1987. The lack of vertical migrations in *Daphnia*: the effect of homogenously distributed food. *Hydrobiologia* 148: 175–181.

Post, J.R. and McQueen, D.J. 1987. The impact of planktivorous fish on the structure of plankton community. *Freshwater Biology* 17:79–89

Pourriot, R. 1983. Influence selective de la predation sur la structure et la dynamique de zooplancton d'eau douce. *Acta Oecologica* 4: 13–25.

Riessen, H.P. 1984. The other side of cyclomorphosis: Why *Daphnia* lose their helmets. *Limnology and Oceanography* 29: 1123–1127.

Riessen, H.P. 1985. Exploitation of prey seasonality by a planktonic predator. *Canadian Journal of Zoology* 63: 1729–1732.

Richards, R.C., Goldman, C.R., Frantz, T.C., and Wickwire, R. 1975. Where have all the *Daphnia* gone? The decline of a major cladoceran in Lake Tahoe. *Internationale Vereinigung für theoretische und angewandte Limnologie, Verhandlungen* 19: 835–842.

Ringelberg, J. 1980. Introductory remarks: Causal and teleological aspects of diurnal vertical migration, pp. 65–68, in Kerfoot, W.C. (editor), *Evolution and Ecology of Zooplankton Communities*. University Press of New England, Hanover, New Hampshire.

Rosenthal, H. and Hempel, G. 1970. Experimental studies in feeding and food requirements of herring larvae (*Clupea harengus* L.), pp. 344–364, in Teele, H. (editor), *Marine Food Chains*. Oliver and Boyd, Edinburgh.

Sandstrom, O. 1980. Selective feeding by Baltic herring. *Hydrobiologia* 69: 199–207.

Scavia, D., Fahnenstiel, G.L., Evans, M.S., Jude, D.J., and Lehman, J.T. 1986. Influence of salmonid predation and weather on long-term water quality trends in Lake Michigan. *Canadian Journal of Fisheries and Aquatic Sciences* 43: 435–443.

Seitz, A. 1977. Die Bedeutung von Umweltfactoren, Konkurenz und Rauber-Beute-Beziehungen für die Koexistenz drier Daphnienarten. *Ph.D. Thesis, Universität München*, 160 pp.

Shapiro, J. and Wright, D.I. 1984. Lake restoration by biomanipulation: Round Lake, Minnesota, the first two years. *Freshwater Biology* 14: 371–383.

Sommer, U., Gliwicz, Z.M., Lampert, W., and Duncan, A. 1986. The PEG-model of seasonal succession of planktonic events in fresh waters. *Archiv für Hydrobiologie* 106: 433–471.

Stasiak, I. 1981. Reconstruction of history of Tatra lakes zooplankton from sediment cores. *M. Sc. Thesis, University of Warsaw*, 40 pp.

Stearns, S.C. 1976. Life-history tactics: a review of the ideas. *Quarterly Review of Biology* 51: 3–47.

Stemberger, R.S. and Evans, M.S. 1984. Rotifer seasonal succession and copepod predation in Lake Michigan. *Journal of Great Lakes Research* 10: 417–428.

Stenson, J.A.E. 1978. Relations between vertebrate and invertebrate zooplankton predators in some arctic lakes. *Astarte* 11: 21–26.

Stenson, J.A.E. 1987. Variation in capsule size of *Holopedium gibberum* (Zaddach): a response to invertebrate predation. *Ecology* 68: 928–934.

Stenson, J.A.E., Bohlin, T., Henrickson, L., Nilsson, B.I., Nyman, H.G., Oscarson, H.G., and Larsson, P. 1978. Effects of fish removal from a small lake. *Internationale Vereinigung für theoretische und angewandte Limnologie, Verhandlungen* 20: 794–801.

Stich, H.B. and Lampert, W. 1981. Predator evasion as an explanation of diurnal vertical migration by zooplankton. *Nature* 293: 396–398.

Stich, H.B. and Lampert, W. 1984. Growth and reproduction of migrating and non-migrating *Daphnia* species under simulated food and temperature conditions of diurnal vertical migrations. *Oecologia* (Berlin) 61: 192–196.

Strickler, J.R. and Twombly, S. 1975. Reynolds number, diapause and predatory copepods. *Internationale Vereinigung für theoretische und angewandte Limnologie, Verhandlungen* 19: 2943–2950.

Stross, R.G. 1973. Zooplankton reproduction and water blooms, pp. 467–478, in Glass, G.A. (editor), *Bioassay Techniques and Environmental Chemistry*. Environmental Protection Agency, Duluth, Minnesota.

Swift, M.C. and Fedorenko, A.Y. 1975. Some aspects of prey capture by *Chaoborus* larvae. *Limnology and Oceanography* 20: 418–425.

Swift, M.C. and Forward, R.B. Jr. 1981. *Chaoborus* prey capture efficiency in the light and dark. *Limnology and Oceanography* 26: 416–466.

Szlauer, L. 1965. The refuge ability of plankton animals before plankton eating animals. *Polskie Archiwum Hydrobiologii* 13: 89–95.

Szlauer, L. 1968. Investigations upon ability in plankton *Crustacea* to escape the net. *Polskie Archiwum Hydrobiologii* 16: 78–86.

Tessier, A.J. 1986. Comparative population regulation of two planktonic Cladocera (*Holopedium gibberum* and *Daphnia catawba*). *Ecology* 67: 285–302.

Thorp, J.H. 1986. Two distinct roles for predators in freshwater assemblages. *Oikos* 47: 75–82.

Threlkeld, S.T. 1979. The midsummer dynamics of two *Daphnia* species in Wintergreen Lake, Michigan. *Ecology* 60: 165–179.

Threlkeld, S.T. 1981. The recolonization of Lake Tahoe by *Bosmina longirostris*: Evaluating the importance of reduced *Mysis relicta* population. *Limnology and Oceanography* 26: 433–444.

Tschumi, P.A., Bangerter, B., and Zbaren, D. 1982. Zehn Jahre limnologische Forschung am Bielersee (1972–1982). *Vierteljahresschrift der Naturforschenden Gesellschaft Zurich* 127: 337–355.

Vanni, M.J. 1986. Fish predation and zooplankton demography: indirect effects. *Ecology* 67: 337–354.

Vanni, M.J. 1987. Effects of food availability and fish predation on a zooplankton community. *Ecological Monographs* 57: 61–88.

Vijverberg, J. and van Densen, W.L.T. 1984. The role of fish in the food web of Tjukemeer, The Netherlands. *Internationale Vereinigung für theoretische und angewandte Limnologie, Verhandlungen* 22: 891–896.

Wiljanen, M. and Holopainen, I.J. 1982. Population density of perch (*Perca fluviatilis* L) at egg, larval and adult stages in the dysoligotrophic Lake Soumunjarvi, Finland. *Annales Zoologici Fennici* 19: 39–46.

Vuorinen, I., Rajasilta, M., and Salo, J. 1983. Selective predation and habitat shift in a copepod species—support for the predation hypothesis. *Oecologia* (Berlin) 59: 62–64.

Ware, D.M. 1973. Risk of epibenthic prey to predationby rainbow trout (*Salmo gairdneri*). *Journal of Fisheries Research Board of Canada* 30: 786–797.

Weglenska, T., Dusoge, K., Ejsmont-Karabin, J., Spodniewska, I., and Zachwieja, J. 1979. Effect of winter kill and changing fish stock on the biocenose of the pond-type Lake Warniak. *Ekologia polska* 27: 39–70.

Weider, L.J. 1984. Spatial heterogeneity of *Daphnia* genotypes: Vertical migration and habitat partitioning. *Limnology and Oceanography* 29: 225–235.

Werner, E.E. and Gilliam, J.F. 1984. The ontogenetic niche and species interactions in size-structured populations. *Annual Review of Ecology and Systematics* 15: 393–425.

Wilson, D.S. 1975. The adequacy of body size as a niche difference. *American Naturalist* 109: 769–784.

Wong, B. and Ward, F.J. 1972. Size selection of *Daphnia pulicaria* by yellow perch (*Perca flavescens*) fry in West Blue Lake, Manitoba. *Journal of Fisheries Research Board of Canada* 29: 1761–1764.

Wright, J.C. 1965. The population dynamics and production of *Daphnia* in Canyon Ferry Reservoir, Montana. *Limnology and Oceanography* 10: 583–590.

Wright, D.I. and O'Brien, W.J. 1982. Differential location of *Chaoborus* larvae and *Daphnia* by fish: The importance of motion and visible size. *American Midland Naturalist* 108: 68–73.

Wright, D.I. and O'Brien, W.J. 1984. The development and field test of tactical model of the planktivorous feeding of white crappie (*Pomoxis annularis*). *Ecological Monographs* 54: 65–98.

Zaret, T.M. 1972a. Predator-prey interaction in a tropical lacustrine ecosystem. *Ecology* 53: 248–257.

Zaret, T.M. 1972b. Predators, invisible prey, and the nature of polymorphism in the Cladocera (class Crustacea). *Limnology and Oceanography* 17: 171–184.

Zaret, T.M. 1978. A predation model of zooplankton community structure. *Internationale Vereinigung für theoretische und angewandte Limnologie, Verhandlungen* 20: 2496–2500.

Zaret, T.M. 1980. *Predation and Freshwater Communities.* Yale University Press, New Haven, 180 pp.

Zaret, T.M. and Kerfoot, W.C. 1975. Fish predation on *Bosmina longirostris*: body-size selection versus visibility selection. *Ecology* 56: 232–237.

Zaret, T.M. and Suffern, J.S. 1976. Vertical migration in zooplankton as a predator-avoidance mechanism. *Limnology and Oceanography* 21: 804–813.

8

Toward an Autecology of Bacterioplankton

Carlos Pedrós-Alió

Institute of Marine Sciences
Higher Council of Scientific Research
Barcelona, Spain

8.1 Introduction

The purpose of this chapter on bacterioplankton is to place these procaryotes within the frame of reference commonly used to study and understand the eucaryotic plankton. The study of bacterioplankton is very recent compared to that of phyto- and zooplankton. In the last few years quantitative evidence of the importance of bacterioplankton in carbon cycling in lakes is accumulating at a fast rate. It now appears that bacterioplankton production constitutes around 30 percent of the total production of phytoplankton (Cole et al., 1988). However, because bacteria are very small and morphologically not very diverse, very little is known about their species composition in situ (Staley, 1980; Atlas, 1983).

Species identification is almost impossible without both the isolation of pure cultures and their biochemical study in the laboratory. Unfortunately, there is evidence that some of the most abundant bacteria in natural environments are hard to isolate in pure cultures (Hoppe, 1976; Sieburth, 1979; Torrella and Morita, 1981; Jones, 1987; Ward et al., 1987). Thus, the species composition of bacterioplankton has not been determined for most natural environments. Furthermore, since knowledge of species composition is necessary before seasonal succession can be examined, it is very difficult to study the succession of bacterioplankton species.

Applying the species concept itself to bacteria is considered to be problematic by many microbial ecologists (Brock, 1987; Jones, 1987). Sonea (1988) has tried to characterize the peculiar organization of bacterial communities by considering them as "adaptable chimeras" in which the classical concept of species loses all meaning. However, it is obvious that each bacterial cell

297

in nature has a certain genotype, i.e., the base sequence of its DNA, both genoforic and plasmidic. A continuum from perfect similarity to large differences among genotypes can be envisioned. Some genotypes may be members of the same clone (identical base sequences). Some will show minor differences that will still allow nucleic acid hybridization. Others will carry out the same main physiological processes with substantially different genotypes. Finally, many genotypes will be so different from each other as to merit inclusion in widely separated branches of the phylogenetic tree (Woese, 1987). Several of those genotypes will be close enough among themselves to produce phenotypes carrying out approximately the same ecological functions in the environment (for example, sulfate reduction). These functional groups can be used to study microbial ecology. The present chapter has been organized as a route from what is presently known (synecology of bacteria) toward what we would like to know in the future (autecology of all the species in a community). A theoretical framework is proposed to guide such a trip, and an attempt will be made to fit diverse and separate bits of evidence into such a framework. Hopefully, the spaces left empty will indicate where more research is needed. This chapter will also examine what kinds of techniques should be used, and what results may be expected from such future work.

We shall start from the current paradigm of the plankton to see what the role of bacteria is. We shall then examine whether this narrow view of bacterial roles is realistic in view of the metabolic diversity of bacteria. Next, the concept of the guild will be introduced as a tool to guide autecological studies of bacterioplankton. We shall show how this concept is more useful than that of species (and of immediate practical use). We shall then collate the evidence and interpret it in light of the guild concept to see what we can learn about seasonal succession of bacterioplankton. Finally, we shall present an example in which detailed knowledge of "true" species succession exists for a specific group of bacteria.

8.2   Paradigm of the Plankton

The basis for division of the plankton into three categories (phyto-, zoo-, and bacterioplankton) is functional. Such a division works fairly well when different trophic modes and life strategies can be clearly assigned to the three different taxonomic groups. In the classical model, phytoplankton consist of phototrophic organisms carrying out oxygenic photosynthesis and having little or no motility. The organisms forming the zooplankton have a high potential for motility and a heterotrophic metabolism based on raptorial feeding on phytoplankton or on other members of the zooplankton. Finally, the bacterioplankton is composed of procaryotic organisms with a heterotrophic metabolism, based on osmotrophy of dead organic matter originating from the other two components of the plankton.

This scheme mimics nicely the classical model for terrestrial ecosystems with plants, animals, and decomposers. In such a model, plants are the dominant component of the community. They determine the general aspect of the landscape (desert, prairie, forest) and, thus, which kinds of animals can survive. Plants represent the main source of energy and carbon to any system and are, therefore, at the base of virtually every food chain without regard for the dominance of herbivores or decomposers. Many of the developments in ecology have been derived from the study of terrestrial plant and animal communities: species composition, species substitution along environmental gradients, competition, succession toward a climax, etc. (Odum, 1971; Krebs, 1985). But it has not been possible to apply these ideas to the third component of the communities: the decomposers.

The same classical model has been applied to the plankton. In this case, different phytoplankton species (nannoplankton, net plankton) determined different zooplankton species (net zooplankton, mucus-net makers, protozoa, etc.). Zooplankton was eaten by fishes and, eventually, all organic matter was decomposed by bacterioplankton (Steele, 1974). Pomeroy (1974) claimed a "new paradigm" in which the role of microbes was more diverse and important than had been realized previously. Since Pomeroy's contribution considerable information has accrued, and important insights have been gained from theoretical developments and experimental data concerning the synecology of bacterioplankton. Some of these insights will be summarized in the following pages.

Many techniques for the study of bacterioplankton are fairly recent and there is still much discussion about their validity. Perhaps because of this uncertainty, there has been a wealth of review articles on methodology (Jones, 1979; van Es and Meyer-Reil, 1982; Overbeck et al., 1984; Staley and Konopka, 1985; Newell et al., 1986; Karl, 1986) reflecting the need felt by microbial ecologists to arrive at an established set of standard procedures.

Techniques notwithstanding, there is general agreement on the following three points:

1. Bacterial abundance in fresh and marine waters oscillates around 10^6 cells ml^{-1} (Hobbie, 1979; Gorlenko et al., 1983). Very oligotrophic systems may have lower numbers, around 10^5 cells ml^{-1}, while eutrophic lakes may easily reach 10^7 cells ml^{-1}. But, within a given system, the cell number only changes by a factor of 5 to 10 year-round.

2. Bacterial heterotrophic production constitutes 20 to 30 percent of primary production (Cole et al., 1988). Thus, a substantial portion of the carbon flux will circulate through bacterioplankton, and a certain degree of complexity is to be expected among bacteria.

3. Bacteria are grazed intensely by ciliates and flagellates (Sherr and Sherr, 1984, 1987; Güde, 1986, Chapter 9 of this book) and to a much lesser degree, by crustacean zooplankton (Pedrós-Alió and Brock, 1983a; Porter, 1984; Rieman, 1985; Børsheim and Anderson, 1987).

Thus, bacterial production may reenter the planktonic food chains via ciliates and flagellates. This "microbial loop" (Azam et al., 1983) confounds the traditional separation of herbivore from detritivore food chains and shows that decomposition is not the only role of bacteria in the plankton.

These three points have been incorporated into the current paradigm of the plankton (Williams, 1981; Pomeroy, 1984; Porter et al., 1985). However, all these studies have considered the bacterioplankton as a black box (synecology). In pelagic habitats the aerobic decomposers have been studied almost to the exclusion of all other groups of bacteria. And most studies have been limited to the epilimnion of lakes. As we shall see, aerobic decomposition is only one of the many roles played by bacterioplankton in lakes and their activities extend all the way down into the bottom sediments. Because of the difficulties in identifying species in the field, it has not been possible to apply ecological concepts, such as diversity, competition, and succession. When these terms are used in microbial ecology they usually only refer to protists (Alexander, 1971; Atlas, 1983), while procaryotes are either not analyzed or only examined in pure culture (van Es et al., 1984; Kuenen et al., 1985). Although these studies are necessary for understanding the potential of microorganisms and have clarified many aspects of competition among bacteria, as long as the extent of the diversity of natural systems remains unknown, pure culture studies cannot be easily extrapolated to field situations.

8.3   Trophic Roles of Bacterioplankton

To consider bacteria only as decomposers of organic matter is an oversimplification that ignores the wealth of chemical reactions carried out by procaryotes and the complexities of their interactions with the organisms of the eucaryotic plankton. In order to grasp the diversity of these processes we shall briefly review a few metabolic concepts.

The essential requirements for a typical growing organism are shown in Figure 8.1A. Most organisms use minute quantities of inorganic nutrients such as nitrogen, phosphorus, sulfur, and others, but large amounts of carbon and energy sources are processed for growth. This growth always requires large amounts of a reduced electron donor and an oxidized electron acceptor. Depending on the compounds used for these purposes, organisms can be classified in three different trophic systems (see, for example, Stolp and Starr, 1981):

1. With respect to the energy source, organisms can be either phototrophs (using light energy) or chemotrophs (using energy from exergonic chemical reactions).

2. With respect to the source of electrons, organisms can be either organotrophs (using electrons from organic matter) or lithotrophs (taking electrons from inorganic compounds such as water, hydrogen or sulfide).

3. Finally, according to the carbon source, organisms are divided into autotrophs (using CO_2) and heterotrophs (using organic matter).

A complete trophic description of an organism, therefore, involves much more information than is usually given, and should include information on the three systems given above. For example, most plants are photolithoautotrophs and all animals are chemoorganoheterotrophs.

For the most part, phytoplankton organisms only use one carbon source (CO_2), only one energy source (visible light), only one electron donor (H_2O), and only one electron acceptor (CO_2), which is reduced to form new organic matter (Figure 8.1B). Therefore, they are photolithoautotrophs. Members of the zooplankton are also specialized in one single trophic mode. They use particulate organic matter (POM) as a carbon source (often in the form of life phytoplankton cells), respiration of organic matter as an energy source, the same organic matter as an electron donor, and oxygen as an electron acceptor; the oxygen is reduced to water (Figure 8.1C). As a result of excretion by zooplankton, both organic matter and inorganic nutrients may be released to the environment. Therefore, they are all chemoorganoheterotrophs.

Thus, both the phyto- and zooplankton are specialized in one trophic mode each. As a result of their trophic activities they process large amounts of organic matter, CO_2, O_2, and H_2O, but no other elements are affected to such a large extent. Bacterioplankton, on the other hand, must be divided into smaller groups, each of which use a particular combination of the energy sources and the electron donors and acceptors available to procaryotes. We shall call such groups "guilds." Some of these guilds are the ones that are most often considered in synecological studies of lakes— the aerobic decomposers. These guilds of bacteria use the same trophic mode as the zooplankton, except that dissolved organic matter (DOM) is usually more important than POM, and they use osmotrophy instead of phagotrophy as the mechanism for acquiring the necessary compounds.

Many other guilds exist in lakes and, moreover, these other guilds

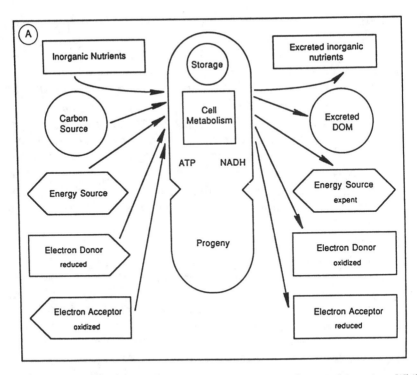

Figure 8.1   A. Essential requirements and waste products of a typical organism. While only small amounts of nutrients are needed, large amounts of carbon sources, energy sources, electron donors, and electron acceptors are processed in the life cycle of organisms. Thus, the cycles of the components of any one of these four requirements have rapid turnover rates and are extremely dynamic. B. Essential requirements and waste products of a typical phytoplankton representative. C. Essential requirements and waste products of a typical zooplankton representative. In both B and C a single type of nutrition is used. Thus, zooplankton and phytoplankton fulfill one single functional role each in the plankton community.

are responsible for many of the most important biogeochemical properties of aquatic systems. For example, the concentrations of nutrients and gasses from the metalimnion down to the sediment are almost exclusively due to the action of various bacterial guilds. Examples of such guilds are shown in Figure 8.2. Many of these organisms are anaerobic and extend life in lakes to the anaerobic hypolimnion, a zone from which both the phyto- and the zooplankton are almost completely excluded. Other bacterio-plankton guilds require the simultaneous presence of oxygen and of reduced inorganic compounds such as ammonia, hydrogen sulfide, and methane. Since these compounds tend to oxidize to nitrate, sulfur, and carbon dioxide, respectively, in the presence of oxygen, they can coexist

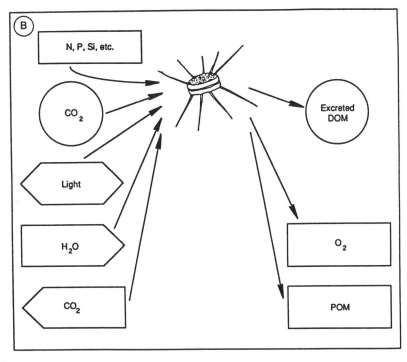

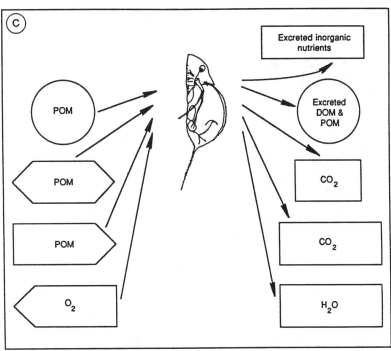

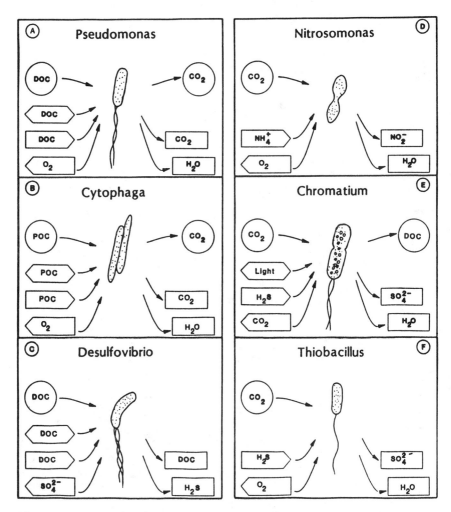

Figure 8.2 Examples of bacterial guilds. A. *Pseudomonas* is a member of a guild of aerobic decomposers using DOM. B. *Cytophaga*, a gliding bacterium, is an example of an aerobic decomposer using mostly POM. C. *Desulfovibrio* is a member of the sulfate reducers, one of the anaerobic guilds. D. *Nitrosomonas* belongs to a chemolithoautotrophic guild requiring oxygen and ammonia. E. *Chromatium* is a photolithoautotroph requiring light, sulfide, and anaerobic conditions. F. *Thiobacillus* is a chemolithoautotroph requiring sulfide and oxygen.

only at interfaces where oxygen is continuously supplied from above and the reduced compound from below.

Fortunately for the microbial ecologists, guilds can be studied as single units (instead of species), and their "succession" with time can be analyzed with respect to environmental parameters, competition, nutrient limitation, etc. This is because guilds were defined as "a group of species that exploit the same class of environmental resources in a similar way" without regard to the taxonomic position of each species (Root, 1967). Therefore, a guild can be studied as 1) a functional unit of the ecosystem or, 2) its internal composition and functioning can be analyzed. In either case guilds are valid units for testing ecological hypotheses.

In the first approach, a community is seen as a collection of guilds. Each one is formed by one or more species, which may or may not be taxonomically related. Each guild corresponds to a given functional role that is important in the structure and dynamics of the community. In this way, the complexity of a natural community, especially that of a microbial community, can be reduced to a handful of roles that are much less than the total number of species present (Root, 1967; Adams, 1985; Terborgh and Robinson, 1986). Relieved of the need to assign each bacterium to a species, the microbiologist can now proceed to study the autecology of bacterial guilds. Because bacterial taxonomy is mostly based on physiology, a bacterial guild will usually correspond to a uniform bacterial taxon.* The relative importance of different guilds in a system can be assessed by physiological or biochemical measurements. The contributions of each guild to the fluxes of energy and matter in different ecosystems can be compared, and the succession of guilds through the seasons can be followed.

In the second approach, the structure of the guild itself is the object of analysis. Guilds are thought to be formed by a core, or organizer, species and several peripheral species. Many species can be interchangeable members of a guild and thus, the composition of the guild may change depending on environmental conditions without any major changes in its overall role in the community. Competition is considered to be intense within a guild, and some differences "in the efficiency with which a portion of a common range of situations can be exploited" must exist among the members of the guild in order to avoid competitive exclusion (Root, 1967). However, this intra-guild approach can only be applied to a few microbial guilds with unique properties (see Section 8.5).

Bacterioplankton guilds   The guild concept has been used with many complex organisms: plants, birds, triclads and spiders, arthropods, reptiles, and even human beings (Root, 1967; Adams, 1985; Terborgh and Robinson, 1986). In most of these cases the vague definition of guild given earlier has been used to constitute the set of species to be studied. In complex organisms this definition is not only vague, but also subjective (Adams, 1985). Despite

---

*When true phylogenetic taxonomy will be applied more extensively to bacteria, however, this may not prove to be true, since some physiologically related bacteria have been shown to be very distant relatives phylogenetically (Woese, 1987).

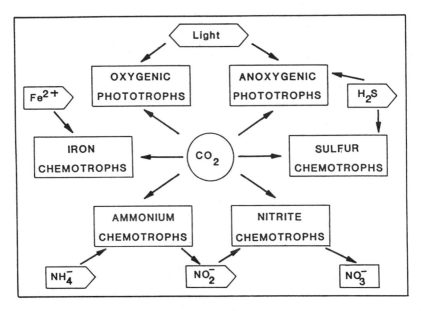

Figure 8.3   The autotrophic guilds and the relationships among them. Algae and cyanobacteria are oxygenic phototrophs. *Chromatium* (Figure 8.2E) is an anoxygenic phototroph. *Thiobacillus* (Figure 8.2F) is a sulfur chemotroph. The nitrogen chemotrophs are divided into two groups: those using ammonia and excreting nitrite (*Nitrosomonas*, Figure 8.2D) and those using nitrite and liberating nitrate (*Nitrobacter*). Some *Thiobacillus* species can be either sulfur or iron chemotrophs. Expected vertical distributions of these guilds in freshwaters is shown in Figure 8.6.

this, we can define bacterial guilds objectively and precisely. For example, one could choose to have one guild for each type of autotrophic metabolism, as shown in Figure 8.3: oxygenic phototrophs, anoxygenic phototrophs, sulfide-oxidizing chemotrophs, ammonium- and nitrite-oxidizing chemotrophs, and iron-oxidizing chemotrophs.*

The contribution of each one of these guilds to CO_2 fixation can then be assessed by using radioisotopes and appropriate inhibitors, such as DCMU, darkness, or N-serve (a compound that inhibits ammonia oxidizers). This division into guilds on the basis of metabolism is particularly useful for our purposes for the following reasons:

1. It emphasizes the metabolic diversity of procaryotes, since only one out of the six guilds (oxygenic phototrophs) includes eucaryotes.

2. It pools all of the organisms that are usually called phytoplankton

---

*See Kelly (1981) for a description of chemolithoautotrophic bacteria.

into one single guild (except for anoxygenic cyanobacteria) and divides a small part of the bacterioplankton, the autotrophs, into several categories. Obviously, if we want to analyze bacterioplankton succession in detail we need this type of subdivision. The finer we divide the closer we get to species succession.

3. It demonstrates that bacterioplankton organisms fulfill many different functional roles in the plankton community.

For these reasons the term "bacterial guild" will be defined as a group of bacteria that use the same energy, carbon, and electron sources and the same electron acceptor, regardless of their taxonomy.

One can expect strong competitive interactions to occur among the species within each guild, because they exploit the same resources using the same metabolic mechanisms. Also, there will be some competition among the different guilds for the same carbon source, and stronger interactions between pairs of guilds using the same source of energy or reducing power (see Figure 8.3). Some of this competition is relieved by spatial differentiation of niches. For example, some chemolithoautotrophs need oxic/anoxic interfaces, but it is unlikely that nitrogen-dependent chemolithoautotrophs can live at interfaces with hydrogen sulfide due to the toxicity of this compound. Sulfur-dependent chemolithoautotrophs, on the contrary, would grow very well only at such interfaces with sulfide. Thus a potential competitive interaction is avoided. Similarly, oxygenic and anoxygenic phototrophic guilds are usually found at different depths, the oxygenic guild above the anoxygenic one.

It is also convenient to divide chemoorganoheterotrophic bacteria into several different guilds. Even if organic matter is their common carbon, energy, and electron source, different guilds use different kinds of organic matter. Chemoorganoheterotrophic bacteria present a much wider range of interaction with phyto- and zooplankton than their role as decomposers suggests. Presumably, different guilds of bacterioplankton have different types of relationships with eucaryotic plankton. Thus, analyzing the relative importance of each relationship through the yearly cycle would provide more information about the succession of bacterioplankton. Some of the possible relationships of bacterioplankton guilds to phytoplankton are summarized in Figure 8.4A (also see Paerl and Kellar, 1978; Cole, 1982; Bird and Kalff, 1987). Their relationships with zooplankton are presented in Figure 8.4B (see Turner and Ferrante, 1979; Pedrós-Alió and Brock, 1983a; van Bruggen et al., 1983). Finally, some of the relationships among different bacterioplankton guilds are summarized in Figure 8.4C (see van Gemerden and Beeftink, 1983; Kuenen et al., 1985; McInerney, 1986; Guerrero et al., 1986, 1987a; Ward and Winfrey, 1985).

In summary, it is possible to define guilds objectively by their carbon

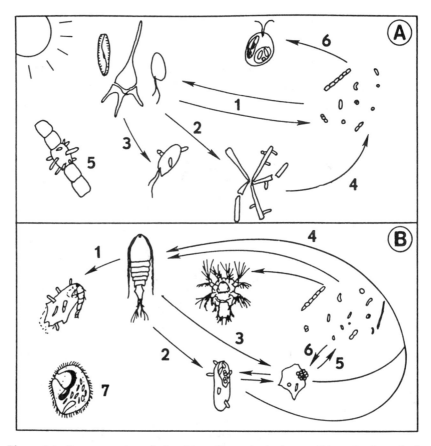

Figure 8.4   Representative relationships of bacterioplankton guilds to A, phytoplankton; B, zooplankton; and C, to other members of the bacterioplankton. The numbers indicate the following relationships: in A. 1, exchange of DOM and nutrients between phytoplankton and bacteria; 2, colonization of dead algae; 3, predation on live algae; 4, liberation of DOM and nutrients from decomposing particles; 5, mutualistic relationship between respiring bacteria and heterocysts of cyanobacteria; 6, feeding on bacteria by mixotrophic algae. In B. 1, colonization of dead zooplankton; 2, decomposition of fecal pellets; 3, formation of POM from remains and excreta of zooplankton and colonization by bacteria; 4, feeding on bacteria by zooplankton; 5 and 6, exchange of nutrients, DOM, and bacterial cells between free-living and attached bacteria; 7, mutualistic relationship between methanogenic bacteria and anaerobic ciliates. In C. 1, competition between different phototrophic bacteria (*Chromatium* and *Chlorobium*); 2, predation by specialized bacteria (*Vampirococcus* and *Daptobacter*) on the phototrophic bacteria; 3, production of low-molecular-weight DOM that can be degraded by anaerobic decomposers (such as *Clostridium*); 4, excretion of photosynthate; 5, DOM and POM coming from the aerobic to the anaerobic parts of the lake; 6 and 7, interspecies hydrogen transfer—acetogenic bacteria convert organic acids produced by the clostridia and others to acetate, hydrogen, and CO_2. In the second step, methanogenic bacteria convert these compounds to methane; 8 and 9, exchange of DOM, nutrients, and bacterial cells between attached and free-living fractions.

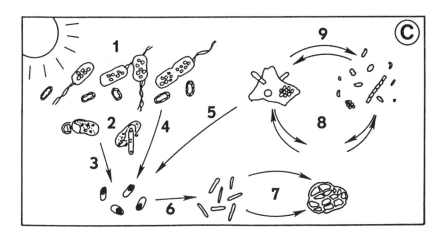

sources, by their energy sources, and by their electron donors and acceptors. The concept of the guild is, therefore, beautifully suited for the study of the procaryotic plankton. Furthermore, it has the advantage of having already been used by many microbial ecologists who measure processes such as sulfide production vs. methane formation. However, it is necessary to have a theoretical model if hypotheses are to be tested. The concept of the guild, with its associated body of ecological theory, provides the required model. Moreover, at present, it may be the only way to approach the problem of succession and community structure among procaryotes.

8.4 Seasonal Succession of Bacterioplankton Guilds

Succession of guilds requiring anaerobic zones   Many of the different trophic modes shown in Figure 8.2 either require a completely anaerobic environment or the simultaneous presence of both reduced and oxidized compounds. Therefore, the appearance of guilds with such metabolic requirements will require the presence of anaerobic zones in lakes. Most lakes undergo stratification of the water column during some time of the year. These stratified periods can result in an anaerobic hypolimnion. In some lakes, which are very well protected from wind and are rich in organic matter, decomposition can be so intense that the whole lake may become anaerobic, even during periods of complete mixing. Finally, meromictic lakes will possess a permanent anaerobic monimolimnion.

This vertical stratification of lakes results in a vertical gradient of redox conditions. Figure 8.5 shows the location of the chemocline in four different kinds of lakes. All the examples are of lakes in the Banyoles karstic area in northeastern Spain (see Guerrero et al., 1987b, for a description of these lakes).

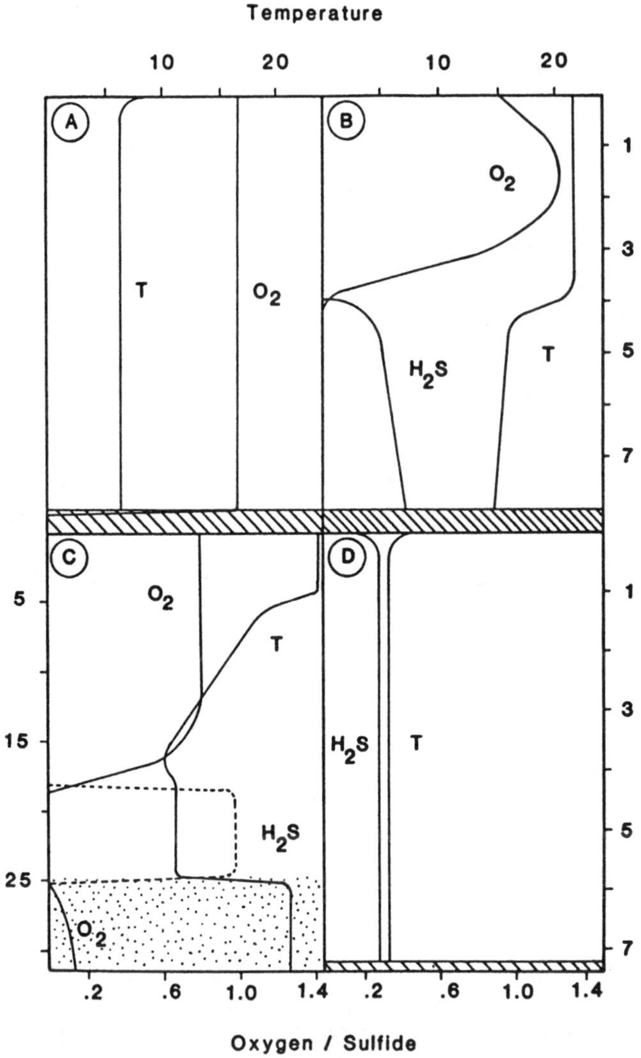

Temperature

Oxygen / Sulfide

Figure 8.5 Location of the chemocline between aerobic and anaerobic zones in lakes. Vertical scales show depth in meters. Upper horizontal scales show temperature in degrees Celsius. Lower horizontal scales indicates oxygen (mg l^{-1}, values have to be multiplied times 10) and sulfide (mM). Hatching indicates sediment. A. Chemocline at the upper-most cm of sediments. The whole water column is aerobic; usually found in mixed lakes (Lake Banyoles, basin I). B. Chemocline at the metalimnion; usually found in stratified lakes (Lake Vilar). C. Complex pattern of chemoclines (Lake Banyoles, basin III). The stippled area indicates loose, fine sediment kept in suspension by water entering through a bottom chimney. This water is aerobic and warmer than hypolimnetic water, but denser due to high amounts of sulfate and sediments. D. Chemocline at the water surface; usually found only in small, holomictic lakes (Lake Cisó).

The chemocline is usually at the sediment-water interface in mixed lakes (Figure 8.5A); at the metalimnion in stratified lakes (Figure 8.5B); in a complex pattern in some lakes (Figure 8.5C); and at the water surface in small holomictic lakes that are protected from wind and very eutrophic (Figure 8.5D). In turn, such vertical redox gradients will determine the chemical form in which the different inorganic compounds used by bacteria (as energy sources or electron donors and acceptors) will be found. Therefore, one can expect a vertical stratification of guilds such as those shown in Figure 8.6 (see also Jones, 1987).

When the chemocline is at the metalimnion, some oxygen diffusion can occur, and organic matter concentrations are usually low. Thus, autotrophic guilds can occupy different layers of the metalimnion over a range of centimeters. On the other hand, when the chemocline is at the sediment-water interface, oxygen diffusion is more difficult, but the organic matter concen-

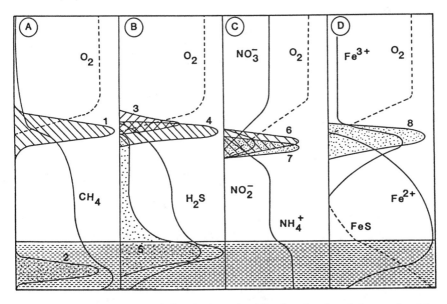

Figure 8.6   Idealized vertical distribution of oxidized and reduced forms of certain chemical compounds in stratified lakes and consequent stratification of bacterial guilds utilizing such compounds. Sediment is indicated by dashed lines. A. Guilds using one-carbon compounds: 1, methylotrophs; 2, methanogens. B. Guilds using sulfur compounds: 3, sulfur chemolithoautotrophs (*Thiobacillus* type); 4, anoxygenic photo-lithoautotrophs (*Chromatium*); and 5, sulfate-reducing chemoorganoheterotrophs (*Desulfovibrio*). C. Guilds using nitrogen compounds: 6, nitrite chemolithoautotrophs; 7, ammonium-dependent chemolithoautotrophs (*Nitrosomonas*). Nitrate- or nitrite-respiring chemoorganoheterotrophs may be found at the interface. D. 8, iron-dependent chemolithoautotrophic guild obtaining energy from the oxidation of ferrous to ferric iron.

tration is high due to the settling of matter from the whole water column. Therefore, heterotrophic, anaerobic guilds dominate here and also stratify vertically over a range of millimeters. A whole series has not been described, but partial establishment of such vertically stratified guilds is a well-known phenomenon (Sorokin, 1970; Gorlenko et al., 1983; Ward et al., 1987).

The presence or absence of such metabolically defined guilds depends on stratification, that is, on the yearly cycle of solar radiation and the physical characteristics of the lake. An abundance of organic matter will favor chemoorganoheterotrophic guilds, while low concentrations of organic matter will favor the presence of chemo- or photolithoautotrophic guilds. All these guilds are important in both the carbon cycle and in the cycles of other elements such as iron, sulfur, or nitrogen.

In the carbon cycle, autotrophic bacterial guilds fix CO_2, which may constitute an additional input of organic matter to the system. These organisms frequently form dense layers (usually made up of two or three species) at pelagic chemoclines. Accumulations of zooplankton feeding on such metalimnetic layers have been described on several occasions (Sorokin, 1970; Gophen et al., 1974; Guerrero et al., 1978; Parkin and Brock, 1981; Caumette et al., 1983). Therefore, the products of these autotrophs can reenter the herbivore food chains. The role of such guilds in the carbon cycle of lakes is analogous to that of phytoplankton. Their quantitative importance in CO_2 fixation, however, is not well known (Biebl and Pfennig, 1979; Gorlenko et al., 1983).

During the winter Lake Cisó is completely anaerobic (see Figure 8.5D) and there is no carbon-dioxide fixation by phytoplankton. But two guilds of autotrophic bacteria are active at the surface layers of the lake (at the chemocline). Chemolithoautotrophic bacteria fix carbon dioxide at the surface using the oxygen diffusing from the air and the sulfide coming from the lake (the *Thiobacillus* type, Figure 8.2F). Photolithoautotrophic bacteria (of the *Chromatium* type, Figure 8.2E) fix carbon dioxide in the upper 75 cm of water using light as a source of energy and hydrogen sulfide as the source of electrons. Estimated contribution is about 66 percent of total fixation for the former and 33 percent for the latter. During stratification, these bacterioplankton guilds are confined to the metalimnion (between 1 and 2 m) where they can fix around 52 percent (chemotrophs) and 20 percent (phototrophs) of total carbon fixation in the lake. A layer of eucaryotic phototrophs appears slightly above this layer, absorbing light and, thus, reducing the supply of light available for the phototrophic bacteria, and providing oxygen through oxygenic photosynthesis. Therefore, this layer should favor the thiobacilli over the phototrophs. An exhaustive account of relationships among sulfur-oxidizing guilds can be found in Kuenen et al. (1985), although most of the work summarized there was done with laboratory cultures and is beyond the scope of this review.

Seasonal succession of these bacterial guilds has rarely been studied.

However, Gorlenko et al. (1983) have summarized work done in lakes in the USSR over many years. A particularly well-studied case is that of methylotrophic bacteria. Rudd and co-workers (Rudd and Hamilton, 1978; Rudd and Taylor, 1980) analyzed the activities of this bacterial guild through the seasons in Lake 227, in the Canadian Precambrian Shield region. The results of their studies on the seasonal distribution and activity of methylotrophic bacteria is shown in Figure 8.7. During summer stratification, the methane-oxidizing activity is limited to the metalimnion. During winter, due to inverse stratification, methane diffuses throughout the lake and methane oxidation occurs at almost all depths, athough at much lower rates due to low winter temperatures. The result of the activities of the methylotrophic guild is that a portion of primary production that could have been lost through methane evasion is recycled by being converted back to CO_2 (a compound that algae and autotrophic bacteria can use) and bacterial biomass (which zooplankton can eat). It has been estimated that 5 percent of primary production follows this path in eutrophic Lake Mendota (Fallon et al., 1980; Brock, 1985) and 9 percent in Lake 227 (Rudd and Hamilton, 1978). Thus, a bacterial guild that is usually ignored is seen to play a quantitatively significant role in the carbon cycle of these lakes.

Reed and Dugan (1978) have prepared fluorescent antibodies to two species of methylotrophic bacteria. The antibodies were used to study the distribution of the two bacterial species in an aquatic environment. It should be easy to combine such enumeration techniques with measurements of activities such as those of Rudd and Hamilton (1978) and follow both through a yearly cycle. Autoradiography with ^{14}C-methane could be added to these measurements to determine how many individuals of each antibody-recognized species are active at any particular time point. This sort of study would provide extremely useful information on annual succession within a guild. The same combination of techniques could be applied to the other guilds shown in Figure 8.6. The pattern in Figure 8.7 is probably similar to that of many other bacterial guilds, and their contributions to cycling of elements in lakes could then be evaluated.

When there is an ample supply of organic matter and a slow supply of oxygen, anaerobic heterotrophic guilds are likely to establish themselves. These organisms use organic matter as their source of energy and reducing power, but transport electrons to compounds other than oxygen: some of them to organic molecules (the fermenters) and some to inorganic oxidized compounds (the anaerobic respirers of iron, nitrate, sulfate, and carbon dioxide). These organisms, by processing dead organic matter and converting it to carbon dioxide and methane, recycle the carbon that otherwise would be lost from the system and would accumulate in the sediments. Their most important contribution to element cycling is the inorganic compounds they alter chemically. Sulfate respirers are frequently the most important guild in freshwater habitats rich in sulfate. They are often limited by the supply of

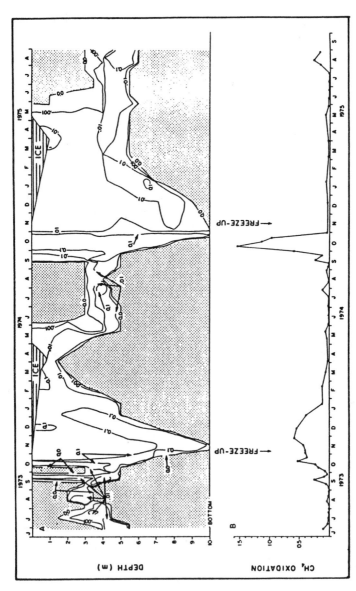

Figure 8.7   Distribution of the methylotrophic guild activity in Lake 227 over a three-year period. Top, depth-time distribution of methane oxidation rates (μmol l⁻¹ h⁻¹); bottom, total rate of methane oxidation in the lake (moles lake⁻¹ day⁻¹ × 10⁻³). The guild is confined to the metalimnion during the summer (as in Figure 8.6A). In October and November, the lake mixes, and the activity becomes distributed throughout the water column. At this time the activity reaches its maximum level. During the winter, the lake becomes stratified again and the guild is distributed in the upper part of the lake (from Rudd and Hamilton, 1978).

organic matter and sulfate (Ingvorsen et al., 1983), but sometimes their production of hydrogen sulfide is large enough to fill the whole hypolimnion with this gas. Sometimes even the whole lake may be permeated with it and since hydrogen sulfide is a toxic gas, its presence causes the disappearance of all eucaryotic life except for a few ciliates (Fenchel et al., 1977; Dyer et al., 1986). Therefore both phytoplankton and zooplankton disappear and the hypolimnion becomes an exclusively procaryotic ecosystem.

The sulfate-reducing guild enters into competition with the methanogenic guild (Ward and Winfrey, 1985; McInerney, 1986) and forms mutualistic relationships with phototrophic bacteria (Pfennig, 1980) and, possibly, with ciliates. The methanogenic guild is the most abundant in the anaerobic sediments of lakes with little sulfate, and can establish symbiotic relationships with ciliates (Van Bruggen et al., 1983). These guilds usually dominate the sediments, but not the plankton, so we shall not deal with them further. An excellent discussion of them can be found in Ward and Winfrey (1985).

Succession of aerobic heterotrophic guilds   The aerobic heterotrophic bacteria perform the role usually denoted as "decomposition" in the epilimnion of lakes. Due to their physical proximity to phytoplankton and zooplankton, these bacteria are the most intimately associated with the eucaryotic plankton. In terrestrial ecosystems subtle differences between feeding modes of organisms reveal adaptations which allow coexistence of many different species. Thus, phytophagous mammals in a Serengeti plain can graze on short, long, or dead grass, browse on low bushes, climb trees for leaves, or reach the upper branches (Krebs, 1985). We can assume that similar differences exist among heterotrophic bacteria. Each group of bacteria will be better adapted than any other one to one of the possible relationships shown in Figure 8.4. Small, free-living bacteria will be adapted to the uptake of low-molecular-weight compounds in very low concentrations. Other bacteria will be better suited to close associations with phytoplankton, quickly using excreted photosynthate. Still others will live attached to phytoplankton cells, perhaps even parasitizing or grazing on them. Finally, other bacteria will establish mutualistic symbioses with phototrophs, such as those between certain bacteria and cyanobacterial heterocysts (Paerl and Kellar, 1978).

We can almost be certain that one unique group of species will be best suited for each one of the relationships described. If the relative importance of these processes could be followed through the seasons, some information on succession would be available. The compounds used as carbon, energy, and electron sources will be different for each group of bacteria. We shall assume that the bacterial partners in each association shown in Figure 8.4 constitute a different guild and that the relative importance of such guilds can be followed by microscopic observation or by measuring the process carried out by each guild. Unfortunately, most studies concentrate on short periods of time, usually during the spring and summer, or use data from only a few sampling dates in each season. Nevertheless, we shall try to summarize whatever is known of seasonal variations for each of these heterotrophic guilds.

Attached and free-living guilds   One way to divide the heterotrophic community of the bacterioplankton into two different guilds is into those using DOM and those using POM. Presumably, POM-using bacteria are attached to their substrate. However, what can be seen as attached bacteria under the microscope corresponds to a much wider group of microorganisms (Pedrós-Alió and Brock, 1983b; Melack, 1985). Nevertheless, attached vs. free-living bacteria present a simple and quantifiable—although crude—way to divide the aerobic heterotrophic component of bacterioplankton. Some studies have been conducted of the seasonal abundance of attached and free-living bacteria (Kirchman, 1983; Pedrós-Alió and Brock, 1983b; Lovell and Konopka, 1985; Simon, 1985). Even though attached bacteria may constitute from an almost insignificant portion up to 50 percent of the bacterioplankton, a seasonal pattern in their abundance is apparent in some studies: attached bacteria are more abundant and active during early summer than during the winter. Presumably, suitable particles are more abundant and loss factors are less intense during the early summer. In some cases, the peak times of attached bacterial abundance are clearly correlated with those of diatoms or cyanobacteria. From the few reports available, it seems that attached bacteria are better adapted to live on phytoplankton cells that are not subjected to zooplankton feeding. This is one of the few instances in which some feature of bacterioplankton succession fits in with the PEG-model (Sommer et al., 1986; see Chapter 1 of this book). During early spring small, fast-growing species of eucaryotic algae dominate the phytoplankton and bloom in early summer (step 1 in the model). A parallel increase in the total number of bacteria has been observed in several lakes (Pedrós-Alió and Brock, 1982; Bell et al., 1983; Simon, 1985). This increase in total bacteria is usually the result of an increase in free-living bacteria, which are metabolically very active as shown by thymidine incorporation rates and other methods.

When zooplankton, especially cladocerans, feed on this spring bloom (steps 2 to 4 in the model), they also eat a substantial part of the bacterioplankton (Pedrós-Alió and Brock, 1983b; Børsheim and Anderson, 1987). As a result, the so-called "clear-water phase" is sometimes paralleled by a minimum in bacterial numbers. In the next step (step 8), usually around midsummer, large-sized species of phytoplankton develop, which are harder for the cladocerans to eat. Attached bacteria may also be more abundant at this time. Two factors are probably responsible for this: a stable and abundant supply of phytoplankton particles and a reduction of the grazing pressure on such particles (Pedrós-Alió and Brock, 1983a; Lovell and Konopka, 1985).

In general, attached bacteria seem to take up selectively more glucose, glutamate (Goulder, 1977; Kirchman and Mitchell, 1982), amino acids (Simon,

1985), and acetate (Pedrós-Alió and Brock, 1983b) than free-living bacteria. On the other hand, they may take up less thymidine (Kirchman, 1983; Lovell and Konopka, 1985), and sulfate (Pedrós-Alió and Brock, 1983b), which suggests less DNA and protein synthesis than free-living bacteria. This relatively high rate of uptake of carbon compounds with slower cell growth and division is consistent with scanning-electron micrographs which show attached bacteria surrounded by thick layers of extracellular materials (Paerl, 1974; Pedrós-Alió and Brock, 1983b). Autoradiography studies (Paerl, 1978) also show that a large portion of the dissolved organic carbon taken up by attached bacteria soon appears in their surrounding mucillaginous materials. Thus, attached bacteria seem to take up proportionally more carbon than they need for growth, and to store the excess as extracellular material. This material may help in adhesion and cohesion of the bacteria to particles, and produces a new sort of environment modified by its own microbiota. An equivalent process has been very well characterized in the case of bacterial plaque formation on teeth (Gibbons and Van Houte, 1975). In a sense, this extracellular material, usually considered to be polysaccharides, is analogous to the wood in a forest, and constitutes nonliving biomass. This material does not contribute to production of new cells, but generates complexity and heterogeneity in the environment, allowing colonization by many different and specialized organisms.

The treatment of the attached bacterioplankton given above is not without dangers. Pooling all attached bacteria together into one guild separated from free-living guilds may be experimentally convenient, but it is certainly an oversimplification. It is obvious that many different types bacteria may be found on particles: passengers on unedible particles, parasites or predators on their prey cells (including algae, zooplankton, and other bacteria), commensals of algal-excreted products, decomposers of zooplankton fecal pellets, and mutualistic bacteria conferring some advantage to the organisms to which they are attached. If a more detailed analysis were intended, all such groups should be considered as separate guilds, performing different functions, and composed of different species. Such a detailed analysis is unwarranted now due to a dearth of data. The guild concept, however, could easily accommodate this type of studies. Clearly, each such guild would follow the seasonal variations of its corresponding type of particles.

A different sort of problem is more serious. If a species is very well adapted to becoming attached, it has to colonize new particles periodically, before it sinks with its particle to the bottom of the lake and before the particle is exhausted as a food source. Thus, some portion of the free-living bacterioplankton must be composed of normally attached cells seeking new particles to colonize and not contributing in any significant way to the free-living bacteria (Pedrós-Alió and Brock, 1983b). There is no simple solution to this problem.

Differentiation becomes even harder when we look at the free-living guilds. A classification according to type of particles colonized is not possible. Whatever characteristics are used, they have to be looked for in the individual bacteria themselves. Morphology is no help, since every physiological group of bacteria is composed of species with several different morphologies (for example, methanogens can be filamentous, rods, cocci, etc.), and many different bacteria are alike in shape. Differences, therefore, have to be sought in the physiology and biochemistry of the bacteria.

Uptake of dissolved organic compounds (DOM) is mediated by enzyme systems with specific kinetics. It is to be expected that bacteria in the plankton will have a variety of different enzymatic systems with diverse kinetic parameters. Some degree of specialization is likely, with some bacteria being better at taking up organic acids, for example, and other bacteria being more efficient at taking up amino acids. We do not know whether aerobic bacterial species, adapted to the use of only one or two carbon substrates, are abundant in nature. There are anaerobic bacteria which only use a few organic acids, such as butyrate, and there are also methylotrophs that can only utilize compounds with a single carbon atom. But most aerobic bacteria isolated up to now are able to use a number of compounds. Some strains of *Pseudomonas* for example, isolated from the field are able to grow on dozens of substances as sole carbon and energy sources. There is some truth in the argument that *Pseudomonas* and the other easily cultivable bacteria may not be the main bacteria active in natural systems. (However, recently, mRNA of *Pseudomonas* species has been found to be very abundant in some environments (M.G. Höfle, Max-Planck Institute for Limnology, Plön, FRC, pers. comm.). The famous Russian bacteriologist Sergei Winogradsky (1949) called such easily cultivable bacteria zymogenous, while the hard-to-isolate, slow-growing organisms were named autochthonous.

More recently, the bacterioplankton has been divided into copiotrophic and oligotrophic bacteria, which resemble Winogradsky's groupings (Sieburth, 1979). Copiotrophic bacteria would be mostly dormant in natural systems with low concentrations of organic matter. When a food source became locally abundant, they would grow rapidly and use it quickly. These copiotrophs would have starvation-survival mechanisms (Kjelleberg et al., 1985) to endure natural conditions and would opportunistically grow only when high concentrations of a substrate were present. Oligotrophic bacteria would not need such starvation-survival mechanisms, because they would be adapted to growth on natural concentrations of DOM, making such mechanisms less necessary. Of course, there must be a certain threshold below which all bacteria have to find a way to survive starvation. Oligotrophic bacteria would divide only a few times when confronted with a rich medium. Some could even form microcolonies (Torrella and Morita, 1981), but they would never grow enough to form visible colonies on plates and be isolated. These bacteria would be active most of the time in natural waters, using the small amounts of DOM available.

Biochemical characterization of guilds   If we accept that some specialization must exist in the enzyme systems that different bacteria possess to
take up different substrates, studying the percentage of the bacterioplankton
active at taking up each substrate (as well as the magnitude of such uptake)
would reflect the importance of each such guild. If such experiments were
performed through a yearly cycle, seasonal succession of the different guilds
could be followed. Many investigations have analyzed the uptake of different
organic compounds by the bacterioplankton, including amino acids (van Es
et al., 1984; Jørgensen, 1987; Simon, 1985; Fuhrman, 1987), acetate (Stanley
and Staley, 1977; Pedrós-Alió and Brock, 1983b), glucose (Overbeck, 1979),
or several compounds simultaneously (Hoppe, 1976; Simon, 1985). For these
studies to be useful for guild succession they must fulfill two conditions: First,
radioactive substrates have to be added in tracer amounts so as not to change
the in situ concentrations; and second, some additional technique, such as
direct counts incorporating autoradiography (Brock and Brock, 1968; Hoppe,
1976; Meyer-Reil, 1978), has to be performed simultaneously to determine
that the percentage of bacteria taking up that substrate is really changing
through the seasons. Otherwise, changes in total uptake could be the result
of changes in concentration or different temperatures, for example.

   If the changes in a guild are to be followed, the number of bacteria taking
up the substrate must be determined. A step in that direction was taken by
Tabor and Neihof (1984). These authors followed by autoradiography the
number of cells able to take up amino acids, thymidine, and acetate for one
year in Chesapeake Bay (Figure 8.8). The three types of bacteria increased
toward late summer and decreased during the winter. Bacteria taking up
amino acids were always more abundant than those taking up thymidine. In
turn, both groups were more abundant than bacteria taking up acetate. Their
autoradiographic method holds promise for autecological studies, especially
if used in conjunction with other direct-microscope techniques (Newell et al.,
1986).

   This approach, however, is even more problematic than the approach
based on examining the attached bacteria. We do not know to what degree
the same species of bacteria are taking up the same substrates. In addition,
we can only assay for substrates that are reasonable metabolic intermediates.
The chemical composition of organic matter can be complex, and the more
specialized guilds may be those using exclusively humic acids, for example.
Perhaps they perform the rate-controlling step in the metabolism of organic
matter. Several workers have attempted to label phytoplankton and use their
excreted radioactive material as food for bacterioplankton (Jones et al., 1983;
Brock and Clyne, 1984). Following these processes throughout the year is
another way to approach succession.

   Recently, a series of biochemical methods has been introduced to approach the problem of bacterial diversity in natural systems (White et al.,
1979; White, 1986). In order to be useful for our purpose, these methods have

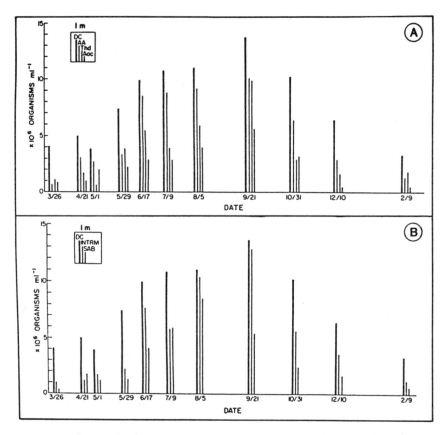

Figure 8.8   One way to examine the succession of aerobic heterotrophic guilds. A. The total number of bacteria (DC), the number of bacteria taking up amino acids (AA), thymidine (Thd), and acetate (Aoc) in Chesapeake Bay. B. The total number of bacteria (DC), of respiring bacteria (INTRM), and of synthetically active bacteria (SAB) in the same waters. All values are the result of different direct-microscopic techniques applied to the aerobic heterotrophic guilds. See Newell et al. (1986) for a detailed description of the techniques (from Tabor and Neihof, 1984).

to rely on direct analysis of natural samples. If isolation of bacterial cultures were necessary prior to use of the technique, we would be back to the hard-to-isolate, native bacteria problem. Fortunately, these biochemical techniques fulfill these requirements and will be described below.

For this method, usually an extraction procedure is applied to natural samples, and a particular macromolecule is assayed by gas-liquid chromatography/mass spectrometry. This technique is very useful for measuring polymers, such as muramic acid, poly-β-hydroxybutyrate, lipopolysaccharides, and other components of bacterial envelopes. However, for the present purpose, the only applicable technique is that of measuring the amount of lipids composing the membrane phospholipids. Since most bacteria share similar phospholipids, the analysis is not going to be of general applicability. Nevertheless, in certain habitats it may be possible to follow a given guild by measuring its "signature" compounds (Ward et al., 1987). For example, lipids of methanogens are certainly different from those of other bacteria (Martz et al., 1983). Something similar may be true for sulfate-reducing bacteria (Edlund et al., 1985). If clear signature compounds can be identified for a given guild, and their concentration per cell is constant, this should be a very convenient way to follow it through the seasons. However, it will probably be impossible to differentiate among species with this technique.

Guilds based on nucleic acid hybridization   All the techniques and approaches discussed so far are indirect attempts to look inside the bacterioplankton "black box," and they will probably only produce blurred, disconnected pictures of its interior. However, this depressing state soon may be greatly improved by the ongoing introduction into ecology of new molecular biological techniques. Pace and colleagues (Lane et al., 1985; Pace et al., 1986; Olsen et al., 1986) have used well-established genetic methods (such as nucleic acid sequencing, rRNA cataloguing, and gene cloning in bacteriophages) and have adapted them for use with natural samples of bacteria. These techniques do not require isolation of microorganisms in pure cultures and, thus, fulfill one of the requirements for use in the study of succession. Since individual nucleic acids are separated, and each type should correspond to a different species, the true diversity of the community can be studied! Moreover, the relative amounts of each nucleic acid type will tell us which species is more abundant in the field.

In this procedure natural assemblages are collected and concentrated. Next, the nucleic acids are extracted and analyzed. Two types of nucleic acids are usually measured: 5S ribosomal RNA (rRNA) and DNA genes coding for 16S rRNA (rDNA). The approach using 5S rRNA has been applied to three different simple ecosystems: a hot spring community (Stahl et al., 1985), the symbionts of animals found in submarine hydrothermal vents (Stahl et al., 1984), and an acid lagoon from a mine-leaching operation system (Olsen et al., 1986). Three species were found in the first community and one symbiont in each animal in the second. By comparing the sequences of the 5S rRNA to those of known organisms, the species from natural samples can be placed into the general phylogenetic tree of bacteria (Woese, 1987). If an organism belongs to any of the already catalogued species, this should be obvious from its position in the tree. If it does not, at least we will know within which group it should fit. Neither the bacteria from the hot spring nor the symbionts from vent animals had been previously isolated in pure culture. Attempts had always failed. Thus, the identity of the organisms was unknown. Using the 5S rRNA technique, these organisms could be placed in the phylogenetic tree without the need to isolate them. The technique is, therefore, extremely appropriate for field studies. Its main limitation derives from the shortness of 5S rRNA. If many species are present in a given system, 5S rRNA does not give enough information to differentiate all of them. It has been estimated that such technique will be useful for environments with up to ten species but not more (Pace et al., 1986; Olsen et al., 1986).

The second technique involves cloning of the DNA from a natural sample and isolating clones whose DNA fragments code for the 16S rRNA. These fragments can then be sequenced and each type of sequence can be quantified. Again, three types of information are available: the number of different 16S rDNA tells us how many species there were in the natural community. The relative amounts reveal which species were more abundant. Finally, sequencing and comparing to the established phylogenetic tree can identify each organism, or at worst, the groups within which each organism falls.

This technique is very time consuming, but several steps easily lend themselves to automation. Once the rDNA gene library has been constructed for a given community, probes can be constructed for each species. Then such probes can be used through in situ hybridization to identify every organism in natural samples. A seasonal study of the bacterioplankton in a lake can then be performed. This would be the best possible way to look for bacterioplankton succession. In fact, it would be more exhaustive and complete than any successional study performed so far with eucaryotes, since no one has determined the diversity of all the animals and plants in any community. Of course, there is no reason why such techniques could not be applied to both procaryotes and eucaryotes simultaneously. If that is achieved we will have comparable seasonal succession sequences and true diversity estimates for the whole plankton community for the very first time.

8.5   Autecology Within Guilds

Approaches to studies of ecology within guilds must take advantage of peculiar traits already present in the organisms or in the environments studied. Alternatively, techniques specific for the system under consideration must be developed. In order to study autecology within a guild, all the important species in the guild must be differentiated in some way. The trivial case would be a guild formed by one single species. A next step would consider guilds whose species can be differentiated morphologically. Finally, some guilds with not much diversity may be amenable to study with a combination of techniques involving immunofluorescence, autoradiography, molecular probes, and others. Several of the possibilities described below were suggested by Brock (1971) in relation to the measurement of in situ growth rates of individual microorganisms.

Monospecific guilds   It is very unusual to find a guild formed by a single species, but in certain extreme environments such as hot acid springs or salinas that may be the case. In fact, there may only be one single guild of organisms in such ecosystems. This simplicity allows fine analysis of the community. Thus, Brock and associates were able to study autecology of *Sulfolobus acidocaldarius* in acid hot springs in Yellowstone National Park. *Sulfolobus* has a characteristic morphology and this permitted microscopic demonstration that it was the only inhabitant of the springs (Mosser et al., 1974a). Through a combination of culture techniques and immunofluorescence, it was possible to show that each spring actually had several different strains of *Sulfolobus* which differed in serological properties and temperature optima (Mosser et al., 1974a; Bohlool and Brock, 1974). In situ growth rates were calculated by several different approaches (Mosser et al., 1974b), and they were found to be lower than potential growth rates. Actually, the spring ponds could be considered to behave as chemostats, with the *Sulfolobus* populations in steady state. The annual cycle was not followed. The constancy of conditions in such springs is well known and, therefore, little seasonal variation is to be expected. Nevertheless, the coexistence of several strains with different temperature optima indicates the possibility of fluctuating conditions. It would be interesting to follow the relative abundances of different strains over long periods of time and see whether small changes in flux or temperature or some other factor allow for succession within the guild.

Guilds with morphologically different species   Species of some bacterial groups show very distinct appearances under the microscope. Examples are purple phototrophic bacteria, cyanobacteria, prosthecate bacteria, methanogens, *Planctomyces*, and others (Starr and Schmidt, 1981). In these cases, cell counts of individual species can be followed through the year. Autoradiography can also be used to ascertain the activity of each species, and specific rates of uptake or production can be determined. One of the best examples of such an approach is that of Stanley and Staley (1977) who examined prosthecate bacteria in a paper-mill waste lagoon.

    Caulobacter or prosthecate bacteria are usually easy to distinguish. Their presence has been reported from many habitats (Stanley and Staley, 1977; Staley et al., 1981), but a complete study of seasonal succession has not been attempted. The most detailed study (Staley et al., 1987) followed the abundance of *Caulobacter* spp. through a yearly cycle in Lake Washington. Unfortunately, the study was done in 1972, when reliable and fast direct count methods had not been developed. Therefore the most-probable-numbers extinction dilution method was used and this, like any other culture method, has its dangers when trying to extrapolate to the natural situation. Staley et al. (1987) found maximal abundances in summer and winter, and minimal numbers in spring and fall. Concentrations were around 10^3 cells ml^{-1}, which would represent about 0.1 percent of the total population, but it was not possible to show that these *Caulobacter* were active in the field. Since this study was done, careful laboratory work has been performed with prosthecate bacteria (Schmidt, 1981). Perhaps a more productive study could be done on these organisms using a combination of the techniques now available.

Like prosthecate bacteria, members of the *Blastocaulis-Planctomyces* group of bacteria (Schmidt and Starr, 1981) are easy to identify during bacterial counts. Many workers have reported the abundances of *Planctomyces*, usually higher during the summer and lower during the winter, as well as higher in the epilimnion and lower in the hypolimnion (Pedrós-Alió and Brock, 1982; Lewis et al., 1986). More detailed studies of *Planctomyces* would seem feasible and rewarding. Members of this group have been found in concentrations from 10^3 ml^{-1} to 10^4 ml^{-1}, which would represent between 0.1 and 1 percent of the total population. Considering their large size relative to other bacteria, their share of the biomass could be larger.

Another group of bacteria in which individual species can be distinguished under the microscope is that of the phototrophic bacteria (Trüper and Pfennig, 1981). Species found coexisting in a given habitat are frequently very different from each other, and a combination of morphology plus pigmentation is usually enough for determination at the species level. This characteristic of the group has been used to analyze competition and seasonal succession in Lake Cisó, and it will be discussed at length later, to illustrate the possibilities of the within-guilds approach.

Other guilds   In most cases, guilds will not be monospecific and species will not be morphologically different. In these cases study of succession becomes more complicated. A lot of work will be necessary just to determine that one has identified all the major species in the guild. One frequently cited technique for autecological studies is immunofluorescence (Bohlool and Schmidt, 1980). This technique involves isolation of pure cultures of the organism of interest, preparation of antibody against it, and chemical coupling of the antibody to a fluorescent dye. The fluorescent antibody can then be used to identify very specifically the organism in natural samples. This method has only been applied a few times to planktonic bacteria (Bohlool and Brock, 1974; Reed and Dugan, 1978; Stanley et al., 1979; Ward, 1984), and never for long enough to say something about succession. If all the members of a guild could be isolated and antibodies could be prepared against all of them, this would be an excellent way to follow their seasonal succession. However, achieving the right specificity of the antibody is not an easy task. Too much or too little antibody specificity may make this technique unworkable.

So far, very interesting information has been obtained with fluorescent antibodies, although it is only marginally relevant to succession. For example, Muyzer et al. (1987) prepared fluorescent antibodies against *Thiobacillus ferrooxidans* and then applied it to coal samples that produced acid from pyrite. They found that *T. ferrooxidans* was practically absent from all of them. It is a humbling thought to realize that so much of the physiological knowledge we have has been obtained with organisms such as *T. ferrooxidans* that do not seem to be important in nature, but grow very well in the laboratory. Another useful piece of information that the immunofluorescence method can produce, is the percentage of the total count corresponding to a given species. Thus, Stanley et al. (1979) found between 10^2 and 10^3 cells ml^{-1} of *Nitrobacter* in the plankton of Lake Ithasca, Minnesota. If the activity of the guild could be measured, and autoradiography were performed simultaneously, an idea of the population dynamics within the guild could be obtained. The combination of autoradiography and immunofluorescence seems to be particularly powerful (Fliermans and Schmidt, 1975; Ward, 1984) for these purposes.

Another possibility concerns the use of nucleic acid probes. Probes with adequate specificity seem easier to prepare than antibodies with the right specificity (Pace et al., 1986; Holben et al., 1988). Probes can be constructed starting with a gene library from the natural population. In this way it is not necessary to isolate pure cultures to begin with. If pure cultures are available, preparing the probes is easier. This approach is already being tested with guilds such as methylotrophic and denitrifying bacteria.

8.6   Succession of Phototrophic Bacteria

The phototrophic bacteria in Lake Cisó offer several advantages as a model of a bacterial guild. Only a few species coexist and they can be morphologically differentiated. Each species is found in extremely high concentrations (around 10^6 cells ml^{-1} each). Their activity can be easily followed by measuring $^{14}CO_2$ incorporation into particulate material, with the appropriate controls—using DCMU to exclude oxygenic photosynthesis and darkness to separate activities of chemolithoautotrophic bacteria. Finally, phototrophic bacteria are present throughout the year, so that the actual species succession can be followed through the seasons without any interruptions.

Lake Cisó is a small, holomictic lake (22 m in diameter, 8 m maximum depth) with water rich in sulfate (from 8 to 10 mM). A complete description of the lake can be found in Guerrero et al. (1985) and Pedrós-Alió et al. (1986). The lake is stratified from March to October. During this period a very sharp stratification of organisms develops in the lake. The epilimnion is very poor in organisms, but a layer of the protists *Cryptomonas phaseolus* (10^5 cells ml^{-1}) and *Coleps hirtus* with endosymbiotic *Chlorella* (10^4 cells ml^{-1}) forms in the upper part of the metalimnion. Slightly below the metalimnion, but overlapping with the previous layer, *Chromatium minus* forms a dense layer (10^6 cells ml^{-1}). *Amoebobacter* M3, another purple phototrophic bacterium has its maximal abundance slightly below *C. minus* (also around 10^6 cells ml^{-1}). And still deeper the two green phototrophic bacteria *Chlorobium limicola* and *Chlorobium phaeobacteroides* appear mixed with the purple bacteria.

After mixing occurs, *C. hirtus* disappears and *C. phaseolus* maintains low populations (around 10^2 cells ml⁻¹), while the phototrophic bacteria are abundant throughout the lake (Pedrós-Alió et al., 1987). During this period, owing to the high production of sulfide in the sediments, the lake is anaerobic and rich in sulfide (around 0.4 mM) except for the uppermost cm.

Each of the phototrophic bacteria found in the lake presents a set of characteristics that separates it from the others. The purple bacteria differ from the green bacteria in being larger (about 50 μm³) instead of small (1 μm³), motile instead of immotile, accumulating sulfur intracellularly instead of depositing it outside the cells, and being more tolerant of microaerophilic conditions. The green bacteria are more tolerant of high sulfide concentrations and are able to use lower light intensities (Trüper and Pfennig, 1981; van Gemerden and Beeftink, 1983). Thus, the purple bacteria are found further up in the water column than the green bacteria and the latter, in turn, are better adapted to the deeper zones of the lake.

In 1977 and 1978 there was a single species of dominant purple bacterium, *C. minus*, and two of green bacteria, *C. limicola* and *C. phaeobacteroides*. The relative abundance of these organisms changed depending on physico-chemical conditions in the lake. High solar radiation and high water flux through the lake favored *C. minus*. The importance of solar radiation for light-limited organisms such as these (Guerrero et al., 1985) seems quite logical. This factor followed the usual annual cycle for temperate areas, with year-to-year variability due to changes in the amount of cloud cover. *C. minus* was more abundant in early summer (12 mg wet weight m⁻²) than during the winter (1 mg wet weight m⁻², Figure 8.9A). The purple bacterium positioned itself in the upper part of the thermocline during stratification, shading the immotile green bacteria underneath. Thus, the latter were more abundant in winter and fall (5 mg wet weight m⁻²) than during the summer (1 mg wet weight m⁻², Figure 8.9A). The green bacteria responded to the shading of *C. minus* by increasing the concentration of their pigments. The amounts of both bacteriochlorophyll and carotenoids were five times larger during stratification than during mixing despite the lower biomass (Pedrós-Alió et al., 1983, Figure 8.9B), but this was not sufficient to trap enough light energy, and the population declined. During mixing periods, motility was of no use to *C. minus*, as it could not maintain its position above the population of green bacteria. Therefore, the better adaptation of the latter to low light intensities and its higher affinities for sulfide (Kuenen et al., 1985; van Gemerden and Beeftink, 1983) allowed these bacteria to dominate the phototrophic guild (see Figure 8.9A). In the mixing period of 1977, heavy rainfall caused a high flux of water through the lake. As a consequence, winter production of sulfide (which depends on surface area of sediments and not on flux) was constantly being flushed out of the lake. Lower sulfide concentrations were found. This allowed dominance of *C. minus* even during mixing in 1977 (Pedrós-Alió et al., 1984). In this example, the outcome of competition between the green and purple

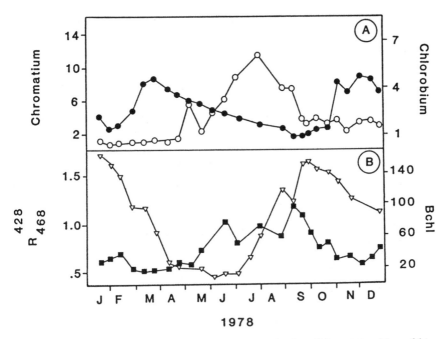

Figure 8.9   Seasonal succession within the anoxigenic photolithoautotrophic guild in Lake Cisó. A, biomass of *Chromatium* (○) and *Chlorobium* (●) during 1978, both in mg fresh weight m⁻². B, specific content of bacteriochlorophyll in *Chlorobium* (■) in μg per mg fresh weight, and ratio between absorbances at 428 and 468 nm (▽). When this ratio is large (winter periods) the green species of *Chlorobium* predominates; when the ratio is low (summer period) the brown species dominates (last curve redrawn from Montesinos et al., 1983, the other curves redrawn from Pedrós-Alió et al., 1983).

bacteria changed throughout the year (and thus produced a seasonal species succession), and was seen to depend on a set of environmental factors which favored one or the other at different times of the year. This caused an alternation in the pattern of dominance of the guild by the two types of phototrophic bacteria.

Despite such alternation, *C. minus* was the "core" species in the guild (see Section 8.3). First, it was the most abundant for most of the time, and second, *C. minus* changed in abundance in direct response to physicochemical conditions in the lake. On the other hand, the number of the green bacteria was dependent on the abundance of *C. minus* (Pedrós-Alió et al., 1983, 1984). In fact, the influence of *C. minus* on the green bacteria had an even more dramatic aspect: the presence of a large population of *C. minus* not only forced higher specific content of pigments and a lower biomass in the green bacteria, but it also determined which one of the two species of *Chlorobium* would be more abundant (Montesinos et al., 1983). The brown-colored species, *C.*

phaeobacteroides, was dominant when *C. minus* was very abundant, while the green-colored species, *C. limicola,* dominated when the number of *C. minus* was low. Montesinos et al. (1983) showed in laboratory experiments that a filter mimicking the spectral properties of *C. minus* caused the same shift in a mixed culture of *Chlorobium* toward the dominance of the brown species. Thus, the quality of the light that filtered through dense layers of *C. minus* was better suited for the antenna pigments of *C. phaeobacteroides.* When *C. minus* was not abundant, *C. limicola* was more efficient and displaced the brown species (see Figure 8.9B). This example illustrates how availability of the energy source (solar radiation); physical determinants such as rainfall, cloud cover, and flux; as well as adaptations of each organism; interact to shift the balance of competition among the different species within the guild.

Superimposed on the annual succession of *C. minus* (with *C. phaeobacteroides*) during stratification and *C. limicola* during mixing, there are longer-period oscillations related to changes in weather and morphology of the lake basin. Lake Cisó is located in an area that is geologically very active. This area is karstic and has many dolines, or pits, and Cisó is one such doline. New sinkholes are being formed constantly. Thus, a new lake appeared not far from Cisó in 1978, and other sinkholes have appeared afterward. Since the studies described above were completed in 1978, Cisó itself has suffered two different sinking episodes, which completely changed its morphology along the northwestern edge (Guerrero et al., 1985). As a concomitant change, sulfide concentration diminished to the present levels: around 0.4 mM during mixing and up to 2 mM in the hypolimnion during stratification. These values were respectively around 1.2 mM and up to 9 mM before the morphological changes took place (Guerrero et al., 1985). As a result, the bacterial photo-trophic guild has shifted its species composition. In addition to the 10^6 cells ml^{-1} of *C. minus,* 10^6 cells ml^{-1} of *Amoebobacter* M3 can now be found. The lower sulfide concentrations have presumably allowed more purple bacteria to develop. Doubling the biomass of such bacteria has diminished tremendously the biomass of green bacteria through more intense shading.

Of course, the two purple bacteria also compete. Again, a combination of adaptive traits and physicochemical changes in the environment results in the coexistence of the two species. *Amoebobacter* M3 is a purple photo-trophic bacterium forming colonies of small, immotile, coccoidal cells surrounded by a thick slime layer and packed with gas vesicles. Within the Chromatiaceae, *Amoebobacter* M3 presents a set of characteristics (immotile, colonial, gas vesicles, slime layer) as different as possible from *C. minus.* This organism has yet to be studied in pure culture in the laboratory, but it seems probable that it will be less tolerant of high sulfide concentrations than *C. minus.* As in the case of *C. minus* versus Chlorobia, the case of *C. minus* versus *Amoebobacter* M3 illustrates the high pressure toward maximal possible differentiation among the members of a guild in order to avoid mutual exclusion.

In Section 8.4 we saw how the bacterial phototrophic guild competed with other autotrophic guilds, both eucaryotic and procaryotic. In this section we have seen how a combination of environmental factors determines the species composition within the guild. Both analyses (within the guild and among guilds) were made possible by a unique set of factors: morphologically or otherwise easily distinguishable species, easily measured activity, high natural abundances, etc. When this system was analyzed, relationships among species and succession appeared to be completely analogous to those found among species of phyto- and zooplankton. Presumably, the same concepts should also be applicable to the heterotrophic bacterial guilds, if their study is approached in the same spirit.

8.7 Concluding Remarks

The present chapter should have been a review of the way in which resource limitation, competition, physical factors, and life-cycle adaptations influence species succession in bacterioplankton as has been done for phyto-and zoo- plankton. Unfortunately, for the reasons given earlier almost nothing is known about such topics. Therefore, there were two possibilities open to the author of this chapter: to review what is known about bacterioplankton, even if it is irrelevant to the study of succession, or to try to collect dispersed pieces of evidence and put them together within a frame of reference adequate to the study of succession. The author decided to follow the second one, with the idea in mind that the guild concept is a good way to place bacteria within the same paradigm as the other members of the plankton. If successful, such an approach could provide 1) a guide to the sort of information needed on bacterioplankton, 2) a frame of reference where researchers could place their experimental results, and 3) a theoretical guide to decide what techniques and studies should be most rewarding. For example, a study of the annual cycle of prosthecate bacteria in a lake acquires a whole new dimension if we think of it as a model of a bacterioplankton guild. With this perspective, we are not only studying a peculiar group of organisms, but we are also testing the way in which general ecological principles apply to bacteria. Moreover, the results obtained should be applicable to other heterotrophic bacterial guilds that are harder to study. Mechanisms should be similar to those of more complex organisms, so that the tremendous amount of knowledge ac- cumulated about phyto- and zooplankton could be used as null hypotheses against which to test bacterioplankton. Certainly, the concept of guild seems more appropriate than that of species for carrying out more detailed studies of bacterial succession. Even if not successful, such an approach should at least provoke criticism that will lead us toward an autecology of bacterio- plankton.

Acknowledgments
I would like to thank J.M. Gasol, H. Güde, A. Konopka, F. Rodá, U. Sommer, and W. Wiebe for their rigorous criticism of earlier versions of this chapter. I am grateful to R. Guerrero and J. Mas for permission to use unpublished results and stimulating discussion.

References

Adams, J. 1985. The identification and interpretation of guild structures in ecological communities. *Journal of Animal Ecology* 54: 43–59.

Alexander, M. 1971. *Microbial Ecology* J. Wiley and Sons, New York. 511 pp.

Atlas, R.M. 1983. Diversity of microbial communities. *Advances in Microbial Ecology* 7: 1–47.

Azam, F., Fenchel, T., Field, J.G., Grey, J.S., Meyer-Reil, L.-A., and Thingstad, F. 1983. The ecological role of water-column microbes in the sea. *Marine Ecology Progress Series* 10: 257–263.

Bell, R.T., Ahlgren, G.M., and Ahlgren, I. 1983. Estimating bacterioplankton production by measuring [³H]thymidine incorporation in a eutrophic Swedish lake. *Applied and Environmental Microbiology* 45: 1709–1721.

Biebl, H. and Pfennig, N. 1979. CO_2 fixation by anaerobic phototrophic bacteria in lakes, a review. *Ergebnisse der Limnologie* 12: 48–58.

Bird, D.F. and Kalff, J. 1987. Algal phagotrophy: Regulating factors and importance relative to photosynthesis in *Dinobryon* (Chrysophyceae). *Limnology and Oceanography* 32: 277–284.

Bohlool, B.B. and Brock, T.D. 1974. Population ecology of *Sulfolobus acidocaldarius*. II. Immunoecological studies. *Archiv für Microbiologie* 97: 181–194.

Bohlool, B.B. and Schmidt, E.L. 1980. The immunofluorescence approach in microbial ecology. *Advances in Microbial Ecology* 4: 203–241.

Børsheim, K.Y. and Anderson, S. 1987. Grazing and food size selection by crustacean zooplankton compared to production of bacteria and phytoplankton in a shallow Norwegian mountain lake. *Journal of Plankton Research* 9: 367–379.

Brock, T.D. 1971. Microbial growth rates in nature. *Bacteriological Reviews* 35: 39–58.

Brock, T.D. 1985. *A Eutrophic Lake: Lake Mendota, Wisconsin.* Springer-Verlag, New York. 308 pp.

Brock, T.D. 1987. The study of microorganisms in situ: progress and problems. *Society for General Microbiology Special Symposium* 41: 1–17.

Brock, M.L. and Brock, T.D. 1968. The application of microautoradiographic techniques to ecological studies. Internationale Vereinigung für Theoretische und Angewandte Limnologie, Mittelungen no. 15, 29 pp.

Brock, T.D. and Clyne, J. 1984. Significance of algal excretory products for growth of epilimnetic bacteria. *Applied and Environmental Microbiology* 47: 731–734.

Caumette, P., Pagano, M., and Saint-Jean, L. 1983. Répartition verticale du phytoplancton, des bactéries et du zooplancton dans un milieu stratifié en Baie de Biétri (Lagune Ebrié, Cote d'Ivoire). Relations trophiques. *Hydrobiologia* 106: 135–148.

Cole, J.J. 1982. Interactions between bacteria and algae in aquatic ecosystems. *Annual Review of Ecology and Systematics* 13: 291–314.

Cole, J.J., Pace, M.L., and Findlay, S. 1988. Prediction of bacterial production in fresh- and saltwater ecosystems: an overview. *Marine Ecology Progress Series* 43: 1–10.

Dyer, B.D., Gaju, N., Pedrós-Alió, C., Esteve, I., and Guerrero, R. 1986. Ciliates from a freshwater sulfuretum. *Biosystems* 19: 127–135.

Edlund, A., Nichols, P.D., Roffley, R., and White, D.C. 1985. Extractable and lipopolysaccharide fatty acid and hydroxy acid profiles from *Desulfovibrio* species. *Journal of Lipid Research* 26: 982–988.

Fallon, R.D., Harrits, S., Hanson, R.S., and Brock, T.D. 1980. The role of methane in internal carbon cycling in Lake Mendota during summer stratification. *Limnology and Oceanography* 25: 357–360.

Fenchel, T., Perry, T., and Thane, A. 1977. Anaerobiosis and symbiosis with bacteria in free-living ciliates. *Journal of Protozoology* 24 : 154–163.

Fliermans, C.B. and Schmidt, E.L. 1975. Autoradiography and immunofluorescence combined for autecological study of single cell activity with *Nitrobacter* as a model system. *Applied Microbiology* 30: 676–684.

Fuhrman, J. 1987. Close coupling between release and uptake of dissolved free amino acids in seawater studied by an isotope dilution approach. *Marine Ecology Progress Series* 37: 45–52.

Gibbons, R.J. and Van Houte, J. 1975. Bacterial adherence in oral microbial ecology. *Annual Review of Microbiology* 29: 19–44.

Gophen, M., Cavari, B.Z., and Berman, T. 1974. Zooplankton feeding on differentially labelled algae and bacteria. *Nature* 247: 393–394.

Gorlenko, V.M., Dubinina, G.A., and Kuznetsov, S.I. 1983. *The Ecology of Aquatic Microorganisms*. E. Schweizerbart'sche Verlagsbuchhandlung (Nägle u. Obermiller), Stuttgart. 252 pp.

Goulder, P. 1977. Attached and free bacteria in an estuary with abundant suspended solids. *Journal of Applied Bacteriology* 43: 399–405.

Güde, H. 1986. Loss processes influencing growth of planktonic bacterial populations in Lake Constance. *Journal of Plankton Research* 8: 795–810.

Guerrero, R., Abellà, C., and Miracle, M.R. 1978. Spatial and temporal distribution of bacteria in a meromictic lake basin: relationships with physicochemical parameters and zooplankton. *Internationale Vereinigung für Theoretische und Angewandte Limnologie. Verhandlungen* 20: 2264–2271.

Guerrero, R., Montesinos, E., Pedrós-Alió, C., Esteve, I., Mas, J., van Gemerden, H., Hofman, P.A., and Bakker, J.F. 1985. Phototrophic sulfur bacteria in two Spanish lakes: vertical distribution and limiting factors. *Limnology and Oceanography* 30: 919–931.

Guerrero, R., Pedrós-Alió, C., Esteve, I., Mas, J., Chase, D., and Margulis, L. 1986. Predatory prokaryotes: predation and primary consumption evolved in bacteria. *Proceedings of the National Academy of Sciences, U.S.A.* 83: 2138–2142.

Guerrero, R., Esteve, I., Pedrós-Alió, C., and Gaju, N. 1987a. Predatory bacteria in prokaryotic communities. *Annals of the New York Academy of Sciences* 503: 238–250.

Guerrero, R., Pedrós-Alió, C., Esteve, I., and Mas, J. 1987b. Communities of phototrophic sulphur bacteria in lakes of the Spanish Mediterranean region. *Acta Academie Aboensis* 47: 125–151.

Hobbie, J.E. 1979. An assessment of quantitative measurement of aquatic microbes. *Ergebnisse der Limnologie* 12: 59–63.

Holben, W.H., Jansson, J.K., Chelm, B.K., and Tiedje, J.M. 1988. DNA probe method for the detection of specific microorganisms in the soil bacterial community. *Applied and Environmental Microbiology* 54: 703–711.

Hoppe, H. 1976. Analysis of actively metabolizing bacterial populations with the autoradiographic method, pp. 179–197, in Reinheimer, G. (editor), *Microbial Ecology of a Brackish Water Environment*. Springer-Verlag, Berlin.

Ingvorsen, K., Zeikus, J.G., and Brock, T.D. 1983. Dynamics of bacterial sulfate reduction in a eutrophic lake. *Applied and Environmental Microbiology* 42: 1029–1036.

Jones, J.G. 1979. *A Guide to Methods for Estimating Microbial Numbers and Biomass in*

Freshwater. Freshwater Biological Association Scientific Publication no. 39, Ambleside, Cumbria. 112 pp.

Jones, J.G., Simon, B.M., and Cunningham, C. 1983. Bacterial uptake of algal extracellular products: an experimental approach. *Journal of Applied Bacteriology* 54: 355–365.

Jones, J.G. 1987. Diversity of freshwater microbiology. *Society for General Microbiology Special Symposium* 41: 235–259.

Jørgensen, N.O.G. 1987. Free amino acids in lakes: concentrations and assimilation rates in relation to phytoplankton and bacterial production. *Limnology and Oceanography* 32: 97–111.

Karl, D.M. 1986. Determination of in situ microbial biomass, viability, metabolism, and growth, pp. 85–176, in Poindexter, J.S. and Leadbetter, E.R. (editors), *Bacteria in Nature*, volume 2. Plenum Press, New York.

Kelly, D.P. 1981. Introduction to the chemolithoautotrophic bacteria, pp. 997–1004, in Starr, M.P., Stolp, H., Trüper, H.G., Balows, A., and Schlegel, H.G. (editors), *The Prokaryotes*, volume 1. Springer-Verlag, Berlin.

Kirchman, D. 1983. The production of bacteria attached to particles suspended in a freshwater pond. *Limnology and Oceanography* 28: 858–872.

Kirchman, D. and Mitchell, R. 1982. Contribution and particle-bound bacteria to total microheterotrophic activity in five ponds and two marshes. *Applied and Environmental Microbiology* 43: 200–209.

Kjelleberg, S., Marshall, K.C., and Hermansson, M. 1985. Oligotrophic and copiotrophic marine bacteria—observations related to attachment. *FEMS Microbiology Ecology* 31: 89–96.

Krebs, C.J. 1985. *Ecology. The Experimental Analysis of Distribution and Abundance*, 3rd edition. Harper and Row, New York.

Kuenen, J.G., Robertson, L.A., and van Gemerden, H. 1985. Microbial interactions among aerobic and anaerobic sulfur-oxidizing bacteria. *Advances in Microbial Ecology* 8: 1–60.

Lane, D.J., Pace, B., Olsen, G.J., Stahl, D.A., Sogin, M.L., and Pace, N.R. 1985. Rapid determination of 16S ribosomal RNA sequences for phylogenetic analyses. *Proceedings of the National Academy of Sciences U.S.A.* 82: 6955–6959.

Lewis, W.M., Frost, T., and Moore, D. 1986. Studies of planktonic bacteria in Lake Valencia, Venezuela. *Archiv für Hydrobiologie* 106: 289–305.

Lowell, C.R. and Konopka, A. 1985. Thymidine incorporation by free-living and particle-bound bacteria in a eutrophic dimictic lake. *Applied and Environmental Microbiology* 49: 501–504.

Martz, R.F., Sebacher, D.K., and White, D.C. 1983. Biomass measurements of methaneforming bacteria in environmental samples. *Journal of Microbiological Methods* 1: 53–61.

McInerney, M.J. 1986. Transient and persistent associations among prokaryotes, pp. 293–338, in Poindexter, J.S. and Leadbetter, E.R. (editors), *Bacteria in Nature*, volume 2. Plenum Press, New York.

Melack, J.M. 1985. Interactions of detrital particulates and plankton. *Hydrobiologia* 125: 209–220.

Meyer-Reil, L.-A. 1978. Autoradiography and epifluorescence microscopy combined for the determination of number and spectrum of actively metabolizing bacteria in natural waters. *Applied and Environmental Microbiology* 36: 506–512.

Montesinos, E., Guerrero, R., Abellà, C., and Esteve, I. 1983. Ecology and physiology of the competition for light between *Chlorobium limicola* and *Chlorobium phaeobacteroides* in natural habitats. *Applied and Environmental Microbiology* 46: 1007–1016.

Mosser, J.L., Mosser, A.G., and Brock, T.D. 1974a. Population ecology of *Sulfolobus acidocaldarius*. I. Temperature strains. *Archiv für Microbiologie* 97: 169–179.

Mosser, J.L., Bohlool, B.B., and Brock, T.D. 1974b. Growth rates of *Sulfolobus acidocaldarius* in nature. *Journal of Bacteriology* 118: 1075–1081.

Muyzer, G., de Bruyn, A.C., Schmedding, D.J.M., Bos, P., Westbroek, P., and Kuenen, G.J. 1987. A combined immunofluorescence-DNA-fluorescence staining technique for enumeration of *Thiobacillus ferrooxidans* in a population of acidophilic bacteria. *Applied and Environmental Microbiology* 53: 660–664.

Newell, S.Y., Fallon, R.D., and Tabor, P.S. 1986. Direct microscopy of natural assemblages, pp. 1–48, in Poindexter, J.S. and Leadbetter, E.R. (editors), *Bacteria in Nature*, volume 2. Plenum Press, New York.

Odum, E.P. 1971. *Fundamentals of Ecology* 3rd edition. Saunders, Philadelphia. 574 pp.

Olsen, G.J., Lane, D.J., Giovannoni, S.J., and Pace, N.R. 1986. Microbial ecology and evolution: a ribosomal RNA approach. *Annual Review of Microbiology* 40: 337–365.

Overbeck, J. 1979. Studies on heterotrophic functions and glucose metabolism of microplankton in Plußsee. *Ergebnisse der Limnologie* 13: 56–76.

Overbeck, J., Höfle, M.G., Krambeck, C., and Witzel, K.-P. (editors). 1984. *Proceedings of the Second Workshop on Measurement of Microbial Activity in the Carbon Cycle of Aquatic Ecosystems*. *Ergebnisse der Limnologie*, volume 19. E. Schweizerbart'sche Verlagsbuchhandlung (Nägele u. Obermiller), Stuttgart.

Pace, N.R., Stahl, D.A., Lane, D.J., and Olsen, G.J. 1986. The analysis of natural microbial populations by ribosomal RNA sequences. *Advances in Microbial Ecology* 9: 1–55.

Paerl, H.W. 1974. Bacterial uptake of dissolved organic matter in relation to detrital aggregation in marine and freshwater systems. *Limnology and Oceanography* 19: 966–972.

Paerl, H.W. 1978. Microbial organic carbon recovery in aquatic systems. *Limnology and Oceanography* 23: 927–935.

Paerl, H.W. and Kellar, P.E. 1978. Significance of bacterial-*Anabaena* (Cyanophyceae) associations with respect to nitrogen fixation in freshwater. *Journal of Phycology* 14: 254–260.

Parkin, T.B. and Brock, T.D. 1981. Photosynthetic bacterial production and carbon mineralization in a meromictic lake. *Archiv für Hydrobiologie* 91: 354–379.

Pedrós-Alió, C. and Brock, T.D. 1982. Assessing biomass and production of bacteria in eutrophic Lake Mendota, Wisconsin. *Applied and Environmental Microbiology* 44: 203–218.

Pedrós-Alió, C. and Brock, T.D. 1983a. The impact of zooplankton feeding on the epilimnetic bacteria of a eutrophic lake. *Freshwater Biology* 13: 227–239.

Pedrós-Alió, C. and Brock, T.D. 1983b. The importance of attachment to particles for planktonic bacteria. *Archiv für Hydrobiologie* 98: 354–379.

Pedrós-Alió, C., Montesinos, E., and Guerrero, R. 1983. Factors determining annual changes in bacterial photosynthetic pigments in holomictic Lake Cisó, Spain. *Applied and Environmental Microbiology* 46: 999–1006.

Pedrós-Alió, C., Abellà, C., and Guerrero, R. 1984. Influence of solar radiation, water flux and competition on biomass of phototrophic bacteria in Lake Cisó, Spain. *Internationale Vereinigung für Theoretische und Angewandte Limnologie, Verhandlungen* 22: 1097–1101.

Pedrós-Alió, C., Gasol, J.M., and Guerrero, R. 1986. Mircobial ecology of sulfurous Lake Cisó, pp. 638–644, in Megusar, F. and Gantar, M. (editors), *Perspectives in Microbial Ecology*. Slovenian Society for Microbiology, Ljubljana, Yugoslavia.

Pedrós-Alió, C., Gasol, J.M., and Guerrero, R. 1987. On the ecology of a *Cryptomonas phaseolus* population forming a metalimnetic bloom in Lake Ció, Spain: Annual distribution and loss factors. *Limnology and Oceanography* 32: 285–298.

Pfennig, N. 1980. Syntrophic mixed cultures and symbiotic consortia with phototrophic bacteria: a review, pp. 127–131, in Gottschalk, G., Pfennig, N., and Werner, H. (editors), *Anaerobes and Anaerobic Infections*. Gustav Fischer, Stuttgart.

Pomeroy, L.R. 1974. The ocean's food web, a changing paradigm. *BioScience* 24: 499–504.

Pomeroy, L.R. 1984. Microbial processes in the sea: diversity in nature and science, pp. 1–33, in Hobbie, J.E. and Williams, P.J.LeB. (editors), *Heterotrophic Activity in the Sea*. Plenum Press, New York.

Porter, K.G. 1984. Natural bacteria as food resources for zooplankton, pp. 340–345, in Klug, M.J. and Reddy, C.A. (editors), *Current Perspectives in Microbial Ecology*. American Society for Microbiology, Washington, D.C.

Porter, K.G., Sherr, E.B., Sherr, B.F., Pace, M., and Sanders, R.W. 1985. Protozoa in planktonic food webs. *Journal of Protozoology* 32: 409–415.

Reed, W.M. and Dugan, P.R. 1978. Distribution of *Methylomonas methanica* and *Methylosinus trichosporium* in Cleveland Harbor as determined by an indirect fluorescent antibody-membrane filter technique. *Applied and Environmental Microbiology* 35: 422–430.

Riemann, B. 1985. Potential importance of fish predation and zooplankton grazing on natural populations of freshwater bacteria. *Applied and Environmental Microbiology* 50: 187–193.

Root, R.B. 1967. The niche exploitation pattern of the blue-gray gnatcatcher. *Ecological Monographs* 37: 317–350.

Rudd, J.W.M. and Hamilton, R.D. 1978. Methane cycling in a eutrophic shield lake and its effects on whole lake metabolism. *Limnology and Oceanography* 23: 337–348.

Rudd, J.W.M. and Taylor, C.D. 1980. Methane cycling in aquatic environments. *Advances in Aquatic Microbiology* 2: 77–150.

Schmidt, J.M. 1981. The genera *Caulobacter* and *Asticcacaulis*, pp. 466–476, in Starr, M.P., Stolp, H., Trüper, H.G., Balows, A., and Schlegel, H.G. (editors), *The Prokaryotes*, volume 1. Springer-Verlag, Berlin.

Schmidt, J.M. and Starr, M.P. 1981. The *Blastocaulis-Planctomyces* group of budding and appendaged bacteria, pp. 496–504, in Starr, M.P., Stolp, H., Trüper, H.G., Balows, A., and Schlegel, H.G. (editors), *The Prokaryotes*, volume 1. Springer-Verlag, Berlin.

Sherr, B.F. and Sherr, E.B. 1984. Role of heterotrophic protozoa in carbon and energy flow in aquatic ecosystems, pp. 412–423, in Klug, M.J. and Reddy, C.A. (editors), *Current Perspectives in Microbial Ecology*. American Society for Microbiology, Washington, D.C.

Sherr, B.F. and Sherr, E.B. 1987. High rates of consumption of bacteria by pelagic ciliates. *Nature* 325: 710–711.

Sieburth, J.M. 1979. *Sea Microbes*. Oxford University Press, New York. 491 pp.

Simon, M. 1985. Specific uptake rates of amino acids by attached and free-living bacteria in a mesotrophic lake. *Applied and Environmental Microbiology* 49: 1254–1259.

Sommer, U., Gliwicz, Z.M., Lampert, W., and Duncan, A. 1986. The PEG-model of seasonal succession of planktonic events in fresh waters. *Archiv für Hydrobiologie* 106: 433–471.

Sonea, S. 1988. A bacterial way of life. *Nature* 331: 216.

Sorokin, Y.I. 1970. Interrelation between sulphur and carbon turnover in meromictic lakes. *Archiv für Hydrobiologie* 66: 391–446.

Stahl, D.A., Lane, D.J., Olsen, G.J., and Pace, N.R. 1984. Analysis of hydrothermal vent-associated symbionts by ribosomal RNA sequences. *Science* 224: 409–411.

Stahl, D.A., Lane, D.J., Olsen, G.J., and Pace, N.R. 1985. Characterization of a Yellowstone hot spring microbial community by 5S ribosomal RNA sequences. *Applied and Environmental Microbiology* 49: 1379–1384.

Staley, J.T. 1980. Diversity of aquatic heterotrophic bacterial communities, pp. 321–322, in Schlessinger, D. (editor), *Microbiology–1980.* American Society for Microbiology, Washington, D.C.

Staley, J.T. and Konopka, A. 1985. Measurement of in situ activities of nonphotosynthetic microorganisms in aquatic and terrestrial habitats. *Annual Review of Microbiology* 39: 321–346.

Staley, J.T., Hirsch, P., and Schmidt, J.M. 1981. Introduction to budding and/or appendaged bacteria, pp. 451–455, in Starr, M.P., Stolp, H., Trüper, H.G., Balows, A., and Schlegel, H.G. (editors), *The Prokaryotes*, volume 1. Springer-Verlag, Berlin.

Staley, J.T., Konopka, A.E., and Dalmasso, J.P. 1987. Spatial and temporal distribution of *Caulobacter* spp. in two mesotrophic lakes. *FEMS Microbiology Ecology* 45: 1–6.

Stanley, P.M. and Staley, J.T. 1977. Acetate uptake by aquatic bacterial communities measured by autoradiography and filterable radioactivity. *Limnology and Oceanography* 22: 26–37.

Stanley, P.M., Gage, M.A., and Schmidt, E.L. 1979. Enumeration of specific populations by immunofluorescence, pp. 46–55, in Costerton, J.W. and Colwell, R.R. (editors), *Native Aquatic Bacteria.* American Society for Testing and Materials Special Technical Publication no. 695, Philadelphia.

Starr, M.P. and Schmidt, J.M. 1981. Prokaryote diversity, pp. 3–42, in Starr, M.P., Stolp, H., Trüper, H.G., Balows, A., and Schlegel, H.G. (editors), *The Prokaryotes*, volume 1. Springer-Verlag, Berlin.

Steele, J.H. 1974. *The Structure of Marine Ecosystems.* Harvard University Press, Cambridge, Massachusetts. 128 pp.

Stolp, H. and Starr, M.P. 1981. Principles of isolation, cultivation, and conservation of bacteria, pp. 133–175, in Starr, M.P., Stolp, H., Trüper, H.G., Balows, A., and Schlegel, H.G. (editors), *The Prokaryotes*, volume 1. Springer-Verlag, Berlin.

Tabor, P.S. and Neihof, R.A. 1984. Direct determination of activities for microorganisms of Chesapeake Bay populations. *Applied and Environmental Microbiology* 48: 1012–1019.

Terborgh, J. and Robinson, S. 1986. Guilds and their utility in ecology, pp. 65–90, in Kikkawa, J. and Anderson, D.J. (editors), *Community Ecology: Pattern and Process.* Blackwell Scientific Publications, Oxford.

Torrella, F. and Morita, R.Y. 1981. Microcultural study of bacterial size changes and microcolony and ultramicrocolony formation by heterotrophic bacteria in seawater. *Applied and Environmental Microbiology* 41: 518–527.

Trüper, H.G. and Pfennig, N. 1981. Characterization and identification of the anoxygenic phototrophic bacteria, pp. 299–312, in Starr, M.P., Stolp, H., Trüper, H.G., Balows, A., and Schlegel, H.G. (editors), *The Prokaryotes*, volume 1. Springer-Verlag, Berlin.

Turner, J.T. and Ferrante, J.G. 1979. Zooplankton fecal pellets in aquatic ecosystems. *BioScience* 29: 670–677.

Van Bruggen, J.J.A., Stumm, C.K., and Vogels, G.D. 1983. Symbiosis of methanogenic bacteria and sapropelic protozoa. *Archives of Microbiology* 136: 89–95.

Van Es, F.B. and Meyer-Reil, L.-A. 1982. Biomass and metabolic activity of heterotrophic marine bacteria. *Advances in Microbial Ecology* 6: 111–170.

Van Es, F.B., Lanbroeck, H.J., and Veldkamp, H. 1984. Microbial ecology: an overview, pp. 1–33, in Codd, G.A. (editor), *Aspects of Microbial Metabolism and Ecology.* Society for General Microbiology Special Publication no. 11, Academic Press, London.

Van Gemerden, H. and Beeftink, H.H. 1983. Ecology of phototrophic bacteria, pp.

146–185, in Ormerod, J.G. (editor), *The Phototrophic Bacteria: Anaerobic Life in the Light.* Blackwell Scientific Publications, Oxford.

Ward, B.B. 1984. Combined autoradiography and immunofluorescence for estimation of single cell activity by ammonium-oxidizing bacteria. *Limnology and Oceanography* 29: 402–410.

Ward, D.M. and Winfrey, M.R. 1985. Interactions between methanogenic and sulfate-reducing bacteria in sediments. *Advances in Aquatic Microbiology* 3: 142–179.

Ward, D.M., Tayne, T.A., Anderson, K.L., and Bateson, M.M. 1987. Community structure and interactions among community members in hot spring cyanobacterial mats. *Society for General Microbiology Special Symposium* 41: 179–210.

White, D.C. 1986. Quantitative physicochemical characterization of bacterial habitats, pp. 177–203, in Poindexter, J.S. and Leadbetter, E.R. (editors), *Bacteria in Nature,* volume 2. Plenum Press, New York.

White, D.C., Bobbie, R.J., Herron, J.S., King, J.D., and Morrison, S.J. 1979. Biochemical measurements of microbial mass and activity from environmental samples, pp. 69–81, in Costerton, J.W. and Colwell, R.R. (editors), *Native Aquatic Bacteria.* American Society for Testing and Materials Special Technical Publication no. 695, Philadelphia.

Williams, P.J. LeB. 1981. Incorporation of microheterotrophic processes into the classical paradigm of the planktonic food web. *Kieler Meeresforschungen* 5: 1–28.

Winogradsky, S. 1949. *Méthode dans la Microbiologie du Sol.* Masson et Cie., Paris.

Woese, C. 1987. Bacterial evolution. *Microbiological Reviews* 51: 221–271.

9

The Role of Grazing on Bacteria in Plankton Succession

Hans Güde

Institute of Lake and Fish Research
Langenargen, Federal Republic of Germany

9.1 Introduction

As has been pointed out in Chapter 1, the planktonic community is especially suitable for testing general ecological theories. Its spatial structure is relatively simple, and many species can be easily cultured and have short generation times. An experimental approach can be effective, particularly with those plankton showing the highest potential growth rates: the bacteria. Indeed, many important biotic interactions such as competition, commensalism, mutualism, predation, and parasitism have been studied using bacteria (Frederickson, 1977). In spite of this wealth of experimental data, however, there is no consensus regarding the control of planktonic bacterial succession. This lack of agreement is not surprising considering the inability of current methodologies to clearly distinguish different populations of bacteria. The information gained from microscopic examination of bacteria is limited, unlike the case for phyto- and zooplankton. Bacterial composition can at best be differentiated according to rough morphologic criteria such as length and width of cells, multicellular or single-celled growth, attachment to surfaces, etc. The indirect approach of analyzing the composition of colonies grown on agar plates is hindered by the large discrepancy between plate counts and direct microscopic counts (Staley and Konopka, 1985). As has been emphasized by Pedrós-Alió (see Chapter 8), promising approaches based on species-specific biomolecular cues may improve this situation in the near future.

Presently, however, this gap in our knowledge limits our attempts to determine the succession of bacteria on a species or population level. Nevertheless, the role played by bacteria in the theater of planktonic succession may be considered on a more general level. Such considerations appear to be justified by our knowledge of the importance of bacteria in the flow of inorganic and organic material through the planktonic community (Staley and Konopka, 1985). Consequently, it must be expected that the succession of the planktonic community is, to a considerable degree, the result of bacteria and their interactions with other components of the planktonic community. Grazing on bacteria is certainly one of the most direct of these interactions. In this chapter, the importance of grazing on bacteria is examined for its influence on both the succession of bacterial populations and the succession of the entire plankton community.

The importance of grazing as a loss factor for bacterioplankton will be discussed as it relates to the structure of the bacterial community. The impact that grazing on bacteria has on the larger planktonic community will be discussed using zoo- and phytoplankton as examples and using the PEG-model developed by Sommer et al. (1986).(Also see Section 1.2)

9.2   Impact of Grazing for Bacterial Populations

Importance of bacterial loss processes   The success of a planktonic bacterial population will ultimately be determined by its ability to acquire nutrients for growth and to minimize losses. Because of the low concentrations of organic substrates in natural waters (observed concentrations of amino acids and sugars usually are at the nanomolar level), one initial attempt to understand heterotrophic bacterial life in aquatic habitats focused on the question of how bacteria could efficiently take up nutrients in this dilute environment. These studies showed that aquatic bacteria have highly efficient uptake systems (e.g., Wright and Hobbie, 1965). Moreover, it was demonstrated by chemostat studies that those bacteria with uptake systems with a high affinity for a limiting substrate will outcompete those with a lower affinity (Veldkamp, 1977; Tempest and Njeissel, 1978). Improved methods for visualizing bacterioplankton, especially by epifluorescence microscopy (Zimmermann and Meyer-Reil, 1974; Hobbie et al., 1977) provided further insight into aquatic bacterial life. It became clear that freely-suspended single cells were more common than aggregated or attached bacteria (Hoppe, 1984). The initial attempts at estimating the growth rates of bacteria were also based on epifluorescence microscopy, either by determining the frequency of dividing cells (e.g., Hagström et al., 1979), or by determining the change of bacterial densities in a given time (e.g., Kirchman et al., 1982; Wright and Coffin, 1984). Estimates of bacterial production calculated from these approaches, together with those obtained from measurements of the incorporation of radioactive substrates into cell material, suggest that bacterial production constitutes a considerable fraction of total planktonic production, which may range between 20 and 50 percent of measured primary production (Van Es and Meyer-Reil, 1982; Cole et al., 1988). If these estimates are realistic, bacterial biomass has the potential for increasing several times within a few days, provided no losses occur. Because bacterial densities are usually rather stable within the time range of several days, bacterial losses must approximately equal bacterial growth.

One of the processes considered responsible for bacterial losses is physiological mortality or autolysis of bacteria (e.g., Riemann et al., 1982). It must be remembered, however, that bacteria, unlike higher multicellular organisms, are not programmed for death. Indeed, contrary to many textbook illustrations, the bacterial growth curve is characterized by a slow decline of bacterial densities after reaching a maximum (Figure 9.1). This trend was demonstrated by starvation experiments where bacterial numbers were observed to even increase initially after exhaustion of substrate (Morita, 1982; Kjelleberg et al., 1987). Long-term stability of bacterial numbers was also demonstrated in lake water cultures free of grazers (e.g., Güde, 1986; Tranvik and Höfle, 1987). Physiological death in bacterial populations must therefore be regarded as resulting mainly from adverse external factors (e.g., rise or decline of pH, depletion of oxygen, or excretion of toxic products). In dense laboratory cultures, the adverse conditions are often created by the bacteria themselves (Mason, 1987). The much lower bacterial densities found in natural conditions can hardly be expected to produce such effects. Given the demonstrated ability of aquatic bacteria for long-term survival, it is suggested that physiological death in natural waters is in most cases too small to offset bacterial growth.

The sinking of bacteria to deeper water layers or to the sediment is another loss factor to consider. For freely-suspended single cells, this factor is negligible (Jassby, 1975), but for the fraction of bacteria attached to larger particles, sinking may be more important (Pedrós-Alió and Brock, 1983). As mentioned above, the majority of bacteria in many lakes consist of free single cells, even though the productivity of attached cells may exceed that of free cells in certain situations (Simon, 1987). Therefore, sinking also only partly explains bacterial losses.

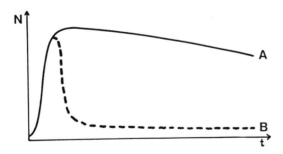

Figure 9.1   Idealized development of bacterial numbers (N) in the absence (A) or the presence (B) of grazing flagellates as are regularly observed in lake water cultures (e.g., Güde, 1986).

Bacterial death can be caused by attack of bacterial predators or by phages. Although there is circumstantial evidence for the action of these factors (e.g., Guerrero et al., 1987), their importance cannot be evaluated at present due to a lack of data. Densities of bacterial populations in natural waters, however, appear to be below the level at which an efficient attack by viruses or bdellovibrios can occur (Shilo, 1984). Moreover, due to the usual host specificity of these organisms, the attack would be directed against only one population and would not explain losses of a broad range of bacterial populations. Because these factors seem to only partially explain bacterial losses, it is possible that grazing may be responsible for the "enigma of bacterial death rates" (Pace, 1988).

Evidence for bacterial grazing   Until now, there have been no established methods for measuring bacterial grazing. The evidence to date has been based on a variety of different approaches, which give indirect or direct indications of the importance of grazing. One indirect approach has been to determine the abundance of potential grazers and estimate grazing rates on the basis of filtering rates found in laboratory experiments. According to this approach, the heterogeneous group of phagotrophic microflagellates is one of the most important groups of grazers in marine and freshwater habitats. Their observed numbers of 10^3–10^4 ml^{-1} would be sufficient to remove substantial amounts (up to 100 percent) of bacterial biomass per day (Fenchel, 1982, 1986; Linley et al., 1983; Riemann, 1985; Güde, 1986, 1988a; Boraas, 1986; Bern, 1987; Scavia et al., 1987; Nagata, 1988). Considerable grazing pressure may also be exerted by crustacean zooplankton, especially by filtering cladocera, which become very abundant during the seasonal, clearwater phases. Although the ability to filter bacteria-size particles is not common to all cladocerans, it has been observed in a number of species (e.g., Geller and Müller, 1981; Brendelberger, 1985; De Mott, 1985; Hessen, 1985b).

Less evidence of grazing exists for the other groups in the planktonic community. Low grazing pressure by ciliates is predicted because observed bacterial numbers are often below threshold levels at which grazing becomes efficient (Fenchel, 1980). On the other hand, there are several studies showing that the contribution of ciliates to grazing on bacteria must not be regarded as negligible (e.g., Pace, 1984; Sherr and Sherr, 1987; Rassoulzadegan et al., 1988). Rotifers and copepodes are considered as inefficient grazers on freely-suspended bacteria (Starkweather, 1980; Porter, 1984; Sterner, Chapter 4). Therefore, grazing by these zooplankton groups probably affects only those bacteria growing as aggregates or attached to particles (e.g., Schoenberg and Maccubin, 1985). Recent evidence indicates that mixotrophic algae may graze on bacteria under natural conditions (Bird and Kalff, 1986; Porter, 1988). The general importance of these organisms, however, cannot yet be evaluated. In summary, available evidence suggests that grazing on bacteria may be primarily by heterotrophic microflagellates and cladocera; only limited information is available on the contribution of other members of the planktonic community.

Further indirect demonstration of grazing is indicated by changes in bacterial numbers after the removal of grazers by prefiltration, dilution of water samples with particle-free lake water, or addition of substances that inhibit eucaryotes (McManus and Fuhrman, 1988). An increase of bacterial numbers results from this reduction or exclusion of grazing activities (e.g., Fuhrman and Azam, 1980; Kirchman et al., 1982; Wright and Coffin, 1984a; Landry et al., 1986; Sherr et al., 1986; Güde, 1986). Moreover, these approaches support the assumption that grazers occurred predominantly in size classes smaller than 12 μm (Wright and Coffin, 1984a; Güde, 1986), because the strongest effects were observed after elimination of organisms smaller than 12 μm. Provided that the observed population changes of bacteria were exclusively due to the absence of grazers, these techniques could be used to quantify grazing rates, with some attention to possible under- or overestimates.

An underestimate may result from 1 μm prefiltration without dilution by sterile lake water. With the elimination of all larger organisms, the direct or indirect producers of growth substrates for bacteria (i.e., phyto- and zooplankton) are also eliminated from the system. The resulting rapid exhaustion of the substrate pool causes bacteria to stop their growth (Güde, 1986; Sherr et al., 1988). Dilution of the water samples with sterile lake water would prevent this rapid exhaustion of substrates. In this case, however, overestimates of bacterial production may occur, because the dilution procedure decreases not only grazing pressure but also bacterial density. The logistic growth model (Wright and Coffin, 1984b), predicts that a given amount of substrate will then be partitioned among fewer bacteria. Consequently, the amount required for maintenance will decrease and the amount left for new production will increase (Sieracki and Sieburth, 1985). This consideration could become especially relevant in situations of prevailing substrate control, which may frequently occur in nature (Güde, in preparation). The use of eucaryotic inhibitors may also be disturbed by artifacts (Sanders and Porter, 1986). Therefore, the indirect approaches may provide a better understanding of grazing effects, but their use for quantification of grazing rates is useful only after careful examination of the assumptions involved.

Because of these intrinsic difficulties, quantitative estimates of bacterial grazing losses should rely on direct determinations of bacterial grazing rates. Radioactively labeled bacteria have been used in many cases for this purpose. It is important, however, to label natural bacterioplankton rather than culture bacteria, because the size of the latter is usually much larger than that of natural bacteria. Because of its exclusive uptake by bacteria (Bern, 1985; Güde, 1988b) and its high specific labeling, ^{3}H-thymidine is now used in many cases as a labeling substrate (e.g., Hollibaugh et al., 1981; Riemann and Bosselmann, 1984; Bjørnsen et al., 1986; Schoenberg and Maccubin, 1985; Güde, 1986;

Bern, 1987; Nagata, 1988). This method allows the determination of grazing rates for those organisms which are easily separated from bacteria, e.g., crustacean zooplankton. This zooplankton group has been shown to contribute significantly to bacterial grazing (e.g., Riemann and Bosselmann, 1984; Riemann, 1985; Bjørnsen et al., 1986), although not as much as the phagotrophic flagellates (Bern, 1987; Güde, 1988a). Difficulties arise when this method is used for very small grazers, because they cannot be easily separated from the bacteria. Thus, this method is not useful for heterotrophic microflagellates, the majority of which may often pass a 3-μm filter (Wright and Coffin, 1984a; Güde, 1986). Although grazing may become apparent by a shift of label from smaller to larger size fractions, its quantification may again be limited by the high number of assumptions involved (Güde, 1986).

Alternatively, grazing may be directly determined by the use of fluorescent beads of bacterial size (e.g., Børsheim, 1984; Cynar and Sieburth, 1986; McManus and Fuhrman, 1986; Bird and Kalff, 1986; Porter, 1988). The clearance rates obtained by these approaches, however, are lower than those reported from experimental studies, suggesting discrimination between bacteria and fluorescent microspheres by the grazers (Sherr, Sherr, and Fallon, 1987; Nygaard et al., 1988). Better results were obtained with fluorescently-labeled natural bacteria (Sherr, Sherr, and Fallon, 1987).

In summary, the methods presently available allow only rough estimates of bacterial grazing losses. The evidence accumulated from indirect and direct approaches, however, gives strong support to the assumption that grazing is one of the major loss factors for bacteria.

Influence of grazing on bacteria   If grazing is one of the major loss factors for bacteria, it may in turn influence the abundance, activities, and composition of bacterial populations. Various responses may be expected, depending on the intensity of substrate supply and grazing. At low substrate supply and low grazing (Figure 9.2A), bacterial densities will be rather constant at a low level. This situation may occur in marine, oligotrophic, offshore waters (e.g., Wright and Coffin, 1984) and in the hypolimnetic, deep-water layers of deep lakes (e.g., Güde et al., 1985; Nagata, 1987). The minimum values consistently observed in these water layers of Lake Constance (Figure 9.3) are similar to minimum values observed in other lakes and culture experiments (e.g., Ammerman et al., 1984; Pace, 1988; Jürgens and Güde, in preparation). This pattern, as well as studies by Fenchel (1982) suggest that there is a threshold density between 10^5 and 10^6 bacteria ml^{-1}, below which grazing on bacteria becomes inefficient. According to the Lotka-Volterra equations, oscillations of bacterial numbers (Figure 9.2B) are to be expected when the supply of substrate is increased (Williams, 1980; Laake et al., 1983). This situation may occur in the epilimnetic water layers of lakes during the vegetation period, although observed fluctuations appear rather irregular, and amplitudes are relatively small (Figure 9.3). Obviously, there are mechanisms that lead to

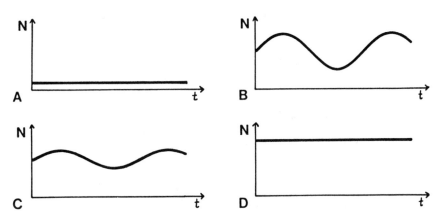

Figure 9.2 Theoretically expected development of bacterial numbers, based on the intensity of substrate supply and grazing. A. Low substrate, grazers absent. B. High substrate, grazers present. C. High substrate, predatory control of grazers. D. High substrate, grazers absent.

the dampening of oscillations. One may be the predatory control of grazers by larger zooplankton (Figure 9.2C), as shown by Wright (1988). High, stable bacterial densities are also expected for high substrates at low grazing (Figure 9.1, Figure 9.2D). Although this situation seems less probable, there are indications for its occasional occurrence (e.g., Güde, 1988a).

Grazers may also influence bacterial metabolic activities. Evidence from many experimental and field studies suggests that grazing leads to a stimulation of bacterial activity and growth (Stout, 1980; Sherr et al., 1982; Sherr and Sherr, 1984; Šimek, 1986; Güde, 1986). Conceptually this stimulation can be explained in a similar way to that of algal grazers on algal growth (Sterner, Chapter 4). As is predicted by the logistic growth model (Wright and Coffin, 1984b; Wright, 1988), bacterial growth will decrease with increasing population densities, because a given supply of substrate has to be partitioned among more bacteria. By keeping bacterial densities at submaximal levels, the same amount of substrate is available to fewer bacteria, which can multiply more rapidly because of the strong dependence of specific growth rate on substrate supply. Thus, grazing can prevent bacteria from reaching self-limiting population densities. Alternatively, it has been observed that grazing results in the release of organic carbon to the environment (e.g., Lampert, 1978; Andersson et al., 1985; Güde, 1988a). This carbon may be utilized by bacteria as a source of additional substrate otherwise not available. Grazers also release inorganic nutrients (e.g., Sherr et al., 1983; Goldman and Caron, 1985; Güde, 1985; Berman et al., 1987; Sherr et al., 1988), that may contribute to enhanced bacterial growth if inorganic nutrients are limited. At present, it cannot be decided which of these explanations is most important.

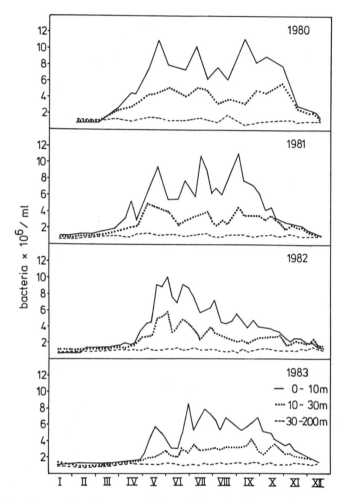

Figure 9.3 Seasonal development of bacterial densities observed throughout the year at a midlake station of Lake Constance. From Güde et al. (1985).

Because they are nonexclusive, they may contribute equally to the observed stimulation.

In either case, these observations are important to answering the question central to the understanding of bacterial succession, i.e., whether and how selection conditions for bacterial populations are changed by grazing activities. In the absence of grazers, the ability to compete for the given substrates will determine the success of the bacterial populations. Under this condition, co-existence of different populations becomes less probable and bacterial diversity will decrease (Figure 9.4A). In the presence of grazers, the selection pressure exerted by competition for substrates can be reduced, because the grazers

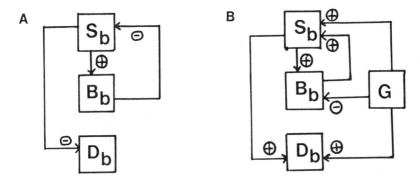

Figure 9.4 Conceptual influence of the presence or the absence of grazers on the substrate supply (S_b), biomass (B_b), and diversity D_b of bacterial populations. A. Grazers absent, less substrate, more competition, less coexistence. B. Grazers present, more substrate, less competition, morphologic selection, increased coexistence.

increase either directly and indirectly, the amount of substrate per cell. On the other hand, the success of a bacterial population will now also depend on its ability to compensate for grazing losses with more rapid growth. With the addition of selection factors, conditions for the coexistence of different species are improved, and bacterial diversity can increase (Figure 9.4B).

A change of selection conditions for bacteria, brought on indirectly by grazing activities, is possible without selective grazing occurring. Grazers, however, may change selection conditions directly by selective grazing (Figure 9.4B). In this case, criteria such as size, shape, taste, etc., will additionally affect the success of a bacterial population. One effect of selective grazing was demonstrated by a series of chemostat experiments that were inoculated by an undefined mixture of natural bacteria (Güde, 1979, 1982). When grazers were absent, bacterial populations always consisted exclusively of free, dispersed single cells. This pattern was independent of the taxonomic composition of the bacteria and the type of substrates limiting bacterial growth. However, drastic changes in taxonomic composition, as well as morphologic structure of the bacteria were regularly observed in the presence of grazing ciliates and zooflagellates. In these cases, populations of filamentous, spiral-shaped or aggregated growth forms increase (Figures 9.5 and 9.6). These observations were originally made with inocula from activated sludge systems, but the same findings were observed in later experiments with algae/bacteria systems and water cultures (Güde, 1985, 1986). Corresponding observations were also made in experimental flagellate/bacteria systems (Van Wambeke and Bianchi, 1985; Caron et al., 1988) and in laboratory investigations on the succession of microbial populations during the decay of plant material (e.g., Newell et al., 1981; Robertson et al., 1981). The uniformity of this pattern suggests that free single cells are selected when competition for

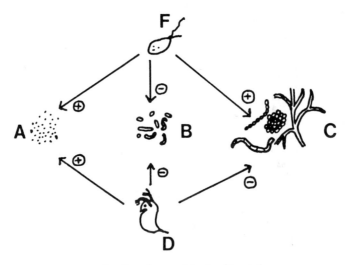

expected size distributions

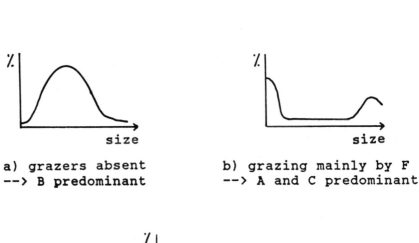

a) grazers absent
--> B predominant

b) grazing mainly by F
--> A and C predominant

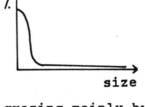

c) grazing mainly by D
--> A predominant

Figure 9.5 Influence of selective grazing by either flagellates (F) or daphnids (D) on the morphologic structure of bacterial population with consideration of three morphologic types. (A: small single-celled bacteria. B: "normal"-sized single-celled bacteria. C: bacteria with filamentous, aggregated or attached growth forms.)

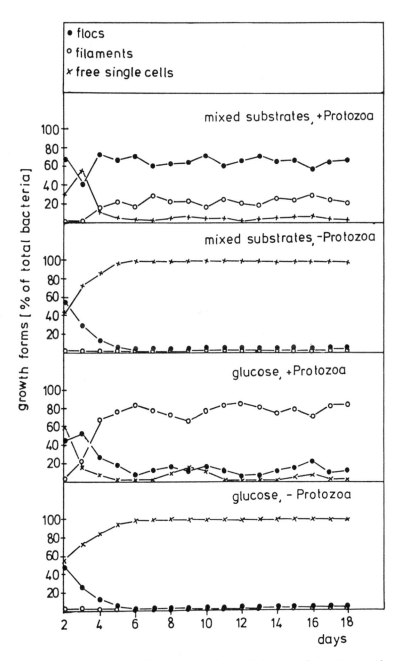

Figure 9.6   Development of different bacterial growth forms in chemostat experiments that were inoculated by a mixture of natural bacteria and protozoa or only by a mixture of bacteria. Either glucose or a mixture of different substrates were offered as limiting substrates. The dilution rate was 0.03 h⁻¹.

substrates is the exclusive selection factor, because this growth form provides an optimum surface/volume ratio and hence a maximum probability of cell/substrate contacts. At the same time, however, these growth forms are most susceptible to protozoan grazing and are therefore increasingly replaced by more complex growth forms with a relative resistance to grazing.

Because of the complex and varying conditions of natural habitats, it is not easy to evaluate the relevance of chemostat findings. It has already been mentioned that planktonic bacteria growing as free single cells are predominate in most lakes. Does this mean that grazing on bacteria is usually of minor importance for selection of bacteria as compared to competition for substrates? First of all, it must be noted that an increase of filamentous and aggregated growth forms could indeed be observed regularly at periods of high densities of zooflagellates in Lake Constance (Güde et al., 1985; Güde, 1988) although free cells were still dominant. Similarly, high densities of flagellates (10^4–10^5 ml^{-1}) and ciliates (10^2–10^3 ml^{-1}) were observed in the eutrophic lakes Schleinsee and Federsee (FRG), a large percentage of which were filaments and aggregates (Güde, unpublished). Additionally, there is increasing evidence that not only large and complex growth forms but also very small cells may represent a refuge from grazing. This conclusion is suggested by the observed shift of the bacterial size spectrum toward the smallest size classes after the introduction of grazers (e.g., Turley, 1985; Ammerman, et al., 1984; Andersson et al., 1986; Kambeck, 1988). Moreover, it must be remembered that grazing pressure is not only exerted by protozoa, but also by crustacean zooplankton, which prefer a much broader size range of food particles than that provided by free single-celled bacteria. Consequently, growth forms that are relatively resistant to protozoan grazing may nevertheless be susceptible to grazing by crustaceans. This suggestion is supported by the regular observation that filamentous and aggregated growth forms disappeared completely at the onset of the clearwater phase in Lake Constance, resulting in a very homogenous bacterial population of exclusively small single cells (Simon, 1987; Güde, 1988a). This predominance of very small cells at the clearwater phase in June is also apparent in a concomitant sharp decrease in bacterial activity occurring in size fractions >1 μm and >3 μm (Güde, 1988b). Finally, it must be emphasized that the morphologic structure of bacterial populations is, of course, not exclusively a result of grazing activities. Small cells may result from starvation (Morita, 1982) and aggregated and attached cells from physicochemical forces (Marshall, 1984).

These observations show that the complex relationships in natural, planktonic communities yield differing responses of bacterial populations to grazing. In spite of this variation, however, a common effect is an increased probability of coexistence of different bacterial populations. Supporting evidence exists for both selective and nonselective grazing. Jost et al. (1973) demonstrated that two bacterial populations coexisted in chemostats in the presence of a ciliate that grazed nonselectively on the bacteria, whereas one population outcompeted the other in the absence of grazers. Güde (1982) provided a corresponding chemostat example for the case of selective grazing. In this case, one population was a superior competitor for substrates but was very susceptible to grazing due to its growth as freely-dispersed single cells. The second population was unable to coexist with the first in the absence of grazers, but was resistant to protozoan grazing because of its flocculent growth. These two populations coexisted in the presence of grazing ciliates, with the single-celled population growing attached to the other, flocculent population.

In summary, there is now sufficient evidence from experimental and field studies supporting the idea that grazing can influence the composition of bacterial populations. Consequently, it must be regarded as an important factor determining the succession of bacterial populations. The most general effect appears to be an enhancement of taxonomic and morphologic diversity. Beyond this general trend, however, our present knowledge does not allow more precise predictions of the specific outcome of the interplay of bacteria with grazers and the planktonic environment.

9.3 Impact of Bacterial Grazing on the Planktonic Community

Even if the ecosystem structure is reduced to five subsystems, and only connections via matter flux are considered (Figure 9.7), one may imagine many possible ways in which grazing on bacteria may influence the other biotic and abiotic subsystems. Of these numerous potential influences, only a few will be considered here. These were selected because a certain amount of information is available on them and because of their assumed importance.

Influence on zooplankton   Grazing on bacteria can influence the quantity, as well as the quality of the zooplankton community. The quantitative influence depends mainly on whether bacteria act as a sink or a link of energy in food webs (Ducklow et al., 1987; Sherr, Sherr, and Albright, 1987; Pomeroy and Wiebe, 1988). The opinion currently favored is that bacteria represent a quantitatively important food source (e.g., Williams, 1981; Azam et al., 1983; Cole et al., 1988). This view is based mainly on estimated high bacterial net productivities and on the assumption of high bacterial growth efficiencies, which would suggest an efficient carbon transfer from primary producers via bacteria to higher trophic levels. As was emphasized by Güde (in press), however, the validity of this conclusion must be questioned. First of all, recent estimates of bacterial growth efficiencies (Höfle and Tranvik, 1986; Bjørnsen, 1986; Nagata, pers. comm.) suggest that 10 percent rather than the formerly assumed 50–80 percent of carbon taken up by bacteria will be transformed into new bacterial biomass. The high estimates were derived indirectly from measurements of the respiration of ^{14}C-substrates, without consideration of

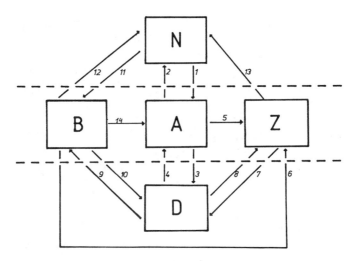

Figure 9.7 Simplified scheme of the matter flow in the planktonic environment, consisting of three living (B, bacteria; A, algae; Z, zooplankton) and two abiotic (N, inorganic nutrients; D, dissolved and particulate dead organic matter or detritus) subsystems.

isotopic equilibrium effects (Güde, 1984; King and Berman, 1984). In contrast, the low estimates rely on observed growth yields. An additional argument against high bacterial productivity is based on surprisingly high estimates of bacterial biomass, which may be comparable in magnitude to algal biomass (Bratbak, 1985; Bjørnsen, 1986; Simon and Tilzer, 1987; Simon and Azam, 1988). Although these findings at first seem to support the quantitative importance of bacterial food, they can hardly be reconciled with the currently suggested doubling times of bacterial biomass in the range of once per day. Given the comparable magnitude of algal and bacterial biomass, and a bacterial growth efficiency of 10 percent, the daily carbon requirements to sustain bacterial growth would then be equivalent to ten times the carbon content of algal biomass. Even if all losses of primary production to compartments other than bacteria were negligible, such high carbon requirements cannot be met by any existing estimate of primary production. Excluding the possibility of a "creatio ex nihilo," the apparent discrepancies must be due either to underestimates of primary production, overestimates of bacterial production, or neglect of alternative (e.g., allochthonous) carbon sources. At present, it cannot be determined which of these causes is more important. Therefore, the final word on the "sink or link" debate has yet to be spoken.

The bacteria represent an additional food niche that potentially enhances the diversity of zooplankton. This qualitative influence is especially apparent for phagotrophic microflagellates, which are now recognized as regular and frequent members of the zooplankton community. These organisms may be mainly dependent on bacterial food, although autotrophic picoplankton are included in their diet (e.g., Nagata, in press). Therefore, a close relationship between the development of these organisms and that of bacteria is expected. The regular appearance of high numbers of zooflagellates after the strong increase of bacterial populations during the algal spring bloom in Lake Constance confirms such a relationship (Güde, 1986, 1988a). Because this connection is less apparent during other seasons, it is suggested that factors other than the supply of bacterial food must also influence the development of the flagellates. Control by higher trophic levels appears to be important, as suggested by the apparent antagonistic development of daphnids and flagellates before, during, and after the clearwater phase in Lake Constance (Güde, 1988a). This pattern suggests that the zooflagellates represent a suitable food source for the daphnids, especially during food limitation in the clearwater phase. Similar effects were published by Riemann (1985) who observed increased densities of zooflagellates and decreased densities of Cladocera in lakes containing planktivorous fishes, as compared to enclosures made free of fish. Thus, the phagotrophic flagellates are controlled alternately or simultaneously by "bottom-up" and "top-down" factors. Initially, there is a predominance of "bottom-up" effects, with an increasing importance of the "top-down" effects as succession proceeds.

Unlike for phagotrophic flagellates, bacteria serve as a facultative rather than an obligate food source for crustacean zooplankton. The extent to which this additional food source can be exploited, however, may vary even with closely related species, as exemplified by *Daphnia galeata* and *D. hyalina*. The strong niche separation of these two species (Stich and Lampert, 1981; Geller, 1985) is also characterized by a different mesh size of the filter apparatus. Only *D. galeata* can filter efficiently on suspended natural bacterioplankton. In any case, one expects that the species able to exploit bacterial food will be selected when the availability of bacterial food exceeds that of algal food. This trend can be expected, for example, in dystrophic humic lakes, where cladocera, with their fine filters that allow efficient grazing on bacteria, were indeed predominant (Hessen, 1985a, b). Additionally, it must be remembered that the bacterial community represents an assemblage of different populations, and as a result, a potential variety of different food niches. Very little, however, is known about the importance of these qualitative differences in bacteria for the development of crustacean zooplankton. There are indications that grazing on bacterial fractions that grow attached to particles or as aggregates can become important for crustacean zooplankton in certain situations (e.g., Pedrós-Alió and Brock, 1983; Schoenberg and Maccubin, 1985).

In summary, bacteria represent a potential food source which can in turn influence zooplankton quantitatively and qualitatively. Although the evidence for the quantitative importance is controversial, the qualitative aspect becomes especially apparent for the zooflagellates, which owe their existence mainly to bacteria. For crustacean zooplankton, the dependence on bacterial food is less pronounced. Its utilization, however, certainly results in a selective advantage to the zooplankton when bacterial food is abundant.

Influence on phytoplankton   Many interactions exist between algae and bacteria (Cole, 1982). Heterotrophic bacteria depend strongly on the algae because primary production supplies, directly or indirectly, the carbon sources for the bacteria. In contrast, phototrophic growth may be independent of bacterial growth, because ideally, only light and inorganic nutrients are required. The situation may change completely, however, if inorganic nutrients become depleted, as frequently occurs in many lakes. In this case, bacteria may stimulate or inhibit algal growth, depending on whether they act as sinks or sources of the required nutrients.

According to the original paradigm, bacteria are considered as mineralizers and act as a nutrient source. This paradigm certainly holds true for carbon. The flow of inorganic carbon from bacteria to the environment (Figure 9.7, Flux 12) always exceeds the reversed flow (Figure 9.7, Flux 11), because the bacterial carbon supply comes almost exclusively from the detritus pool (Figure 9.7, Flux 9) and large parts of it are converted to inorganic carbon via respiration. The old paradigm, however, does not appear to be valid for phosphorus, which is much more frequently depleted in natural waters than carbon. In this case, bacteria must be regarded as sinks that decrease primary production, rather than as sources that stimulate primary production. Conceptually, this pattern can be postulated because the phosphorus demand of algae and bacteria is supplied from the same inorganic-P pool. During growth, the flow from the environment to the cells (Figure 9.7, Fluxes 1 and 11) exceeds the reversed flow (Figure 9.7, Fluxes 2 and 12) in both groups. Consequently, algae and bacteria compete for the same resources and tend to exclude each other. As is now confirmed by many experimental studies, bacteria must be considered as superior competitors for phosphorus (e.g., Rhee, 1972; Currie and Kalff, 1984; Güde, 1985). The result will be an increase of algal P-limitation and a decrease of algal production. As is suggested by Bratbak and Thingstad (1985), a positive feedback mechanism can be induced at this situation. Increased P-limitation leads to increased excretion of algal carbon, resulting in enhanced bacterial growth and increased binding of phosphorus by bacteria, which increases algal P-limitation, etc. Besides these culture experiments, it has been demonstrated by ^{32}P-uptake experiments in the field that bacteria possess efficient P-uptake systems. The majority of the uptake is usually performed by organisms <3 μm (e.g., Currie and Kalff, 1984; Berman, 1985). This effect is most pronounced in Lake Constance during the period of summer epilimnetic P-depletion, when the majority of ^{32}P-uptake occurs in bacteria <3 μm (Güde, unpublished). The importance of bacteria as P-sinks is confirmed by estimates of the percentage of total phosphorus bound by bacteria. Bacteria can bind more than 50 percent of total (i.e., dissolved and particulate) phosphorus during the period of epilimnetic P-depletion in Lake Constance (Jürgens and Güde, in preparation).

Given the ability of bacteria to bind phosphorus, any process leading to the regeneration of this phosphorus will decrease algal P-limitation and contribute to enhanced algal growth. As already emphasized by Johannes (1968), grazing on bacteria must be considered to be of outstanding importance for this regeneration, because grazers receive their phosphorus through food and excrete excess phosphorus into the environment. Because no inorganic nutrients are taken up by the grazers, the matter flow is always unidirectional (Figure 9.7, Flux 13) from the grazers toward the inorganic pools, resulting in a net regeneration of nutrients (e.g., Güde, 1985; Andersen et al., 1986; Berman et al., 1987).

The extent of this regeneration, however, will strongly depend on the C:P ratio, because it becomes less efficient with larger C:P ratios (Anderssen et al., 1986; Jürgens and Güde, in preparation). Because the C:P ratios increase with increasing P-limitation, the desired effect of nutrient regeneration would be weakest when it is most required. Unfortunately, there is little evidence to date on bacterial limitation during epilimnetic P-depletion. It can be assumed, however, that even then, bacteria are predominantly limited by the carbon supply. Moreover, C:P ratios of bacteria grown under P-limitation were shown to be much lower than those observed for algae grown under P-limitation (Dicks and Tempest, 1966; Fuhs et al., 1972; Jürgens and Güde, in preparation). It can be assumed, therefore, that grazing on bacteria would result in an effective regeneration of phosphorus and that this effect would become especially important for algae during periods of epilimnetic P-depletion.

As has been mentioned in Section 9.2, there is a direct link between bacteria and algae, inasmuch as algae themselves may represent the grazers on bacteria (Figure 9.7, Flux 14). Although this type of algal nutrition has long been known, its relevance to bacterial grazing losses has only recently been suggested (Bird and Kalff, 1986; Porter, 1988). Because these findings are comparatively new, however, a realistic evaluation is not yet possible. It has been hypothesized that the main advantage of mixotrophic growth is the supply of additional energy sources, which would be especially useful in conditions of reduced light availability, as occur in deeper water layers. In connection with the role of grazing on bacteria in P-regeneration discussed above, it could be equally important for the P-supply of the algae. Eating the competitors and obtaining at the same time the nutrients which were the source of competition could be regarded as an almost ideal solution of the problems the algae are facing in P-deficient conditions.

In conclusion, there are a variety of algae/bacteria interactions that may be influenced to a certain extent by grazing on bacteria. One of the most important must be the competition between bacteria and algae for phosphorus, because grazing is an efficient mechanism to make bacterial phosphorus available to algae. This grazing may occur directly by phagotrophic algae and/or indirectly by zooplankton. Evidence for the first mechanism is scarce at present, whereas the latter possibility is now well documented by experimental and field studies.

9.4   Grazing on Bacteria in the PEG-model

Field observations from Lake Constance provided a substantial basis for the development of the PEG-model (Sommer et al., 1986), in which, however, bacteria were not considered. Therefore, it may be useful to examine here the participation of bacteria and their grazers in the seasonal succession events. As is shown in Figures 9.3 and 9.8, bacterial densities remain consistently low from January to March. Minimum thymidine incorporation values observed at this time suggest that bacterial populations are controlled mainly by the short supply of organic substrates, because of low primary production and cold temperatures (Güde et al., 1985). Bacterial grazing is also negligible at this time, because low temperatures and the absence of algal and bacterial food prevents abundant growth of zooplankton.

Soon after the beginning of the algal spring bloom, bacteria awake from their "winter dormancy," because of the increased substrate supply from the development of primary producers. The result is an increased density and activity of bacterial populations. During this phase, bacterial growth can be assumed to initially be controlled mainly by abiotic factors such as temperature, as is the starting algal populations (Figure 9.8). Soon after bacteria have reached high densities, however, bacterial growth is probably controlled increasingly by the supply of substrates. The morphologic structure of the bacterial populations is at this time very uniform, consisting almost entirely of small, free single cells, which, because of their optimum surface/volume ratio, allow maximum efficiencies of substrate uptake.

Up to this point, bacterial growth has not been much influenced by grazing. As a consequence of the increased supply of bacterial food, however, phagotrophic microflagellates soon reach high densities, with maximum numbers up to 10^4 ml^{-1}. Thus, the structure of the zooplankton community, which at this time is dominated by small, fast-growing species (step 2 of the PEG-model, see Section 1.2) is now influenced by the availability of bacterial food. Although estimated grazing rates on bacteria are high at this time, bacteria can maintain elevated population densities, because the grazing losses can be offset by high bacterial production. High primary production provides a sustained replenishment of the pool of organic substrates from algal exudates and the release of organic carbon by herbivorous zooplankton.

The result of increased grazing is a reduction in substrate limitation and a stimulation of bacterial growth rates. Thus, control of bacterial populations shifts from substrate regulation to grazer regulation. Selective grazing leads

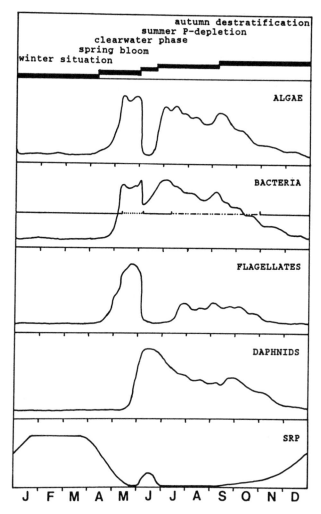

Figure 9.8 Synopsis of the idealized development of algae, bacteria, phagotrophic microflagellates, daphnids and concentration of soluble reactive phosphorus (SRP) during the annual succession observed in Lake Constance. (———) prevailing substrate control; (----) prevailing grazing control; (-----) simultaneous or shortly alternating grazing and substrate control of bacteria.

to an increased morphologic diversity of bacterial populations and a change of the bacterial size structure. The result is a higher percentage of bacterial growth forms with a relative resistance to protozoan grazing (i.e., very large forms such as filaments, flocs, and very small forms).

During the second half of the spring bloom, large herbivorous zooplankton increase, contributing to increased bacterial substrate supply and grazing losses. The substrate supply is increased because organic carbon is released during grazing (Lampert, 1978; Olsen et al., 1986; Güde, 1988) which allows bacterial populations to reach short peaks of production (Simon and Tilzer, 1987). On the other hand, large zooplankton contribute to grazing losses of bacteria. The nature of this grazing loss is different from that of protozoans, because it comprises a much larger spectrum of food sizes. Phagotrophic flagellates fall into this food spectrum but can maintain initially high levels as long as their food supply is high from high bacterial productivity.

The situation becomes completely changed at the subsequent clearwater phase, by which time the zooplankton have strongly reduced algal biomass (step 5 of the PEG-model, see Section 1.2). This reduction of primary producers leads to a reduction of bacterial productivity (see Figure 9.8). Grazing losses can no longer be offset, and bacterial densities decrease. Increasing food limitation of daphnics leads to a maximum exploitation of all potential food sources, including large size classes of bacteria and phagotrophic microflagellates. As a consequence, the morphologic structure of bacterial populations is changed from a more heterogeneous toward an extremely uniform population of very small cells (Güde, 1988a). Additionally, the population densities of phagotrophic flagellates are drastically reduced.

Because of this elimination of efficient bacterial grazers by the daphnids, combined with the limited ability of daphnids to graze on the now predominating very small bacteria, grazing pressure on bacteria is strongly reduced. Bacteria are again under strong substrate control. The lack of grazing pressure allows the bacteria to increase slowly toward a maximum at the end of the clearwater phase. This maximum coincides with the development of the algal summer populations and the depletion of epilimnetic phosphorus (steps 8 and 9 of the PEG-model, see Section 1.2). The beginning of algal P-limitation also initiates the period at which bacteria compete most strongly with algae for phosphorus. Because of their strong competitive abilities, bacteria can bind considerable amounts of available phosphorus, contributing to increased P-limitation of algae. In this situation, grazing on bacteria contributes to a reduction of algal P-limitation, because it results in increased regeneration of bacterial phosphorus. This regeneration can occur because grazing pressure on bacteria increases as the flagellates increase after the summer decline of daphnids.

During the subsequent summer and autumn months, the events controlling bacterial populations and their influence on the total planktonic community are determined by the balance of the different effects described above. The outcome may vary, depending on which of the multiple interactions predominate. For example, grazing on bacteria can be intensified at periods of increased fish predation, which results in decreased numbers of daphnids and increased densities of flagellates. The opposite would be expected when fish predation is low. In any case, bacterial populations will continue to be affected, simultaneously or alternately, by the availability of organic substrates and by grazing, unless the flow of organic carbon from primary production has become so low that it allows only the maintenance of a very reduced bacterial population. Low bacterial numbers reduce the efficiency of grazing and the grazers also decline to their winter minimum.

Obviously, the changes occurring during the annual succession of plankton described above should be accompanied by changes in the species composition of the bacterial community. Such changes are indeed indicated by changes of the composition of bacterial colonies on agar plates (Güde et al., 1985). Further information on this question, however, is lacking because of the methodological constraints mentioned in the Introduction.

9.5 Concluding Remarks

When the role of bacteria in aquatic ecosystems was first being studied, comparatively little attention was directed toward grazing. During the past years, however, intensive research activities have focused on the bacteria/grazer relationships. Due to these efforts, there are now strong indications that grazing on bacteria can be an important factor for the seasonal succession of bacteria, as well as for the total planktonic community. This does not, of course, mean that the bacterial community is always and exclusively controlled by grazing. Moreover, the obvious gaps of our current knowledge make it clear that we are still far from a comprehensive understanding of bacterial succession in the plankton community. The following conclusions may help to stimulate future research efforts to fill these gaps.

1. Understanding of bacterial succession requires a qualitative analysis of bacterial populations. This is difficult because of insufficient methodologies. Methodological progress is urgently needed to resolve the question of the composition of natural bacterial communities.

2. Evidence from field and experimental studies suggests that grazing is one of the major loss factors for bacteria. Phagotrophic microflagellates and filtering cladocera have been proven to be efficient grazers on bacteria. Additionally, ciliates and phagotrophic algae have been implicated as contributors to bacterial grazing losses.

3. Bacterial populations are alternately or simultaneously controlled by substrate supply ("bottom-up") or grazing ("top-down"). The intensity of grazing depends on bacterial densities (which in turn depend on the substrate supply) and on the predatory control of the grazers.

4. Grazing changes the selection conditions for bacteria indirectly

and directly. Indirectly, grazers reduce nutrient limitation by keeping bacterial densities low and thus increasing the amount of substrate available per cell. This effect can be increased by a partial regeneration of organic carbon taken up in the food by the grazers. Selective grazing directly influences the selection conditions, resulting in an enrichment of grazing-resistant bacteria. Morphology is one of the most important criteria for grazing resistance, with very small and very large growth forms being the most resistant. Both the direct and indirect changes of selection conditions tend to increase the taxonomic and morphologic diversity of bacterial populations.

5. Grazing on bacteria can also influence other members of the planktonic community. The most obvious effect is exerted on the structure of the zooplankton community, because bacteria represent an additional or alternative food source. The quantitative importance of this food source is thought to be high. Because of recently determined low bacterial growth efficiencies, however, and the resulting energetic considerations, bacterial food seems to be quantitatively less important than suggested by current estimates of bacterial production. In any case, bacterial food potentially enhances the ecological niches occupied by zooplankton populations. The resulting increase of the diversity of the zooplankton community is exemplified by bacterivorous flagellates.

6. The algal community may be influenced by grazing on bacteria, because bacteria compete efficiently with algae for phosphorus and can bind substantial amounts of available phosphorus. As a result, P-limitation of algae increases, and algal productivity decreases at periods of external P-depletion. Grazing on bacteria counterbalances these effects because it provides regeneration of bacterial phosphorus. This regeneration contributes to a reduction of algal P-limitation and a stimulation of algal productivity.

Acknowledgments
I owe thanks to U. Sommer for encouragement in writing this contribution and assistance in the preparation of the manuscript. I am further indebted to K. Jürgens, C. Pedrós-Alió, R. Sterner, H.B. Stich, and S. Medford for critical discussions and help in improving the manuscript. The study was supported by the Deutsche Forschungsgemeinschaft (SFB 248).

References

Ammerman, J.W., Fuhrman, J.A., Hagström, A., and Azam, F. 1984. Bacterioplankton growth in seawater. 1. Growth kinetics and cellular characteristics in seawater cultures. *Marine Ecology Progress Series* 18: 31–39.

Andersen, O.K., Goldman, J.C., Caron, D.A., and Dennett, M.R. 1986. Nutrient cycling in a marine microflagellate food chain. III. Phosphorus dynamics. *Marine Ecology Progress Series* 31: 47–55.

Andersson, A., Larsson, U., and Hagström, A. 1986. Size selective predation by a microflagellate on pelagic bacteria. *Marine Ecology Progress Series* 33: 51–57.

Andersson, A., Lee, C., Azam, F., and Hagström, A. 1985. Release of amino acids and inorganic nutrients by heterotrophic marine microflagellates. *Marine Ecology Progress Series* 23: 99–106.

Azam, F., Fenchel, T., Field, J.G., Gray, J.S., Meyer-Reil, L.A., and Thingstad, F. 1983. The ecological role of water column microbes in the sea. *Marine Ecology Progress Series* 10: 257–263.

Berman, T. 1985. Uptake of ^{32}P-orthophosphate by algae and bacteria in Lake Kinneret. *Journal of Plankton Research* 7: 71–84.

Berman, T., Nawrocki, M., Taylor, G.T., and Karl, D.M. 1987. Nutrient fluxes between bacteria, bacterivorous nannoplanktic protists and algae. *Marine Microbial Food Webs* 2: 69–82.

Bern, L. 1985. Autoradiographic studies of (methyl-^{3}H)thymidine incorporation in cyanobacterium (*Microcystis wesenbergii*)-bacteria associations and in selected algae and bacteria. *Applied and Environmental Microbiology* 49: 232–233.

Bern, L. 1987. Zooplankton grazing on (methyl-^{3}H)thymidine-labelled natural particle assemblages: determination of filtering rates and food selectivity. *Freshwater Biology* 17: 151–159.

Bird, D.F. and Kalff, J. 1987. Algal phagotrophy: Regulating factors and importance relative to photosynthesis in *Dinobryon* (Chrysophyceae). *Limnology and Oceanography* 32: 277–284.

Bjørnsen, P.K. 1986. Automatic determination of bacterioplankton biomass by image analysis. *Applied and Environmental Microbiology* 51: 1199–1204.

Bjørnsen, P.K., Larsen, J.P., Geertz-Hansen, O., and Olesen, M. 1986. A field technique for the determination of zooplankton grazing on natural bacterioplankton. *Freshwater Biology* 16: 245–253.

Børaas, M.B., 1986. Relationship between densities and sizes of phagotrophic microflagellates and cyanobacteria in Lake Michigan. *IV. International Congress on Ecology*. Syracuse, New York, August 10–16. Abstracts p. 96.

Børsheim, K.Y. 1984. Clearance rates of bacteria-sized particles by freshwater ciliates, measured with mono-disperse fluorescent latex beads. *Oecologia* 63: 286–288.

Børsheim, K.Y. and Olsen, Y. 1984. Grazing activities by *Daphnia pulex* on natural populations of bacteria and algae. *Verhandlungen Internationaler Verein für Theoretische und Angewandte Limnologie* 22: 644–648.

Bratbak, G. 1985. Bacterial biovolume and biomass estimations. *Applied and Environmental Microbiology* 49: 1488–1493.

Bratbak, G. and Thingstad, T.F. 1985. Phytoplankton-bacteria interactions: an apparent paradox? Analysis of a model system with both competition and commensalism. *Marine Ecology Progress Series* 25: 23–30.

Brendelberger, H. 1985. Filter mesh-size and retention efficiency for small particles: comparative studies with cladocera. *Archiv für Hydrobiologie Beiheft Ergebnisse der Limnologie* 21: 135–146.

Caron, D.A., Goldman, J.C., and Dennet, M.R. 1988. Experimental demonstration of the roles of bacteria and bacterivorous protozoa in nutrient cycles. *Hydrobiologia* 159: 27–40.

Cole, J.J. 1982. Interactions between bacteria and algae in aquatic ecosystems. *Annual Review of Ecology and Systematics* 13: 291–314.

Cole, J.J., Findlay, S., and Pace, M.L. 1988. Bacterial production in fresh and saltwater ecosystems: a cross overview. *Marine Ecology Progress Series* 43: 1–10.

Currie, D.J. and Kalff, J. 1984. A comparison of the abilities of freshwater algae and bacteria to acquire and retain phosphorus. *Limnology and Oceanography* 29: 298–310.

Cynar, F.J. and Sieburth, J.McN. 1986. Unambiguous detection and improved quantification of phagotrophy in apochlorotic nanoflagellates using fluorescent microspheres and concomitant phase contrast and epifluorescence microscopy. *Marine Ecology Progress Series* 32: 61–70.

De Mott, W.R. 1985. Relation between filter mesh-size, feeding mode and capture efficiency for cladocera feeding on ultrafine particles. *Archiv für Hydrobiologie Beiheft Ergebnisse der Limnologie* 21: 125–134.

Dicks, J.W. and Tempest, D.W. 1966. The influence of temperature and growth rate on the quantitative relationship between potassium, magnesium, phosphorus and ribonucleic acid of *Aerobacter aerogenes* growing in a chemostat. *Journal of General Microbiology* 45: 547–557.

Ducklow, H.W., Purdie, D.A., Williams, P.leB., and Davies, J.M. 1986. Bacterioplankton: A sink for carbon in a coastal marine plankton community. *Science* 232: 865–867.

Fenchel, T. 1980. Suspension feeding in ciliates protozoa: feeding rates and their ecological significance. *Microbial Ecology* 6: 1–12.

Fenchel, T. 1982. Ecology of heterotrophic flagellates. II. Bioenergetics and growth. *Marine Ecology Progress Series* 8: 225–231.

Fenchel, T. 1986. The ecology of heterotrophic microflagellates. *Advances in Microbial Ecology* 9: 57–97.

Frederickson, A.G. 1977. Behavior of mixed cultures of microorganisms. *Annual Review of Microbiology* 31: 63–87.

Fuhrman, J.A. and Azam, F. 1980. Bacterioplankton secondary production estimates for coastal waters of British Columbia, Antarctica and California. *Applied and Environmental Microbiology* 39: 1085–1095.

Fuhs, G.W., Demmerle, S.D., Canelli, E., and Chen, M. 1972. Characterization of phosphorus-limited plankton algae (with reflection of the limiting nutrient concept). *American Society of Limnology and Oceanography Special Symposium* 1: 113–133.

Geller, W. 1985. Production, food utilization and losses of two coexisting *Daphnia* species. *Archiv für Hydrobiologie Beiheft Ergebnisse der Limnologie* 21: 67–79.

Geller, W. and Müller, H. 1981. The filtration apparatus of Cladocera: filter mesh-sizes and their implications on food selectivity. *Oecologia (Berlin)* 49: 316–321.

Goldman, J.C. and Caron, D.A. 1985. Experimental studies on an omnivorous microflagellate: Implications for grazing and nutrient regeneration in the marine food chain. *Deep Sea Research* 21: 899–915.

Güde, H. 1979. Grazing by protozoa as selection factor for activated sludge bacteria. *Microbial Ecology* 5: 225–237.

Güde, H. 1982. Interactions between floc-forming and nonfloc-forming bacterial populations of activated sludge. *Current Microbiology* 7: 347–350.

Güde, H. 1984. Test for validity of different radioisotope activity measurements by microbial pure and mixed cultures. *Archiv für Hydrobiologie Beiheft Ergebnisse der Limnologie* 19: 257–266.

Güde, H. 1985. Influence of phagotrophic processes on the regeneration of nutrients in two-stage continuous culture systems. *Microbial Ecology* 11: 193–204.

Güde, H. 1986. Loss processes influencing growth of planktonic bacterial populations in Lake Constance. *Journal of Plankton Research* 8: 795–810.

Güde, H. 1988a. Influence of crustacean zooplankton on bacterial populations in Lake Constance. *Hydrobiologia* 159: 63–73.

Güde, H. 1988b. Incorporation of ^{14}C-glucose, ^{14}C-amino acids and ^{3}H-thymidine by different size fractions of aquatic microorganisms. *Archiv für Hydrobiologie Beiheft Ergebnisse der Limnologie* 31: 61–69.

Güde, H. In press. Bacterial production and flow of organic matter in Lake Constance, in: Tilzer, M.M. and Serruya, C. (editors), *Ecological Structure and Function in Large Lakes*. Science Tech Publishers, Madison.

Güde, H., Haibel, B., and Müller, H. 1985. Development of planktonic bacterial populations in Lake Constance (Bodensee-Obersee). *Archiv für Hydrobiologie* 105: 59–77.

Guerrero, R., Esteve, I., Pedrós-Alió, C., and Ganju, N. 1987. Predatory bacteria in procaryotic communities. *Annals of the New York Academy of Sciences* 503: 238–250.

Hagström, A., Larsson, U., Hörstedt, P., and Normark, S. 1979. Frequency of dividing cells, a new approach to the determination of bacterial growth rates in aquatic environments. *Applied and Environmental Microbiology* 37: 805–812.

Hessen, D.O. 1985a. Filtering structures and particle size selection in coexisting Cladocera. *Oecologia (Berlin)* 66: 368–372.

Hessen, D.O. 1985b. The relation between bacterial carbon and dissolved humic compounds in oligotrophic lakes. *FEMS Microbiology Ecology* 31: 215–223.

Hollibaugh, J.T., Fuhrman, J.A., and Azam, F. 1981. Radioactive labeling of natural assemblages of bacterioplankton for use in trophic studies. *Limnology and Oceanography* 19: 995–998.

Hobbie, J.E., Daley, R.J., and Jasper, S. 1977. Use of Nuclepore filters for counting bacteria by epifluorescence microscopy. *Applied and Environmental Microbiology* 33: 1225–1228.

Hoppe, H.G. 1984. Attachment of bacteria: advantage or disadvantage for survival in the aquatic environment, pp. 283–301, in Marshall, K.C. (editor), *Microbial Adhesion and Aggregation*. Springer Verlag Berlin, Heidelberg, New York, Tokyo.

Jassby, A.D. 1975. The ecological significance of sinking to planktonic bacteria. *Canadian Journal of Microbiology* 21: 270–274.

Johannes, R.E. 1968. Nutrient regeneration in lakes and oceans. *Advances in Microbiology of the Sea* 1: 203–213.

King, G.M. and Berman, T. 1984. Potential effects of isotopic dilution on apparent respiration in ^{14}C-heterotrophy experiments. *Marine Ecology Progress Series* 19: 175–180.

Kirchman, D., Ducklow, H.W., and Mitchell, R. 1982. Estimates of bacterial growth from changes in uptake rates and biomass. *Applied and Environmental Microbiology* 44: 1296–1307.

Kjelleberg, S., Hermansson, M., Marden, P., and Jones, G.W. 1987. The transient phase between growth and nongrowth of heterotrophic bacteria, with emphasis on the marine environment. *Annual Review of Microbiology* 41: 25–49.

Krambeck, C. 1988. Control of bacterioplankton structures by grazing and nutrient supply during the decline of an algal bloom. *Verhandlungen der Internationalen Vereinigung für Theoretische und Angewandte Limnologie* 23: 496–502.

Laake, M., Dahle, A.B., Eberlein, K., and Rein, K. 1983. A modeling approach to the interplay of carbohydrates, bacteria and nonpigmented flagellates in a controlled ecosystem experiment with *Skeletonema costatum*. *Marine Ecology Progress Series* 14: 71–79.

Lampert, W. 1978. Release of dissolved organic carbon by grazing zooplankton. *Limnology and Oceanography* 23: 195–218.

Lampert, W., Fleckner, W., Rai, H., and Taylor, B. 1986. Phytoplankton control by grazing zooplankton: a study on the spring clearwater phase. *Limnology and Oceanography* 31: 478–490.

Landry, M.R., Hass, L.W., and Fagerness, V.L. 1984. Dynamics of microbial plankton communities: Experiments in Kaneohe Bay, Hawaii. *Marine Ecology Progress Series* 16: 127–133.

Linley, E.A.S., Newell, R.C., and Lucas, M.I. 1983. Quantitative relationships between phytoplankton, bacteria and heterotrophic microflagellates in shelf waters. *Marine Ecology Progress Series* 12: 77–89.

Marshall, K.C. (editor). *Microbial Adhesion and Aggregation*. Springer Verlag Berlin, Heidelberg, New York, Tokyo.

Mason, C.A., Hamer, G., and Bryers, J.D. 1986. The death and lysis of microorganisms in environmental processes. *FEMS Microbiology Reviews* 39: 373–401.

McManus, G.B. and Fuhrman, J.A. 1986. Bacterivory in seawater studies with the use of inert fluorescent particles. *Limnology and Oceanography* 31: 420–426.

McManus, G.B. and Fuhrman, J.A. 1988. Control of marine bacterioplankton populations: Measurement and significance of grazing. *Hydrobiologia* 159: 51–62.

Morita, R.Y. 1982. Starvation survival of heterotrophs. *Advances in Microbial Ecology* 6: 171–198.

Nagata, T. 1988. Production rate of planktonic bacteria in the north basin of Lake Biwa. *Applied and Environmental Microbiology* 53: 2878–2882.

Nagata, T. In press. Bacterial production and their contribution to the grazer food chain, in Tilzer, M.M. and Serruya, C. (editors), *Ecological Structure and Function in Large Lakes*. Science Tech Publishers, Madison.

Newell, R.C., Lucas, M.I., and Linley, E.A.S. 1981. Rate of degradation and efficiency of conversion of phytoplankton debris by marine microorganisms. *Marine Ecology Progress Series* 6: 123–136.

Nygaard, K., Borsheim, K.Y., and Thingstad, T.F. 1988. Grazing rates on bacteria by marine heterotrophic microflagellates compared to uptake rates of bacterial-sized monodisperse fluorescent latex beads. *Marine Ecology Progress Series* 44: 159–165.

Olsen, Y., Varum, M.M., and Jensen, A. 1986. Some characteristics of the carbon compounds released by *Daphnia*. *Journal of Plankton Research* 8: 505–518.

Pace, M.L. 1982. Planktonic ciliates: their distribution, abundance and relationship to microbial resources in a monomictic lake. *Canadian Journal of Fisheries and Aquatic Sciences* 39: 1106–1116.

Pace, M.L. 1988. Bacterial mortality and the fate of bacterial production. *Hydrobiologia* 159: 41–49.

Pedrós-Alió, C. and Brock, T.D. 1983a. The importance of attachment to particles for planktonic bacteria. *Archiv für Hydrobiologie* 98: 354–379.

Pedrós-Alió, C. and Brock, T.D. 1983b. The impact of zooplankton feeding on the epilimnetic bacteria of a eutrophic lake. *Freshwater Biology* 13: 227–239.

Pomeroy, L.R. and Wiebe, W.J. 1988. Energetics of microbial food webs. *Hydrobiologia* 159: 7–18.

Porter, K.G. 1988. Phagotrophic algae in microbial food webs. *Hydrobiologia* 159: 89–97.

Porter, K.G. 1984. Natural bacteria as food resources for zooplankton, pp. 340–345, in Klug, M.J. and Reddy, C.A. (editors), *Current Perspectives in Microbial Ecology*. American Society for Microbiology, Washington, D.C.

Rassoulzadegan, F., Laval-Peuto, M., and Sheldon, R.W. 1988. Partitioning of the food ration of marine ciliates between pico- and nannoplankton. *Hydrobiologia* 159: 75–88.

Rhee, G.-Y. 1972. Competition between an alga and an aquatic bacterium for phosphate. *Limnology and Oceanography* 17: 505–514.

Riemann, B. 1985. Potential importance of fish predation and zooplankton grazing on natural populations of freshwater bacteria. *Applied and Environmental Microbiology* 50: 187–193.

Riemann, B. and Bosselmann, S. 1984. *Daphnia* grazing on natural populations of lake bacteria. *Verhandlungen des Internationalen Vereins für Theoretische und Angewandte Limnologie* 22: 795–799.

Riemann, B., Sondergaard, M., Schierup, H.H., Bosselmann, S., Christensen, G., Hansen, J., and Nielsen, B. 1982. Carbon metabolism during a spring diatom bloom in the eutrophic lake Mosso. *Internationale Revue der gesamten Hydrobiologie* 67: 145–185.

Robertson, M.L., Mills, A.L., and Ziemann, J.C. 1982. Microbial synthesis of detritus-like particulates from dissolved organic carbon released by tropical seagrasses. *Marine Ecology Progress Series* 7: 279–285.

Sanders, R.W. and Porter, K.G. 1986. Use of metabolic inhibitors to estimate proto-zooplankton grazing and bacterial production in a monomictic eutrophic lake with an anaerobic hypolimnion. *Applied and Environmental Microbiology* 52: 101–107.

Scavia, D. and Laird, A. 1987. Bacterioplankton in Lake Michigan: Dynamics, controls, and significance to carbon flux. *Limnology and Oceanography* 32: 1017–1033.

Schoenberg, S.A. and Maccubin, A.E. 1985. Relative feeding rates on free and particle bound bacteria by freshwater macrozooplankton. *Limnology and Oceanography* 30: 1084–1090.

Sherr, B.F. and Sherr, E.B. 1984. Role of heterotrophic protozoa in carbon and energy flow in aquatic ecosystems, pp. 412–423, in Klug, M.J. and Reddy, C.A. (editors), *Current Perspectives in Microbial Ecology*. American Society for Microbiology, Washington, D.C.

Sherr, B.F. and Sherr, E.B. 1987. High rates of consumption of bacteria by pelagic ciliates. *Nature* 325: 710–711.

Sherr, B.F., Sherr, E.B., and Berman, T. 1982. Decomposition of organic detritus: a selective role for microflagellate protozooa. *Limnology and Oceanography* 27: 765–769.

Sherr, B.F., Sherr, E.B., and Berman, T. 1983. Grazing, growth and ammonium excretion rates of a heterotrophic microflagellate fed with four species of bacteria. *Applied and Environmental Microbiology* 45: 1196–1201.

Sherr, B.F., Sherr, E.B., and Albright, L.J. 1987. Bacteria: Link or sink? *Nature* 235: 88.

Sherr, B.F., Sherr, E.B., and Fallon, R.D. 1987. Use of monodispersed fluorescently labeled bacteria to estimate in situ protozoan bacterivory. *Applied and Environmental Microbiology* 53: 958–965.

Sherr, B.F., Sherr, E.B., Andrew, T.L., Fallon, R.D., and Newell, S.Y. 1986. Trophic interactions between heterotrophic protozoa and bacterioplankton in estuarine water analyzed with selective metabolic inhibitors. *Marine Ecology Progress Series* 32: 169–179.

Sherr, B.F., Sherr, E.B., and Hopkinson, C.S. 1988. Trophic interactions with pelagic microbial communities: Indications of feedback regulation of carbon flow. *Hydrobiologia* 159: 19–26.

Shilo, M. 1984. *Bdellovibrio* as predator, pp. 334–339, in Klug, M.J. and Reddy, C.A. (editors), *Current Perspectives in Microbial Ecology*. American Society for Microbiology, Washington, D.C.

Sieracki, M.E. and Sieburth, J.McN. 1985. Factors controlling the periodic fluctuation in total planktonic bacterial populations in the upper ocean: Comparison of nutrient, sunlight and predation effects. *Marine Microbial Food Webs* 1: 35–40.

Simek, K. 1986. Bacterial activity in a reservoir determined by autoradiography and its relationships to phyto- and zooplankton. *Internationale Revue der Gesamten Hydrobiologie* 71: 593–612.

Simon, M. 1987. The contribution of small and large free-living and attached bacteria to the organic matter metabolism of Lake Constance. *Limnology and Oceanography* 39: 591–607.

Simon, M. and Tilzer, M.M. 1987. Bacterial responses to seasonal primary production and phytoplankton biomass in Lake Constance. *Journal of Plankton Research* 9: 535–552.

Sommer, U., Gliwicz, Z.M., Lampert, W., and Duncan, A. 1986. The PEG-model of seasonal succession of planktonic events in freshwaters. *Archiv für Hydrobiologie* 106: 433–471.

Staley, J.T. and Konopka, A. 1985. Measurements of in situ activities of nonphotosynthetic microorganisms in aquatic and terrestrial habitats. *Annual Review of Microbiology* 39: 321–346.

Starkweather, P.L. 1980. Aspects of the feeding behavior and trophic ecology of suspension-feeding rotifers. *Hydrobiologia* 73: 63–72.

Stich, H.-B. and Lampert, W. 1981. Predator evasion as an explanation of diurnal vertical migration by zooplankton. *Nature* 243: 396–398.

Stout, J.D. 1980. The role of protozoa in nutrient cycling and energy flow. *Advances in Microbial Ecology* 4: 1–50.

Tempest, D.W. and Neijssel, O.M. 1978. Ecophysiological aspects of microbial growth in aerobic nutrient limited environments. *Advances in Microbial Ecology* 2: 105–154.

Tranvik, L.J. and Höfle, M.G. 1987. Bacterial growth on dissolved organic carbon from humic and clear watgers in mixed cultures. *Applied and Environmental Microbiology* 52: 684–692.

Turley, C.M., Newell, R.C., and Robins, D.B. 1986. Survival strategies of two small marine ciliates in regulating bacterial community structure under experimental conditions. *Marine Ecology Progress Series* 33: 59–70.

Van Es, F.B. and Meyer-Reil, L.A. 1982. Biomass and metabolic activity of heterotrophic marine bacteria. *Advances in Microbial Ecology* 6: 111–170.

Veldkamp, H. 1977. Ecological studies with the chemostat. *Advances in Microbial Ecology* 1: 59–89.

Wambeke, F.Van and Bianchi, M.A. 1985. Bacterial biomass production and ammonium regeneration in Mediterranian seawater supplemented with amino acids. 3. Nitrogen flux through heterotrophic microplankton food chains. *Marine Ecology Progress Series* 23: 117–128.

Williams, F.M. 1980. On understanding predator-prey interactions, pp. 349–375, in Ellwood, D.C., Hedger, J.N., Latham, M.J., Lynch, J.M., and Slater, J.H. (editors), *Contemporary Microbial Ecology*. Academic Press, London.

Williams, P.J.le B. 1981. Incorporation of heterotrophic processes into the classical paradigm of the planktonic food web. *Kieler Meeresforschungen/Sonderheft* 5: 1–28.

Wright, R.T. 1988. A model for short-term control of the bacterioplankton by substrate and grazing. *Hydrobiologia* 159: 111–117.

Wright, R.T. and Coffin, R.B. 1984a. Measuring microzooplankton grazing on planktonic marine bacteria by its impact on bacterial production. *Microbial Ecology* 10: 137–149.

Wright, R.T. and Coffin, R.B. 1984b. Factors affecting bacterioplankton density and production in salt marsh estuaries, pp. 485–494, in Klug, M.J. and Reddy, C.A. (editors), *Current Perspectives in Microbial Ecology*. American Society for Microbiology, Washington, D.C.

Wright, R.T. and Hobbie, J.E. 1965. The uptake of organic solutes in lake water. *Limnology and Oceanography* 10: 22–28.

Zimmermann, R. and Meyer-Reil, L.-A. 1974. A new method for fluorescence staining of bacterial populations on membrane filters. *Kieler Meeresforschungen* 30: 24–27.

Index